Injectable Dispersed Systems

Formulation, Processing, and Performance

DRUGS AND THE PHARMACEUTICAL SCIENCES

Executive Editor

James Swarbrick

PharmaceuTech, Inc.
Pinehurst, North Carolina

Advisory Board

DRUGS AND THE PHARMACEUTICAL SCIENCES
A Series of Textbooks and Monographs

Injectable Dispersed Systems

Formulation, Processing, and Performance

Diane J. Burgess

University of Connecticut
Storrs-Mansfield, Connecticut, U.S.A.

CRC Press
Taylor & Francis Group
Boca Raton London New York

CRC Press is an imprint of the
Taylor & Francis Group, an **informa** business

A TAYLOR & FRANCIS BOOK

CRC Press
Taylor & Francis Group
6000 Broken Sound Parkway NW, Suite 300
Boca Raton, FL 33487-2742

First issued in paperback 2019

ISBN-13: 978-0-8493-3699-7 (hbk)
ISBN-13: 978-0-367-39282-6 (pbk)

Library of Congress Cataloging-in-Publication Data

Catalog record is available from the Library of Congress

---

*This book is dedicated to my parents
Violet Isabel Burgess and George Gartly Burgess.*

Preface

With the increasing number of biopharmaceutical products, the emerging market for gene therapeutics, and the high proportion of small molecule new drug candidates that have very poor solubility, the need for parenteral dispersed system pharmaceuticals is growing rapidly. This book serves as a current in-depth text for the design and manufacturing of parenteral dispersed systems. The fundamental physicochemical and biopharmaceutical principles governing dispersed systems are covered together with design, processing, product performance, characterization, quality assurance, and regulatory concerns. A unique and critically important element of this work is the inclusion of practical case studies together with didactic discussions. This approach allows the illustration of the application of dispersed systems technology to current formulation and processing problems and, therefore, this will be a useful reference text for industrial research and development scientists and will help them in making choices of appropriate dosage forms and consequent formulation strategies for these dosage forms. Quality control and assurance as well as regulatory aspects that are essential to parenteral dispersed system product development are discussed in detail. This book also tackles current issues of in vitro testing of controlled release parenterals as well as the development of in vitro and in vivo relationships for these dosage forms.

This work is equally relevant to industrial and academic pharmaceutical scientists. The text is written in a way that the different chapters and case studies can be read independently, although the reader is often referred to other sections of the book for more in-depth information on specific topics. The case studies provide the reader with real problems that have been faced and solved by pharmaceutical scientists and serve as excellent examples for industrial scientists as well as for academics. This text will not only serve as a practical guide for pharmaceutical scientists involved in the research and development of parenteral dosage forms, but will also be a resource for scientists new to this field. The fundamental aspects together with the practical case studies make this an excellent textbook for graduate education.

The book is laid out as follows: Section (I) Basic Principles; Section (II) Dosage Forms; Section (III) Case Studies; and Section (IV) Quality Assurance and Regulation. The basic principles section includes physicochemical and biopharmaceutical principles, characterization and analysis and in vitro and in vivo release testing and correlation of in vitro and in vivo release data. The dosage forms covered in Section II are suspensions, emulsions, liposomes, and microspheres. These chapters detail design and manufacturing and a rationale for selection as well as any specific considerations for the individual parenteral dosage forms. Some formulation and processing aspects are common to all dosage forms and these are discussed in the basic principles chapters or the reader is referred to the appropriate chapter or case study. The dosage form chapters are followed by a case study section where nine case studies are presented that address: biopharmaceutical aspects of controlled release parenteral dosage forms; liposome formulation, design and product development; emulsion formulation, scale up and sterilization; microspheres formulation and processing as well as microsphere in vitro and in vivo release studies; and development and scale up of a nanocrystalline suspension. The final section of the book covers quality assurance and regulatory aspects as well as an FDA perspective.

Diane J. Burgess

Contents

Acknowledgments

I wish to express my sincere gratitude to all the contributors to this work. Their patience and perseverance throughout this process is greatly appreciated. I wish to acknowledge Dr. Paula Jo Stout who was involved in the initial stages of the writing of this book. I would also like to say a big thank you to Mr. Jean-Louis Raton who encouraged me to make it to the finish line and always with a big smile.

Contributors

Jill P. Adler-Moore Department of Biological Sciences, California State Polytechnic University–Pomona, Pomona, California, U.S.A.

Michael J. Akers Pharmaceutical Research and Development, Baxter Pharmaceutical Solutions LLC, Bloomington, Indiana, U.S.A.

Thomas Berger Pharmaceutical Research & Development, Hospira, Inc., Lake Forest, Illinois, U.S.A.

Gayle A. Brazeau Departments of Pharmacy Practice and Pharmaceutical Sciences, School of Pharmacy and Pharmaceutical Sciences, University at Buffalo, State University of New York, Buffalo, New York, U.S.A.

Diane J. Burgess Department of Pharmaceutical Sciences, School of Pharmacy, University of Connecticut, Storrs, Connecticut, U.S.A.

Mei-Ling Chen   Office of Pharmaceutical Science, Center for Drug Evaluation and Research, Food and Drug Administration, Rockville, Maryland, U.S.A.

N. Chidambaram   Senior Scientist, Research & Development, Banner Pharmacaps Inc., High Point, North Carolina, U.S.A.

Brian C. Clark   Pharmaceutical and Analytical R&D, AstraZeneca, Macclesfield, U.K.

Paul A. Dickinson   Pharmaceutical and Analytical R&D, AstraZeneca, Macclesfield, U.K.

Colm Farrell   GloboMax Division of ICON plc, Hanover, Maryland, U.S.A.

Anthony J. Hickey   School of Pharmacy, University of North Carolina, Chapel Hill, North Carolina, U.S.A.

Jim Jiao   Pharmaceutical Research and Development, Pfizer Global Research and Development, Pfizer Inc., Groton, Connecticut, U.S.A.

F. Kadir   Postacademic Education for Pharmacists, Bunnik, Utrecht, The Netherlands

Robert W. Lee   Elan Drug Delivery, Inc., King of Prussia, Pennsylvania, U.S.A.

Robert T. Lyons   Allergan, Inc., Irvine, California, U.S.A.

Frank J. Martin   ALZA Corporation, Mountain View, California, U.S.A.

Bernie Mikrut   Pharmaceutical Research & Development, Hospira, Inc., Lake Forest, Illinois, U.S.A.

Steven L. Nail   Lilly Research Labs, Lilly Corporate Center, Indianapolis, Indiana, U.S.A.

C. Oussoren Department of Pharmaceutics, Utrecht Institute for Pharmaceutical Sciences, Utrecht, The Netherlands

Siddhesh D. Patil Department of Pharmaceutical Sciences, School of Pharmacy, University of Connecticut, Storrs, Connecticut, U.S.A.

Richard T. Proffitt RichPro Associates, Lincoln, California, U.S.A.

Ian T. Pyrah Safety Assessment, AstraZeneca, Macclesfield, U.K.

Theresa Shepard GloboMax Division of ICON plc, Hanover, Maryland, U.S.A.

James P. Simpson Regulatory and Government Affairs, Zimmer, Inc., Warsaw, Indiana, U.S.A.

Mary P. Stickelmeyer Lilly Research Labs, Lilly Corporate Center, Indianapolis, Indiana, U.S.A.

H. Talsma Department of Pharmaceutics, Utrecht Institute for Pharmaceutical Sciences, Utrecht, The Netherlands

Mark A. Tracy Formulation Development, Alkermes, Inc., Cambridge, Massachusetts, U.S.A.

David Young GloboMax Division of ICON plc, Hanover, Maryland, U.S.A.

J. Zuidema Department of Pharmaceutics, Utrecht Institute for Pharmaceutical Sciences, Utrecht, The Netherlands

1

Physical Stability of Dispersed Systems

DIANE J. BURGESS

Department of Pharmaceutical Sciences,
School of Pharmacy, University of Connecticut,
Storrs, Connecticut, U.S.A.

1. INTRODUCTION AND THEORY

Injectable dispersed systems (emulsions, suspensions, liposomes, and microspheres) have unique properties, that are related to their size, interfacial area, and dispersion state. The physicochemical principles governing their behavior include thermodynamics, interfacial chemistry, and mass transport. The stability of these dosage forms is a major issue and is a function of thermodynamics, interfacial chemistry, and particle size. Drug release from such systems is governed by mass transport principles, interfacial chemistry, and size.

Principles of thermodynamics and interfacial chemistry as applied to dispersed systems are detailed in this chapter. Although the principles of particle size are discussed here, they are reviewed in greater detail in the chapter by Jiao and Burgess on characterization. Due to the unique factors associated with release of drugs from the different dispersed system dosage forms mass transport issues are addressed in the individual dosage form (suspensions, emulsions, liposomes, and microspheres) chapters.

Injectable dispersed systems are often colloidal in nature and therefore the principles of colloidal chemistry are also reviewed here. Dispersed systems for intravenous (i.v.) administration are almost always colloids, since their particle size is restricted to $\leq 1 \mu m$ to avoid problems associated with capillary blockage that can occur with larger particles. Dispersed systems administered via other parenteral routes can be much larger and their size is restricted by performance criteria (such as drug release rates, biopharmaceutical considerations, and potential for irritation) and needle size (larger needles are required for larger particles and can result in more painful injections).

2. COLLOID AND INTERFACIAL CHEMISTRY

Colloids are systems containing at least two components, in any state of matter, one dispersed in the other, in which the dispersed component consists of large molecules or small particles. These systems possess characteristic properties that are related mainly to the dimensions of the dispersed phase. The colloidal size range is approximately $1 nm$ to $1 \mu m$ and is set by the following lower and upper limits: The particles or molecules must be large relative to the molecular dimensions of the dispersion media so that the dispersion media can be assigned continuous properties; and they must be sufficiently small so that thermal forces dominate gravitational forces and they remain suspended. To qualify as a colloid, only one of the dimensions of the particles must be within this size range. For example, colloidal behavior is observed in fibers in which only two dimensions are in the colloidal size range. There are no sharp boundaries between colloidal and non-colloidal systems, especially at the upper size range. For example, an emulsion system may display colloidal properties, yet the average droplet size may be larger than 1 μm.

2.1. Classification of Colloids

Based on interaction between the dispersed and continuous phases, colloidal systems are classified into three groups: (i) *lyophilic* or solvent "loving" colloids, the dispersed phase is dissolved in the continuous phase; (ii) *lyophobic* or solvent "hating" colloids, the dispersed phase is insoluble in the continuous phase; and (iii) *association* colloids, the dispersed phase molecules are soluble in the continuous phase and spontaneously "self-assemble" or "associate" to form aggregates in the colloidal size range.

This book focuses mainly on lyophobic systems: emulsions, suspensions, and microspheres. Liposomes may be classified as association colloids, although larger liposomes can be outside the colloidal range. In some instances, liposomes are surface-treated and/or polymerized rendering them irreversible; they are then considered lyophobic colloids. Often, lyophobic nanospheres, microspheres, liposomes, and emulsions are surface-treated with hydrophilic polymers to improve their stability and/or to avoid/delay interaction with the reticular endothelial system following i.v. injection.

2.1.1. Lyophilic Colloids

The dispersed phase usually consists of soluble macromolecules, such as proteins and carbohydrates. These are true solutions and are best treated as a single phase system from a thermodynamic viewpoint. The dispersed phase has a significant contribution to the properties of the dispersion medium and introduces an extra degree of freedom to the system. Lyophilic colloidal solutions are thermodynamically stable and form spontaneously on adding the solute to the solvent. There is a reduction in the Gibbs free energy (ΔG) on dispersion of a lyophilic colloid. ΔG is related to the interfacial area (A), the interfacial tension (γ), and the entropy of the system (ΔS):

$$\Delta G = \gamma \Delta A - T\Delta S$$

where T is the absolute temperature.

The solute/solvent interaction is usually sufficient to break up the dispersed phase. In addition, there is an increase in the entropy of the solute on dispersion and this is generally greater than any decrease in solvent entropy. The interfacial tension (γ) is negligible if the solute has a high affinity for the solvent; thus, the $\gamma \Delta A$ term approximates zero. The shape of macromolecular colloids will vary depending on their affinity for the solvent. The shape of these therapeutics is important as it can affect their activity. Proteins will take on elongated configurations in solvents for which they have a high affinity and will tend to decrease their total area of contact with solvents for which they have little affinity. Since the molecular dimensions of protein molecules are large compared to those of the solvent, the protein effectively has an "interface" with the solvent. Proteins contain both hydrophilic and hydrophobic moieties and consequently shape changes can result in different moieties being exposed to the solvent. Following from this are physical instability and aggregation problems associated with protein solutions. The use of a solvent for which the protein has a high affinity can reduce these problems, as can the addition of surfactants that adsorb onto the protein and thus alter its "interface" with and affinity for the solvent. (Refer to Sec. 3 in this Chapter.)

2.1.2. Lyophobic Colloids

The dispersed phase consists of tiny particles that are distributed more or less uniformly throughout the solvent. The dispersed phase and the dispersion medium may consist of solids, liquids, or gases and are two-phase or multiphase systems with a distinct interfacial region. As a consequence of poor dispersed phase–dispersion media interactions, lyophobic colloids are thermodynamically unstable and tend to aggregate. The ΔG increases when a lyophobic material is dispersed and the greater the extent of dispersion, the greater the total surface area exposed, and, hence, the greater the increase in the free energy of the system. When a particle is broken down, work is required to separate the pieces against the forces of attraction between them (ΔW). The resultant increase in free energy is proportional to the area of new surface created (A):

$$\Delta G = \Delta W = 2\gamma A$$

Molecules that were originally bulk molecules become surface molecules and take on different configurations and energies. An increase in free energy arises from the difference between the intermolecular forces experienced by surface and bulk molecules. Lyophobic colloids are aggregatively unstable and can remain dispersed in a medium only if the surface is treated to cause a strong repulsion between the particles. Such treated colloids are thermodynamically unstable yet are kinetically stable since aggregation can be prevented for long periods.

Emulsions

Emulsion systems can be considered a subcategory of lyophobic colloids. Their preparation requires an energy input, such as ultrasonication, homogenization, or high-speed stirring. The droplets formed are spherical, provided that the interfacial tension is positive and sufficiently large. Spontaneous emulsification may occur if a surfactant or surfactant system is present at a sufficient concentration to lower the interfacial tension almost to zero. Spontaneously forming emulsions usually have very small particle size ($<100\,\mathrm{nm}$) and are referred to as microemulsions.

2.1.3. Association Colloids

Association colloids are aggregates or "associations" of amphipathic surfactant molecules. Surfactants are soluble in the solvent, and their molecular dimensions are below the

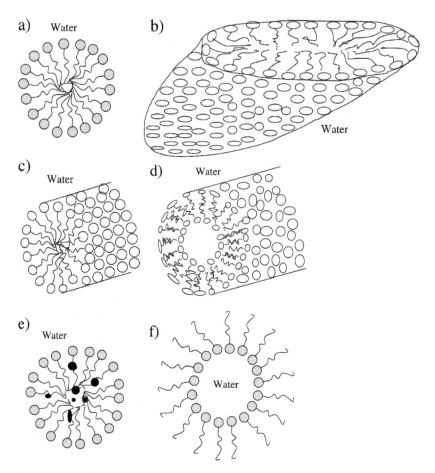

Figure 1 Different types of association colloids (micelles): a) spherical micelle; b) disc-shaped micelle; c) cylindrical micelle; d) micro-
tubular micelle; e) lipid soluble drug in a swollen micelle; f) inverted
micelle.

colloidal size range. However, when present in solution at
concentrations above the critical micelle concentration, these
molecules tend to form association colloids (also known as
micelles) (Fig. 1). Surfactants have a region that has a high
affinity for the medium (lyophilic), and a region that has a
low affinity for the medium (lyophobic). These molecules
adsorb at interfaces to reduce the energy between their
lyophobic region(s) and the medium. On micellization, the

surfactants associate such that their lyophobic regions are in the interior, their lyophilic regions are at the surface, and the solvent is excluded from the hydrophobic core. Not all surfactants form micelles since a subtle balance between the lyophilic and lyophobic portions of the surfactant molecule is required. A charged, zwitterionic, or bulky oxygen-containing hydrophilic group is required to form micelles in aqueous media (1). These moieties are able to undergo significant hydrogen bonding and dipole interactions with water to stabilize the micelles. Micelle formation in strongly hydrogen-bonded solvents is very similar to that in water (1). Micellization occurs spontaneously, depending on the lyophilic–lyophobic balance of the surfactant, surfactant concentration, and temperature.

At room temperature, micellization of surfactants in aqueous media is driven by entropy. The hydrophobic part of the surfactants induces a degree of structuring of water, which disturbs the hydrogen bond pattern and causes a significant decrease in the entropy of the water. This is known as the "hydrophobic" effect. The effect of temperature and pressure on micellization is dependent on surfactant properties (1,2). Ionic surfactants generally exhibit a "Krafft" point. In these systems, micelles form only at temperatures above a certain critical temperature, the Krafft temperature. This is a consequence of a marked increase in surfactant solubility at this temperature, whereas, non-ionic surfactants tend to aggregate and phase separate above a certain temperature (the cloud point) (3).

Micellar shape (Fig. 1) is concentration dependent. At low concentrations, micelles are usually spherical with well-defined aggregation numbers, and are monodisperse. At high concentrations, micellar shape becomes distorted and cylindrical rods or flattened discs form (4). At very high concentrations, the surfactants arrange as liquid crystals (5–7). Surfactants can also form two-dimensional membranes or bilayers separating two aqueous regions under specific conditions. If this bilayer is continuous and encloses an aqueous region, then vesicles result, which are known as liposomes. Micelles are used pharmaceutically to solubilize insoluble drug substances. The drug is solubilized within the micellar core (Fig. 1).

Micellar solubilization allows the preparation of water-insoluble drugs within aqueous vehicles (8). This is advantageous, particularly for i.v. delivery of water-insoluble drugs. Entrapment within a micellar system may increase the stability of poorly stable drug substances and can enhance drug bioavailability (9). Liposomes can have hydrophobic and hydrophilic regions and therefore can be used to solubilize both water soluble and water insoluble drugs.

2.2. Properties of Colloids

Characteristic properties of colloids include: particle size and shape, scattering of radiation, and kinetic properties.

2.2.1. Particle Size and Shape

The colloidal size range is approximately 1 nm to 1 μm and most colloidal systems are heterodisperse. Solid dispersions usually consist of particles of very irregular shape. Particles produced by dispersion methods have shapes that depend partly on the natural cleavage planes of the crystals and partly on any points of weakness (imperfections) within the crystals. The shape of solid dispersions produced by condensation methods depends on the rate of growth of the different crystal faces. Treatments of particle shape are given by Beddow (10), Allen (11), and Shutton (12). Both liquid and solid dispersions often have wide size distributions. (Refer to the chapters by Chidambaram and Burgess and by Nail and Stickelmeyer in this book for methods of preparation of liquid and solid dispersions, respectively. Refer to the chapter by Jiao and Burgess for details on particle size analysis.)

2.2.2. Scattering of Radiation

The scattering of a narrow beam of light by a colloidal system to form a visible cone of scattered light is known as the Faraday–Tyndall effect. The reader is referred to the text of Heimenz (13) for a full account of light scattering by colloidal systems. Electromagnetic radiation induces oscillating dipoles in the material and these act as secondary sources of emission of scattered radiation. The scattered light has the same wavelength (λ) as the incident light. The intensity of the scattered light depends on the intensity of the original light, the polarizability of the material, the size and shape of the material, and the angle of observation. The scattering intensity increases with increase in particle radius, reaching a maximum, and then decreasing. This maximum in scattering intensity coincides with the colloidal size range. Small particles ($\lambda/20 \geq$ particle radius) act as point sources of scattered light. In larger particles, different regions of the same particle may behave as scattering centers and these multiple scattering centers interfere with one another, either constructively or destructively. As particle size increases, the number of scattering centers increases, and the resultant destructive interference causes a reduction in the intensity of the scattered light.

2.2.3 Kinetic Properties

The kinetic properties of colloids are characterized by slow diffusion and usually negligible sedimentation under gravity.

2.3. Thermal Motion

Colloidal particles display a zigzag-type movement as a result of random collisions with the molecules of the suspending medium, other particles, and the walls of the containing vessel. The distance moved by a particle in a given period of time is related to the kinetic energy of the particle and the viscous friction of the medium. The thermal motion of colloidal particles was first observed by the English botanist Robert Brown and is referred to as Brownian motion (14). As a result of Brownian motion, colloidal particles diffuse from regions of high concentration to regions of lower concentration until the concentration is uniform throughout. Gravitational forces, which cause particles to sediment, and Brownian motion oppose one another. Brownian forces are stronger than gravitational forces for particles in the colloidal size range and, therefore, colloids tend to remain suspended.

3. THERMODYNAMICS OF DISPERSED SYSTEMS

The fundamental thermodynamic property influencing the formation and breakdown of dispersed systems is interfacial tension (refer above).

3.1. Colloids

Lyophilic and association colloids are thermodynamically stable, since these are either large molecules or associations of molecules that are in solution. However, lyophobic colloids are thermodynamically unstable. The instability of lyophobic colloids has a major influence on their formulation and performance and they must be rendered kinetically stable for a period that constitutes an acceptable shelf-life for the product. The Derjaguin–Landau and Verwey–Overbeek (DLVO) theory describes the interaction between particles of a lyophobic colloid. This theory is reviewed in the texts of Hunter (15) and Heimenz (16) and is based on the assumption that the van der Waals interactions (attractive forces) and the electrostatic interactions (repulsive forces) can be treated separately and then combined to obtain the overall effect of both of these forces on the particles. Although this is an oversimplification, it is the easiest way to understand the complex interactions that are occurring between colloidal particles.

3.1.1. Attractive Forces

The van der Waals attractive forces between two particles are considered to result from dipole–dipole interactions and are proportional to $1/H^6$, where H is the separation distance between the particles (17). Consequently, the attractive forces operate over very short distances and the value of these attractive forces is greater the smaller the interparticle distance. The Hamaker summation method is used to calculate the total van der Waals interaction energy (V_A), assuming that the interactions between individual molecules in two colloidal particles can be added together to obtain the total interaction and that these interactions are not affected by the presence of all the other molecules. The dispersion theory of Lifshitz (18) overcomes the assumptions made in the London theory and is based on the idea that the attractive interaction between particles is propagated as an electromagnetic wave over distances that are large compared with atomic dimensions. In the Lifshitz theory, the colloidal particles are considered to be made up of many local oscillating dipoles that continuously radiate energy. These dipoles are also continuously absorbing energy from the electromagnetic fields generated by all the surrounding particles. The reader is referred to the text by Hunter (15) for a full treatment of this theory.

3.1.2. Repulsive Forces

The repulsive forces between lyophobic colloids are electrostatic in nature and are a consequence of the charge carried by the particles. All lyophobic colloidal particles acquire a surface charge when dispersed in an electrolyte solution: by adsorption of ions from the solution; by ionization of ionizable groups; or by selective ion dissolution from the particle surface. These charged particles attract ions of the opposite charge (counter ions), some of which become tightly bound to the particle surface. The counter ions are also pulled away from the particle surfaces by the bulk solution as a result of thermal motion. A diffuse layer of ions builds up as a consequence of these two opposing effects so that at a distance from the particle surface, it appears to be electrically neutral (Fig. 2). The tightly bound and the diffuse layers are known collectively as the electrical double layer. The electrical double layer is responsible for the repulsive interaction between two colloidal particles. When two colloidal particles come in close proximity of one another, their electrical double layers overlap. This causes a free energy change and an increase in osmotic pressure as a result of the accumulation of ions between the particles. This repulsive interaction (V_R) decreases exponentially with increase in the distance between the particles.

The DLVO theory combines the attractive and repulsive interactions between lyophobic colloids to explain the

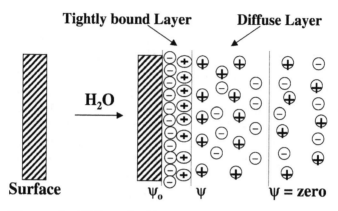

Figure 2 Diffuse double layer of ions on the particle surface.

aggregative instability of two particles at any given separation distance. The two opposing forces are summed, as shown diagrammatically in Fig. 3, where V_T is the summation of V_A and V_R ($V_T = V_A + V_R$). The van der Waals attractive forces dominate at both large and small separation distances. At

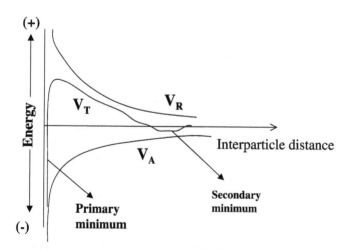

Figure 3 Potential energy diagram for two spherical particles. V_A is vander Waals attraction energy; V_R is repulsive energy; V_T is total potential energy obtained by summation of V_A and V_R ($V_T = V_A + V_R$).

very small distances, van der Waals attraction increases markedly, resulting in a deep attractive well (known as the primary minimum). However, this well is not infinitely deep due to the very steep short range repulsion between the atoms on each surface (15). The DLVO theory can be applied in a broad sense to most lyophobic colloidal systems. This theory is also applied to particles and droplets outside the colloidal size range and is discussed below for liquid and solid dispersion stability in the thermodynamics section. (The reader is referred to Chapter 6 in this book for a treatment of the DLVO theory with respect to coarse suspension.

3.1.3. Stabilization of Lyophobic Colloids

Lyophobic colloids can be kinetically stabilized by electrostatic and polymeric methods. Electrostatic stabilization results from charge–charge repulsion, as discussed above. Polymeric stabilization is achieved by steric stabilization upon adsorption of macromolecules (lyophilic colloids) at the surface of a lyophobic colloid (19). The lyophilic colloids must extend from the surface of the particles over a distance comparable to, or greater than, the distance over which van der Waals attraction is effective. Therefore, a molecular weight of at least a few kDa is required and the lyophilic colloids must be present at a sufficiently high concentration so that they saturate the surfaces of the lyophobic particles. The colloidal particles will then repel one another as a result of volume restriction and osmotic pressure effects, as illustrated in Fig. 4. Lifshitz (18) gives a thermodynamic account of polymeric stabilization. A polyelectrolyte may stabilize a lyophobic colloid by a combination of steric and electrostatic stabilization (electro-steric stabilization).

At low polymer concentrations (and hence low particle surface coverage), lyophilic colloids tend to have multiple points of contact on the lyophobic particle surface and lie along the surface rather than extend from it. Segments of individual molecules may adsorb onto more than one lyophobic colloid and hence bridging flocculation may occur (Fig. 5), causing particle aggregation rather than repulsion.

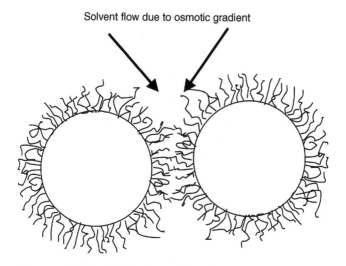

Figure 4   Stabilization of lyophobic colloidal particles by volume restriction and osmotic pressure effects.

3.2.   Thermodynamics of Dispersed Systems Dosage Forms

For the purpose of thermodynamics, dispersed systems can be considered either as liquid dispersions (emulsions) or as solid dispersions (suspensions, liposomes, nano- and microspheres).

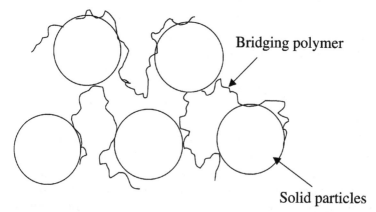

Figure 5   Bridging flocculation by adsorption of polymer chains on lyophobic colloid particles.

The term suspension is used in this section to represent suspensions, liposomes, nano- and microspheres.

3.2.1. Emulsions

When two immiscible liquids or liquids with very limited mutual solubility are agitated together, they fail to dissolve. Immiscibility arises since the cohesive forces between the molecules of the individual liquids are greater than the adhesive forces between different liquids. Consequently, the force experienced by molecules at the interface between the immiscible liquids is imbalanced compared to the force experienced by molecules in the bulk (Fig. 6). This imbalance in the force is manifested as interfacial tension (γ). When the dispersed phase is broken into droplets, the interfacial area (A) of this liquid becomes large compared to that of the bulk liquid and the surface free energy is increased by an amount $\gamma\Delta A$. According to the Gibbs equation (20), increase in interfacial free energy causes thermodynamic instability:

$$\Delta G = \gamma\Delta A - T\Delta S$$

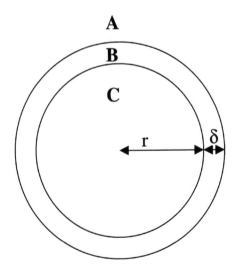

Figure 6 Pressure gradient across a droplet: A, continuous phase; B, thin film with thickness δ; C, dispersed phase; r, radius.

where G is the interfacial free energy, T is the absolute temperature, and S is the configurational entropy of the system. S approximates to the configurational entropy of the dispersed phase since the configurational entropy of the continuous phase approximates to zero.

In most cases G_{form} (formation) is positive. This implies that G_{break} (breakdown) is negative, i.e., emulsions are thermodynamically unstable or metastable. Therefore, there is a tendency to decrease the interfacial area and hence reduce the free energy of the system. This can be achieved by droplet size growth and phase separation. Droplet size growth is attributed to both coalescence and Ostwald ripening. Coalescence is the process by which two droplets converge to form one large droplet. Ostwald ripening is dependent on the dispersed phase molecules having limited solubility in the dispersion media. As a consequence of differences in surface to volume ratios, the net effect is that dispersed phase molecules diffuse from small droplets into the continuous phase and merge with large droplets. Thereby small droplets become smaller and large droplets become larger.

Large amounts of energy are required in emulsion formulation to create the interface since the energy of adhesion between the molecules of the dispersed phase must be broken. Additionally, energy is required due to the high Laplace pressure gradient across the droplets (21) (Fig. 6). The pressure (P) is always greater on the concave side of a curved interface and is inversely proportional to the radius of the spherical droplet

$$\Delta P = \gamma(1/r_1 + 1/r_2)$$

where the curvature of the droplet is defined by the inner and outer radii $r_1 + r_2$.

3.2.2. Suspensions

Suspensions are dispersions of insoluble solid particles in a liquid medium. As with liquid dispersions, work must be done to reduce solids into small particles and disperse them in the continuous medium. The bonds between the molecules in the solid particles must be broken exposing a new interfacial area, with associated interfacial tension since the interfacial adhesive forces are much less than the cohesive forces. The smaller the particles, the larger the interfacial area and free energy, and hence the more thermodynamically unstable the system. Both liquid and solid dispersed systems tend to approach a stable state by reduction in interfacial tension and/or area. Suspensions aggregate rather than coalesce. Like liquid dispersions, they are subject to Ostwald ripening.

3.2.3. Mechanisms of Emulsion and Suspension Destabilization

Instability in emulsion systems is manifested in the following ways: creaming, flocculation, coalescence, Ostwald ripening, phase inversion, and cracking. Instability in suspension systems is manifested as: sedimentation, flocculation, aggregation, Ostwald ripening, and caking.

Creaming /Sedimentation

Creaming/sedimentation occurs as a result of an external force such as gravitation, centrifugation, and electricity. There is no change in droplet/particle size or size distribution. The dispersed phase droplets/particles either rise to the top of the system (cream), if their density is lower than that of the continuous phase or they sediment if their density is higher. Droplet/particle size and size distribution are important as well as the density difference between the two phases and the viscosity of the continuous phase. These parameters are related in Stokes' law:

$$\nu = 2\Delta\rho g a^2 / 9\eta_o$$

where ν is the rate of creaming/sedimentation, $\Delta\rho$ is the density difference between the two phases, a is the hydrodynamic radius of the droplets or particles, and η_o is the viscosity of the continuous phase.

Stokes' law applies to non-interacting dilute dispersions of uniform radius and must be modified for other systems. The structure of the sedimented or creamed layer (that is, the distances between the droplets/particles) is dependent on a balance between the external (gravitational, centrifugal, or electric) field and the interdroplet/particle forces (van der Waals, electrostatic, and/or steric forces associated with an interfacial polymer/surfactant layer). Packing is further complicated by the following factors: polydispersity, deformation (emulsions), flocculation, and coalescence (emulsions)/caking (suspensions). In emulsion systems, the more rigid the interfacial film (refer below to section on emulsion stabilization) the greater the resistance to deformation from spherical geometry. Deformation is also dependent on droplet size, the greater the droplet size the greater the tendency for deformation to occur. Often droplets distort into polyhedral cells resembling foam structures with networks of more or less planar thin films of one liquid separating cells of the other liquid.

A creamed/sedimented system redisperses on shaking, however since the droplets/particles are in close proximity to one another, there is the possibility of flocculation and/or coalescence of the creamed droplets (emulsions) and flocculation and/or caking (suspensions). The stability of the creamed layer to coalescence depends on the stability of the thin film between the droplets to rupture (refer below to the section on the Marangoni effect). Flocculated systems cream/sediment faster as a result of the large size of the floccs. These systems are also easier to redisperse since flocculation occurs at the secondary minimum and therefore involves weaker forces than aggregation at the primary minimum (Fig. 3).

Flocculation

Flocculation is the build up of aggregates (floccs), resulting from weak attractive forces between droplets. The identities of the individual droplets/particles are retained and on shaking the floccs redisperse and there is no change in droplet/particle size or size distribution. The kinetics of flocculation are controlled by diffusion of the droplets/ particles. An equilibrium exists between flocculated droplets/ particles and singlets. In a strongly flocculated system, all the droplets/particles are flocculated. A critical dispersed phase volume exists below which particles are thermodynamically stable with respect to flocculation.

In the flocculation process, emulsion droplets come into contact and are separated by a thin liquid film. The interfacial tensions of each of the two interfaces forming this film are lower than that of an isolated interface:

$$\gamma(h) = \gamma(\infty) + G_i(h)/2$$

where h is the thickness of the film, and $G_i(h)$ is the interfacial free energy of the two droplets across the film.

In the case where only van der Waals attractive forces are operating, then $G_i(h)$ will be negative and $\gamma(h)$ will always be less than $\gamma(\infty)$. Therefore, the film separating the two droplets will spontaneously thin resulting in rupture and coalescence. In the case where repulsive forces are present then the free energy may not fall continuously with decrease in distance between the droplets. According to the DLVO theory sufficient kinetic energy is required to surmount the free energy barrier to coalescence, i.e., droplets must collide with sufficient force (Fig. 3). Flocculation occurs at relatively large inter-droplet distances, where van der Waals attractive forces are larger than the repulsive forces. However, the van der Waals forces are not strong enough to cause coalescence of emulsion droplets and irreversible aggregation of suspensions. Emulsions and suspensions are often formulated to flocculate at the secondary minimum, since this is a relatively stable state. There is a free energy barrier to flocculation and if this is large enough flocculation does not occur.

Coalescence/Caking

Coalescence occurs in flocculated and creamed emulsion systems. Caking occurs in non-flocculated suspension systems that have sedimented. These phenomena result in a change in the particle size and size distribution and involve the elimination of the thin liquid film of continuous phase which separates two particles. The particles then merge forming one large particle. Separation of the dispersed phase is known as cracking in the case of emulsions and caking in the case of suspensions. The limiting state is complete separation into two phases. In the case of emulsion droplets the film separating the droplets ruptures at a specific spot in the lamella due to thinning. Coalescence behavior is dependent on the nature of forces acting across the film and on the kinetic aspects associated with local fluctuations in film thickness (thermal or mechanical). Refer below to a description of the Marangoni effect which opposes film rupture. The kinetics of coalescence are controlled by molecular diffusion of liquid out of the thin film between the droplets. The rate of coalescence is also proportional to the number of contacts between droplets in an aggregate.

The pressure of the packed particles upon one another is sufficient to drive sedimented particles close enough together so that van der Waals attractive forces dominate and the sedimented particles lose their individual identity and can no longer be redispersed on shaking. Sedimented flocculated particles usually do not cake as the particles are aggregated at the secondary minimum (DLVO theory, Fig. 3), where the distance between the particles is sufficient to prevent caking. In order for caking to occur, the particles must approach one another at very short distances—the primary minimum. Flocculation is therefore utilized as a means of stabilizing suspensions to caking.

Ostwald Ripening

Ostwald ripening occurs in dispersed systems where the two phases have a finite mutual solubility (most systems) and the dispersed phase particles are polydisperse. This process results in a change in droplet size and size distribution. Oswald ripening is associated with the difference in the chemical potential between particles of different size. Dispersed phase molecules diffuse from the surface of smaller particles into the continuous medium and join with larger particles. The particles tend to one size and ultimately the formation of one large particle. The effect on droplet solubilities is described by the Kelvin equation:

$$\ln C_\alpha / C_\beta = 2\gamma V / RT (1/\alpha - 1/\beta)$$

where C_α and C_β are the saturation solubilities of the dispersed phase at the surface of the dispersed phase droplets of radii α and β, respectively, V is the molar volume of the dispersed phase, γ is the interfacial tension, R is the ideal gas constant, and T is the absolute temperature. According to this equation, small droplets have a greater solubility than larger droplets due to the effect of curvature on the surface free energy. As a consequence, small droplets become smaller and larger droplets become larger. A discussion of Ostwald ripening and emulsion stability is presented by Jiao and Burgess (22).

Phase Inversion

Phase inversion is the process whereby the dispersed and continuous phases of emulsion systems invert suddenly. A W/O emulsion inverts to an O/W emulsion or vice versa and results in a change in droplet size and size distribution. Phase inversion usually occurs on change in temperature, change in phase ratios, or the addition of a new component and normally follows droplet flocculation and coalescence. Difficulty in packing of the dispersed phase droplets drives phase inversion above a certain volume fraction. An example of a component that can result in phase inversion is the addition of a second emulsifier with a different hydrophilic–lipophilic balance (HLB) value. The second emulsifier can change the overall HLB of the system to a value that favors the opposite type of emulsion.

The various mechanisms involved in emulsion destabilization are inter-related. For example, coalescence usually follows from flocculation and creaming.

3.2.4. Forces Involved in Emulsion
Destabilization

Long-range external field forces operate in emulsion and suspension systems as well as interparticle forces and intermolecular forces. Long-range forces include gravity, centrifugal or applied electrostatic forces and these affect creaming and flocculation. Examples of interparticle field forces are the electrical double layer, dispersion forces, hydrodynamic forces, and diffusional (entropic) forces associated with droplet concentration gradients within the emulsion system. These forces determine whether creaming and sedimentation lead to flocculation and in turn whether flocculation leads to coalescence of droplets and coagulation of particles. Intermolecular forces include diffusion and molecular entropies and are concerned with coalescence, Oswald ripening, and phase inversion.

3.2.5. Factors Affecting Emulsion and Suspension
Breakdown and Stabilization

Factors affecting stability include particle size, size distribution, phase volume ratio, interfacial tension, presence of charge barriers, presence of steric barriers, and the presence of mechanical barriers (film strength).

The smaller the initial droplet size the more stable the emulsion. The narrower the droplet size distribution the more stable the emulsion, since this minimizes contact between creamed and flocculated droplets and therefore reduces the potential for coalescence. The smaller the phase volume ratio of the dispersed phase the more stable the emulsion or suspension as this results in smaller droplet size and reduces the potential for collision. The lower the interfacial tension the more stable the emulsion, as explained by the Gibbs equation. Low interfacial tension results in the formation of droplets with small mean diameters and narrow size distributions. Interfacial charge barriers give rise to electrostatic repulsion between droplets and hence stabilize dispersed systems. Steric barriers result in stabilization by volume restriction and osmotic effects. The stronger the mechanical barrier the greater the resistance to coalescence following flocculation, creaming, and collision.

The Marangoni Effect

A film of continuous phase is present around the dispersed phase droplets and this is very important with respect to emulsion stability (20). If no surfactant is present, this film is very unstable and drains away rapidly under the influence of gravitational forces. In the presence of surfactant, an interfacial tension gradient is created as the film drains. Consequently, surfactant molecules diffuse to the interface of the draining film in an attempt to restore the interfacial tension and in doing so they drag liquid with them restoring the film. This liquid motion is caused by tangential stress and is known as the Marangoni effect. The Marangoni effect helps to stabilize emulsions by restoring the interfacial film at a thinned area on a droplet surface and by forcing liquid into the gaps between approaching droplets and thereby driving them apart.

The difference between the local interfacial tension at a point on the surface of a droplet and the equilibrium value is the surface dilational modulus:

$$E = d\gamma/d(\ln A)$$

where E depends on the nature and amount of surfactant, the rate of transport of surfactant to the interface, and the rate of expansion or contraction of the film. In the case of a thin film, transport toward the interface is very rapid since diffusion over a short distance is rapid and consequently the nature and amount of surfactant are the determining factors. The Gibbs elasticity of a thin film (E_f) equals twice the surface dilational modulus and is given by

$$E_f = -\frac{2dy/d(\ln\Gamma)}{1 + (1/2)h\,dm/d\Gamma}$$

where Γ is the surface excess concentration of surfactant, h is the film thickness, and m is surfactant molar concentration.

The thinner the film the higher the E_f value and consequently the greater the resistance to drainage. Therefore, during stretching the thinnest part of the film will have the greatest resistance to stretching as long as sufficient surfactant is present, since surfactant will diffuse rapidly to that point to restore the interfacial tension. On the contrary, if surfactant concentration is low, then E_f will be lowest for the thinnest part of the film and the film will be unstable. A thin film being stretched will break as soon as it exceeds a certain critical value (γ_{cr}). γ_{cr} will be reached more rapidly for systems with low surfactant concentration that are being stretched rapidly.

This has implications in emulsification as well as in emulsion stability, as it should be easier to form emulsions using slow stirring speeds in the initial part of the processing when droplet formation is rapid.

Strong Marangoni effects can result in distortion of the interface and shredding of droplets when interfacial tension is low. Negative interfacial tension values result in droplet shredding and the spontaneous formation of microemulsions.

3.2.6. Emulsion and Suspension Stabilizing Agents

Emulsifying and suspending agents work by reducing interfacial tension and/or forming an interfacial film barrier to droplet coalescence/aggregation of solids and to Ostwald ripening. Stabilizing agents are usually classified into four groups: inorganic electrolytes, surfactants, macromolecules, and solid particles (20).

Inorganic electrolytes, such as potassium thiocyanate stabilize dispersed systems by imparting a charge at the interface. Electrostatic repulsion reduces the rate of coalescence/aggregation, according to the DLVO theory, refer above.

Surfactants are molecules that contain both hydrophilic and lipophilic groups and consequently have an affinity for interfaces. Surfactants are classified according to the HLB system. This is an arbitrary scale of values that serves to estimate the relative hydrophilicity/lipophilicity of surfactants. According to this system, surfactants with HLB values of approximately 3–8 are effective in stabilizing W/O emulsions and surfactants with HLB values of approximately 8–16 are effective in stabilizing O/W emulsions. In addition to the HLB value, the phase volumes and surfactant concentrations must be taken into account in determination of the type of emulsion that will form in a given system.

Surfactants are added to emulsion and suspension systems to reduce interfacial tension and consequently increase stability. In emulsion systems, surfactants also reduce the initial droplet size and size distribution; draw a liquid film between droplets in areas where film thinning may have occurred (according to the Marangoni effect); impart steric stabilization and in the case of charged surfactants give rise to charge stabilization. In suspension systems, surfactants reduce interfacial tension, decrease the contact angle between the particles and the liquid continuous phase; impart steric stabilization and in the case of charged surfactants result in charge stabilization.

Some surfactant systems form liquid crystalline phases at the emulsion droplet interface (20–22). These phases increase emulsion stability as a consequence of their high interfacial viscoelasticity and the decreased van der Waals force of attraction between droplets since the van der Waals energy distance is increased. At high concentrations, surfactants increase viscosity forming a semisolid emulsion with enhanced stability since droplet movement and molecular diffusion are limited (20). This is known as self-bodying action.

Specific intermolecular interactions can occur between two interfacially adsorbed surfactants, such as cetyl sulfate and cholesterol. This results in enhanced stability by reducing interfacial tension, increasing interfacial elasticity (film strength), which prevents lateral displacement of molecules in thin films and increased stability.

The phase inversion temperature (PIT) is the temperature at which O/W emulsions invert to W/O or vice versa. Emulsion PIT is influenced by surfactant HLB and emulsion droplets are less stable to coalescence at temperatures close to the PIT. Emulsion stability is enhanced when the storage temperature is significantly below the PIT (20–65°C).

Macromolecules such as proteins and certain synthetic polymers are similar to surfactants in that they are interfacially active due to a multiplicity of hydrophilic and lipophilic groups. Interfacial adsorption of macromolecules results in a gain in entropy due to a loss of structured water from the macromolecular surface. Macromolecules reduce interfacial tension, form mechanical interfacial barriers and in some cases charge barriers to droplet coalescence. The mechanism of steric stabilization is due to extension of the macromolecules into the continuous phase. This keeps the emulsion droplets and suspended particles apart by volume restriction and osmotic effects (Fig. 4). Consequently, the droplets and particles are far enough apart that van der Waals attractive forces are not effective. Volume restriction explains the free energy of repulsion by a reduction in the configurational entropy of the macromolecular chains as two droplets approach close to one another. The osmotic effect is due to an increase in the local osmotic pressure and free energy. The osmotic pressure drives continuous phase into the region between the droplets or particles and hence keeps them apart. The polymer must fully cover the dispersed phase surface as bare patches will lead to coagulation. Thicker interfacial films enhance the steric stabilization effect. The polymer must be firmly anchored to the interface. Copolymers with hydrophilic and lipophilic moieties are best with respect to anchoring. The continuous phase should provide a good solvent environment for the polymer to optimize the stabilization effects.

Finely divided solids can adsorb at the emulsion droplet interfaces and form a mechanical barrier to droplet coalescence (23). The type of emulsion formed depends on which phase wets the particles to the greater extent. If they are wetted more by water then they will form a W/O emulsion and if they are wetted more by oil they will form an O/W emulsion.

3.2.7. Predicting Emulsion Stability

Accelerated testing of dispersed systems is used to predict emulsion stability. However, the validity of accelerated tests is questionable as these tests often involve conditions that the product would not normally be subjected to and consequently the stability may be totally different. The level of stress that the system is subjected to must be moderate, sufficient to increase the rate(s) of the natural destabilization mechanism(s) but not so high that the destabilization mechanism(s) are altered (24). Abuse of a dispersed system is useless in predicting stability. Accelerated testing methods involve elevated temperature, freeze/thaw cycling, ultracentrifugation, and vibration.

Thermatropic stress is commonly used in accelerated testing including elevated temperature, low temperature, and freeze/thaw cycling.

Elevated temperature increases the Brownian motion of the droplets and hence increases the probability and force of droplet collision and consequently flocculation and coalescence are enhanced. Molecular diffusion increases with temperature and the mutual solubility of the two phases is also likely to increase, increasing Ostwald ripening. Elevated temperature will decrease the viscosity of the continuous phase of most systems increasing creaming/sedimentation, as predicted by Stokes' law. Therefore, the physical stability of dispersed systems is expected to be lower at elevated temperatures.

In addition to enhancing the rate of emulsion destabilization by the normal mechanisms as outlined above, elevated temperatures can result in several destabilizing reactions (24). Emulsion stability is influenced by the cloud point temperature and the PIT. These two phenomena are related to the solubility of non-ionic surfactants. The cloud point is the temperature above which a particular surfactant self associates as a result of loss of water of hydration. The surfactant separates as a precipitate or gel and is no longer available to stabilize the emulsion. The PIT is the temperature at which an emulsion inverts from an O/W to a W/O emulsion or vice versa, depending on the relative solubility of the surfactant in the two phases (24). Emulsions stabilized by hydrated gel (liquid crystalline) phases exhibit a transition temperature above which the liquid crystalline phase melts destabilizing the emulsion (24). It is pointless to investigate emulsion stability above these temperatures as the emulsions break rapidly and this is unrelated to emulsion stability under normal conditions. Heat may also affect the chemical stability of the system resulting in increased rates of hydrolysis of the surfactant or any incorporated drug and thus complicating the assessment of emulsion physical stability.

Considering the above thermal effects high temperature testing of emulsions must be conducted with care. In general, emulsion breakdown at temperatures of 75°C and above cannot be used as a predictor of stability under normal conditions. Elevated temperature studies should be conducted at moderate temperatures. Rapid emulsion breakdown at 45–50°C or below indicates that the emulsion is highly unstable and should be reformulated.

Low Temperature Studies

There is usually little information to be gained from studies at low temperature, as low temperature increases viscosity and retards the various instability mechanisms. It is therefore rare for emulsions to fail under refrigeration. Exceptions are formulations including certain lipids that form unusual crystalline metamorphic phases at low temperatures which result in emulsion destabilization.

Freeze/Thaw Cycling

The rate of freezing and thawing must be carefully controlled for the test to be useful (24). The freezing rate must be slow so that large ice crystals can form. These crystals may alter emulsion droplet shape and/or penetrate into them destabilizing their interfacial film. Thawing should be performed slowly without agitation. Freeze/thaw cycling should be repeated several times and after the final thawing the temperature should be increased to 35–45°C and the emulsion compared with a similar sample that had not been frozen. Freeze/thaw testing is important as emulsions may encounter freezing temperatures during transportation and storage. Therefore, this is a good predictive test for emulsions that are likely to experience these conditions.

Other accelerated testing methods for emulsions include centrifugation and vibrational stress.

Centrifugation

This method is applicable primarily to fluid emulsions in which sedimentation or creaming is the primary cause of instability. Centrifugation testing under severe conditions (ultracentrifugation) is pointless and has little or no predictive value. The forces encountered in ultracentrifugation are such that steric and electrostatic repulsive barriers around emulsion droplets are usually destroyed (24). This is unrepresentative of the normal conditions experienced by emulsion droplets.

Slow centrifugation 500–5000 rpm can be a useful test as the rate of creaming can be increased without destroying the steric and repulsive barriers (24). This test is particularly useful for emulsions that are normally slow rising (semisolids). The creaming rates of these systems can be speeded up to allow a useful evaluation in the space of a few hours.

Vibrational Testing

As applies to the accelerated stability tests discussed above, extreme stress should be avoided as this will lead to rapid breakdown that is not predictive of the stability of the system (24). Shaking should be conducted at 10–100 cycles/min. This level of vibration simulates the stress of shipping.

This type of test is particularly useful for viscous semisolid emulsions, since droplet interaction is necessary for the various instability mechanisms and these are very limited in viscous systems. Agitation allows the droplets to approach one another with sufficient kinetic energy to effect coalescence and speed up the natural destabilization mechanisms. Shaking may lead to heavy build up of dispersed phase droplets on the container walls and ultimate phase separation.

Interfacial properties (charge, rheology, and tension) can also be used as predictors of emulsion stability (25–34). The interfacial charge on the emulsion droplets gives a direct measurement of the electrostatic barrier to coalescence. Interfacial rheology measures the elasticity of the interfacial film and hence the mechanical barrier to droplet coalescence. Interfacial tension is related to emulsion stability through the Gibbs equation, the lower the interfacial tension the more condensed the interfacial film and the lower the possibility of film drainage and rupture. The relative importance of these interfacial properties has been investigated by Burgess et al. (25–34). They have shown that depending on the predominant mechanism of emulsion destabilization, one or more of these factors can be used to predict emulsion stability. It is therefore important to investigate all of these parameters.

3.2.8. Methods of Emulsion Stability Testing

There is no single universally applicable criterion for quantitative characterization of the rate of de-emulsification (24). It is not possible to determine a specific reaction rate constant for this process as many different reactions often occur simultaneously. It is a common practice to employ several stability tests measuring different parameters to assess the overall stability.

Visual

Visual observations are used to assess sedimentation and creaming (24). The separated layers can be analyzed for content.

Rheological Tests

These are used to determine the viscoelastic nature of a semisolid emulsion and hence the susceptibility to destabilization by creaming, flocculation, and coalescence. Care must be taken as the shearing involved may cause particle size changes and/or destroy the rheological structure of the emulsion.

Particle Size Analysis

Emulsion stability can be defined as the rate of change in mean droplet size and size distribution. Change in particle size or number change is indicative of the strength of forces resisting coalescence and phase separation. According to King and Mukherjee (35), the only precise method for the determination of emulsion stability involves size frequency analysis of the dispersed phase droplets as a function of aging time. The main parameters used for the estimation of emulsion stability are the mean droplet size and the polydispersity index. Droplet size and size distribution analysis can be conducted using photon correlation spectroscopy, light extinction and blockage counting, electronic pulse counting, hydrodynamic chromatography, field flow fractionation, and microscopy. However, there are problems associated with the various methods of particle sizing. With the exception of microscopy, these methods fail to differentiate between floccs and individual droplets. Microscopic tests can be highly subjective and are tedious as a sufficiently large number of particles must be counted. Dilution is usually required prior to particle size analysis and this can cause stability problems. Bulk and interfacial surfactant may re-equilibrate on dilution resulting in loss of surfactant from the interfacial film and destabilization of the emulsion.

Electrical Techniques

Conductivity, dielectric behavior, and zeta potential have all been investigated (24). These methods are not very satisfactory. Changes in dielectric constants have been measured, however these are usually associated with chemical changes in the emulsion such as the hydrolysis of components (36). Changes in zeta potential have correlated with physical stability in cases where reduction in zeta potential has been a consequence of desorption of ionic emulsifiers from droplet interfaces (37,38).

3.2.9. Controls for Emulsion Stability Tests

In some cases, testing is best conducted by comparison with an emulsion of known acceptable shelf life. The stability test is conducted on both emulsion systems and extrapolations are made to predict the shelf life of the unknown emulsion. Other tests use an initial parameter of the emulsion for comparison. For example, the initial droplet size is used to compare particle size growth with time. Burgess and Yoon (26) developed a method of assessing the relative stability of various emulsion systems, based on comparisons with initial measurements. Particle size analysis was conducted at various time intervals until the emulsion cracked. Three to five parameters were then selected (such as initial mean diameter, maximum mean diameter, time to reach maximum mean diameter, and time until phase separation was observed) and the emulsions were given a comparative score for each parameter. The highest scores were given to the apparently most stable emulsion for each parameter. For example, the one that took longest to reach its maximum mean diameter. The scores for each parameter were added and the system with the highest overall score was considered the most stable.

Stability comparisons between different emulsions should be conducted on equivalent systems, that is systems with equal droplet size distributions, equivalent amounts of adsorbed emulsifier and after equilibrium has been reached. Although it may not be possible to obtain exactly equivalent systems, the mechanical processing of the emulsions can be manipulated to improve the match between the emulsions.

3.3. Multiple Emulsions

Multiple emulsions consist of at least three phases and are formed by dispersion of an O/W emulsion in oil to form an O/W/O emulsion or by dispersion of a W/O emulsion in water to form a W/O/W emulsion (39). These complex systems generally have greater stability problems than simple emulsions. Mechanisms of instability specific to multiple emulsions (expressed for W/O/W systems) are as follows:

- Coalescence of inner aqueous droplets.
- Coalescence of multiple droplets.
- Rupture of oil layer on the surface of the internal droplets resulting in loss of internal droplets.
- Transport of water through the oil layer.
- Migration of surfactant from one interface to another.

The multiple droplets must be large so that the inner phase can be incorporated. This large droplet size adds to the instability of the multiple droplets. The amount of inner phase incorporated is also significant, as crowding will lead to rapid coalescence of the inner droplets and rupture of the oil layer. Emulsifier concentrations and ratios are important as phase inversion may occur in multiple emulsion systems as a result of migration of emulsifiers between the two interfaces and consequent alteration of the emulsion HLB value (40,41). Emulsions with a higher concentration of the primary emulsifier are generally more stable (33).

3.3.1. Criteria to Assess Multiple Emulsion Stability

1. *Marker method.* A marker such as NaCl is incorporated in the inner aqueous phase and the percentage released into the external aqueous phase with time is measured (39). The marker is released on rupture of the oil layer separating the two aqueous phases.
2. *Microscopy.* This is an important method, especially for assessment of the internal aqueous droplets. Flocculation and coalescence of the internal droplets cannot be directly measured by any other means. Microscopy can be tedious and subjective. Random samples must be prepared and the size of the multiple, internal and any simple droplets must be determined. It can be difficult to determine whether the multiple droplets contain internal droplets, particularly if they are very small or if they take up a large proportion of the multiple droplets. It may also be difficult to observe the internal droplets due to reflected light from the surface of the oil droplets. Freeze etching electron microscopy is used to avoid problems associated with very small internal droplets. Video microscopy is a useful technique to observe changes in emulsion stability with time.

4. *Viscometry.* This is not a quantitative method, however measurement of continuous phase viscosity changes can be useful to assess multiple droplet rupture (42). Viscosity usually decreases with time as a result of rupture of the oil layer. This is complicated as other factors can affect viscosity, such as incorporated drugs and excipients, in particular polymers. The release of these agents into the continuous phase can result in an increase in viscosity.

5. *Particle size analysis of multiple droplets.* Particle sizing methods as described above are utilized. A problem in assessing the size of multiple droplets is that possible increases in droplet diameter due to coalescence may be offset by decreases in droplet size due to shrinkage or loss of internal droplets from the multiple droplets. In addition, with the exception of microscopy, it is impossible to distinguish between multiple and simple droplets.

REFERENCES

1. Kresheck GC. Surfactants. In: Franks F, ed. Water: A Comprehensive Treatise. New York: Plenum Press, 1975:95–167.

2. Lindman B, Wennerstrom H. Topics in Current Chemistry. Vol. 87. Berlin: Springer, 1980:1–83.

3. Leja J. Surface Chemistry of Froth Flotation. New York: Plenum Press, 1982:284–286.

4. Tanford C. The Hydrophobic Effect. Formation of Micelles and Biological Membranes. 2nd ed. New York: John Wiley & Sons, 1980.

5. Mittal KL, ed. Micellization, Solubilization and Microemulsions. Vols. 1 and 2. New York: Plenum Press, 1977.

6. Everett DH. Basic Principles of Colloid Science. London: Royal Society of Chemistry, 1987:153–166.

7. Tadros ThF, ed. Surfactants. London: Academic Press, 1984.

8. Mukerjee P, Mysels KJ. Critical Micelle Concentrations of Aqueous Surfactant Systems. NSRDS-NBS 36. Washington, DC: National Bureau of Standards, US Government Printing Office, 1971.

9. Elworthy PH, Florence AT, MacFarlane CB. Solubilization by Surface Active Agents. London: Chapman and Hall, 1968.

10. Beddow JK. Particulate Science and Technology. New York: Chemical Publishing, 1980.

11. Allen T. Particle size measurement. In: Williams JC, ed. The Powder Technology Series. London: Chapman and Hall, 1975.

12. Shutton HM. Flow properties of powders and role of surface character. In: Parfitt GD, Sing KS, eds. Characterization of Powder Surfaces. Chapter 3. London: Academic Press, 1976, 107–158.

13. Heimenz PC. Principles of Colloid and Surface Chemistry. 3rd ed. New York: Marcel Dekker, Inc., 1997:193–224.

14. Everett DH. Basic Principles of Colloid Science. London: Royal Society of Chemistry, Paperbacks, 1987:76–108.

15. Hunter RJ. Foundations of Colloid Science. Vol. I. Oxford: Clarendon Press, 1987:395–493.

16. Heimenz PC. Principles of Colloid and Surface Chemistry. 3rd ed. New York: Marcel Dekker, Inc., 1997:499–531.

17. London F. Theory and systematic of molecular forces. Z Phys 1930; 63:245–279.

18. Lifshitz EM. The theory of molecular attractive forces between solids. Sov Phys JETP 1956; 2:73.

19. Napper DH. Polymeric Stabilization of Colloidal Dispersions. New York: Academic Press, 1983.

20. Tadros TF, Vincent B. Emulsion stability. In: Becher P, ed. Encyclopedia of Emulsion Technology. Vol. 1. Basic Theory. New York: Dekker, 1983:130–278.

21. Tadros TF, Vincent B. Liquid/liquid interfaces. In: Becher P, ed. Encyclopedia of Emulsion Technology. Vol. 1. Basic Theory. New York: Dekker, 1983:1–51.

22. Jiao J, Burgess DJ. Ostwald ripening of water-in-hydrocarbon emulsions. J Colloid Interface Sci 2003; 264(2):506–516.

23. Kitchener JA, Mussellwhite RP. The theory of stability of emulsions. In: Sherman P, ed. Emulsion Science. London: Academic Press, 1968:77–125.

24. Rieger MM. Stability testing of macromolecules. Cosmet Toil 1991; 106:59–69.

25. Burgess DJ, Yoon JK. Interfacial tension studies on surfactant systems at the aqueous/perfluorocarbon interface. Colloids Surfaces B Biointerfaces 1993; 1:283–293.

26. Burgess DJ, Yoon JK. Influence of interfacial properties on perfluorocarbon/aqueous emulsion stability. Colloids Surfaces B Biointerfaces 1995; 4:297–308.

27. Burgess DJ, Sahin ON. Interfacial rheology of β-casein solutions. In: Herb CA, Prud'homme RK, eds. Structure and Flow in Surfactant Solutions. ACS Symposium Series, 1994: 380–396.

28. Burgess DJ, Sahin NO. Influence of protein emulsifier interfacial properties on oil-in-water emulsion stability. Pharm Dev Technol 1998; 3(1):1–9.

29. Burgess DJ, Sahin NO. Interfacial rheological and tension properties of protein films. J Colloid Interface Sci 1997; 189(1): 74–82.

30. Yoon JK, Burgess DJ. Interfacial properties as stability predictors of lecithin stabilized perfluorocarbon emulsions. Pharm Dev Technol 1996; 1(4):325–333.

31. Burgess DJ, Sahin ON. Influence of protein emulsifier interfacial properties on oil-in-water emulsion stability. Pharm Dev Technol 1998; 3:21.

32. Opawale FO, Burgess DJ. Influence of interfacial properties of lipophilic surfactants on water-in-oil emulsion stability. J Colloid Interface Sci 1998; 197:142–150.

33. Opawale O, Burgess DJ. Influence of interfacial rheological properties of mixed emulsifier films on W/O/W emulsion stability. J Pharm Pharmacol 1998; 50:965–973.

34. Jiao J, Rhodes DG, Burgess DJ. Multiple emulsion stability: pressure balance and interfacial film strength. J Colloid Interface Sci 2002; 250:444–450.

35. King AT, Mukherjee LN. The stability of emulsions: Part I. Soap-stabilized emulsions. J Soc Chem Ind 1939; 58:243.

36. Reddy BR, Dorle AK. Stability testing of o/w emulsions through dielectric constant-1. Cosmet Toil 1984; 99:67–72.

37. Rambhau D, Phadhe DS, Dorle AK. Evaluation of o/w emulsion stability through zeta potential-1. J Soc Cosmet Chem 1977; 28:183–196.

38. Riddick TM. Control of emulsion stability through zeta potential. Am Perfumer Cosmet 1970; 85:31–36.

39. Florence AT, Whitehill D. The formulation and stability of multiple emulsions. Int J Pharm 1982; 11:277–308.

40. Magdassi S, Frenkel M, Garti N, Kasan R. Multiple emulsion II: HLB shift caused by emulsifier migration to external interface. J Colloid Interface Sci 1984; 97:374–379.

41. Matsumoto S, Kita Y, Yonezawa D. An attempt at preparing water-in-oil-in-water multiple-phase emulsions. J Colloid Interface Sci 1976; 57:353–361.

42. Kita Y, Matsumoto S, Yonezawa D. Viscometric method for estimating the stability of w/o/w-type multiple emulsions. J Colloid Interface Sci 1977; 62:87–94.

2

Biopharmaceutical Principles of Injectable Dispersed Systems

C. OUSSOREN, H. TALSMA, and J. ZUIDEMA
Department of Pharmaceutics, Utrecht Institute for Pharmaceutical Sciences, Utrecht, The Netherlands

F. KADIR
Postacademic Education for Pharmacists, Bunnik, Utrecht, The Netherlands

1. INTRODUCTION

Parenteral routes of administration are often used in cases of hampered non-parenteral bioavailability and severe disease states. Several important therapeutic groups such as antibiotics, anti-asthmatics, anti-convulsics, anxiolytics, and analgesics are administered by parenteral routes. The i.v. route of administration results in a rapid increase of drug concentration in the blood. This route of administration has advantages over local parenteral administration in cases

of emergency, when fast action is essential. Local administration includes subcutaneous (s.c.) and intramuscular
(i.m.) injection and direct injection into target organs such
as the eyes, central nervous system, and heart. When the
drug is intended for local effect, for prolonged and
sustained drug release, or for drug delivery to lymph nodes,
s.c. and i.m. administration is preferred. Clinical advantages
of s.c. and i.m. drug administration may include avoidance of
high systemic peak drug concentrations, less frequent
administration, and fewer problems with compatibility of
the injection components compared to i.v. administration.

Dispersed systems for parenteral use may be developed if
a drug cannot be dissolved and administered in solution or for
the purposes of controlled or sustained drug release, drug
targeting, or reduction of toxic effects to sensitive tissues.
Examples of dispersed systems are emulsions, suspensions,
liposome dispersions, microspheres, or other carriers. Table
1 gives an overview of the size ranges of dispersed systems
that are in use for parenteral administration.

An attractive characteristic of injectable dispersed
systems is their ability to act as sustained release systems.
Pharmaceutical research and development of injectable dispersed systems as sustained release systems following local
administration is mainly focused on altering the pharmacokinetics of the drug so that the therapeutic agent is
delivered over an extended period of time.

This chapter focuses on biopharmaceutical principles of
absorption of locally injected drug formulations. Particular
emphasis is given to the s.c. and i.m. routes as they are most
commonly used. To gain an understanding of the biopharmaceutical principles involved, the general features and basic
kinetics of the s.c. and i.m. absorption processes are
discussed for conventional injectable formulations (such as
solutions and suspensions in aqueous and oily vehicles).
Absorption of drugs formulated in advanced drug carrier systems (such as liposomes and microspheres) is then discussed
in the light of the basic biopharmaceutical principles.

Table 1 Overview Dispersed Systems

Dispersed system	Definition	Particulate material	Applications	Route of administration
Emulsions	Oil in water emulsion	Oil droplet	Parenteral nutrition, administration of liquid oily drugs	i.v., s.c., i.m.
Suspensions	Solid particles in aqueous vehicle	Drugs not soluble in water	Sustained release	s.c., i.m.
Microspheres	Polymeric particles	Biodegradable polymers	Sustained release, site specific delivery	s.c., i.m.
Liposomes	Lipid-based vesicles	Phospholipids	Reduction of drug toxicity, site specific delivery, sustained release	i.v., s.c., i.m., intra-CSF
Nanoparticles	Polymeric particles	Biodegradable polymers		i.v., s.c., i.m.

2. DRUG ABSORPTION FROM CONVENTIONAL FORMULATIONS

2.1. Drug Absorption from Aqueous Vehicles

2.1.1. Variation in Absorption

It is often thought that drugs are rapidly and completely absorbed from the s.c. and i.m. injection site particularly when formulated in aqueous systems. However, it appears that the rate as well as the extent of drug absorption is often erratic and variable (1). Several factors may account for the variability in absorption rate and extent. Differences in physiological parameters such as drainage and blood flow due to muscle activity, inflammation, and physiological reactions to the injection trauma are possible explanations (2,3). For example, after i.m. injection, the absorption of artelinic acid, a water-soluble derivative of the anti-malaria drug artemisinin, is found to be very variable (Fig. 1) (4). Hydrophilic

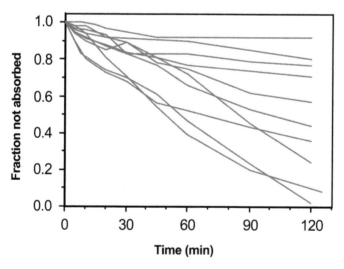

Figure 1 Fraction of the injected dose remaining to be absorbed after i.m. injection of sodium artelinate (20 mg/kg) aqueous solution in rabbits. Each line represents the data of an individual animal ($n = 10$). (From Ref. 4.)

compounds with rather low molecular weights (such as artelinic acid) are preferentially absorbed by the paracellular
route. This transport capacity is influenced by several factors
including muscle activity, inflammation, and flow of the tissue
fluid and, therefore, absorption by this route is generally very
variable (2). Another cause of variable drug absorption is
injection technique. Intra- and intermuscular injections,
within and between the muscle fibrils, respectively (5), may
result in different absorption profiles. This would result in
a bimodal distribution in the absorption rate; however,
experimental evidence for bimodality is lacking in the literature. Another likely explanation, which is not definitely proven, is a variation in the shape of the depot. The shape may
vary from spherical to almost needle-shaped in different subjects. These differences depend on local cohesion between the
muscle components and the tendency for these to be torn open
by the injection procedure. Differences in the shape of the
depot cause differences in the contact area between the depot
and the surrounding tissue, the effective permeation area,
and thus in the absorption rate.

2.1.2. Extent of Absorption

Hydrophilic drugs in solution injected i.m. and s.c. are generally rapidly absorbed from a local depot. However, complete
absorption in a time relevant for therapy does not always
occur (6–9). Absorption only takes place as long as enough
vehicle or essential elements of the vehicle are present to keep
the drug in solution or to drive the absorption process. After
the vehicle has been absorbed, the absorption, rate of the drug
decreases rapidly due to precipitation at the injection site.
Particularly, salts with an alkaline or acidic reaction have
the potential to precipitate after injection due to the neutralizing or buffer capacity of the tissue fluids. This has been
briefly mentioned in the literature for quinidine hydrochloride (9) and is clearly illustrated in a study using human
volunteers by Kostenbauder et al. with i.m. injected phenytoin (10). Relatively high concentrations of phenytoin in solution require a relatively high pH (pH 11 or higher) as well as

co-solvents and/or complexing agents. After i.m. injection, phenytoin is slowly absorbed over a period of approximately 5 days. After 40 hr, 20% of the drug remained unabsorbed. Precipitation and slow redissolution of the drug by tissue fluids at the injection site may explain the slow absorption. The authors developed a mathematical model based on this concept. The observed drug concentration curves in plasma fitted well with this model. Precipitation is probably a result of the pH neutrali-zing effect of the tissue components and possible relatively rapid absorption of one or more of the essential solvent components (10).

Slow absorption due to precipitation is a potential risk for reduced bioavailability as shown for i.m. injection of phenobarbital. After i.m. injection in the deltoid muscle of children, phenobarbital appeared to be completely absorbed, however bioavailability was only 80% compared to oral administration in adults (11). A lack of stability of phenobarbital at the injection site (amide bond hydrolysis) was proposed as an explanation for the incomplete bioavailability. Another possible explanation is precipitation of phenobarbital at the site of injection, since this formulation is very similar to the phenytoin formulation, described above. The detection limit of the analytical method is also very important since slow adsorption of precipitated drug may result in very low and clinically irrelevant concentrations which may not be detected.

Prevention of precipitation of salts and thus enhancement of absorption may be achieved by the use of cosolvents as demonstrated in a study on the effect of propylene glycol on the absorption of benzimidazole hydrochloride (12). It appeared that propylene glycol, which apparently is absorbed slower than water, may prevent, at least partly, precipitation of the free base (or free acid) in certain circumstances. This may enhance the absorption rate of the drugs in question.

2.1.3. Physicochemical Characteristics of Drugs

Physicochemical factors such as the lipophilicity of the drug are important factors that determine bioavailability. Hydrophilic drugs are usually absorbed completely. In contrast, aqueous solutions and suspensions of relatively lipophilic drugs are often absorbed incompletely within a therapeutically relevant time. The influence of drug lipophilicity on absorption rate is illustrated by the difference in absorption rate between midazolam and diazepam. Midazolam is absorbed significantly faster following i.m. injection than the more lipophilic diazepam (13).

β-Blocking agents are an ideal group of compounds for investigation of the effect of drug lipophilicity and release from i.m. and s.c. injection sites since they have similar molecular weights and pK_a values, but differ markedly in lipophilicity. Studies in pigs using crossover experiments have been published with β-blocking agents with varying lipophilicity: atenolol, metoprolol, alprenolol, propranolol, and carazolol (14,15). Figures representing the fraction remaining at the site of injection after s.c. and i.m. injection using i.v. data as references showed biphasic declines; a rapid first phase and a very slow second phase (Fig. 2) (15). Initial release rates appeared to be negatively correlated with drug lipophilicity expressed as fat-buffer partition coefficients, especially after injection into the s.c. fat layers, also called the intra-adipose layers (Fig. 3) (14). The more lipophilic the compound the lower the release rate and the bioavailability. The most hydrophilic compound, atenolol, was the only one which was completely absorbed or bioavailable within 8 hr after i.m. injection and within 24 hr after s.c. injection. The relatively high absorption of relatively hydrophilic drugs is a result of fast transition of the drug into the hydrophilic tissue fluid. More lipophilic drugs, which transit slower into the aqueous phase, are absorbed slower. As a result of the slower absorption process, the aqueous vehicle may be absorbed before drug absorption is complete, which consequently results in reduced bioavailability as discussed above (14,15).

In contrast to what one would expect based on lipophilicity, propanolol was better and faster absorbed from the i.m. injection site (15). This may be related to local tissue irritation by the drug. Propranolol is known to have local irritating properties which may improve blood perfusion in the muscles and account for the deviant behavior after i.m. injection. Absorption of propranolol after s.c. administration was as

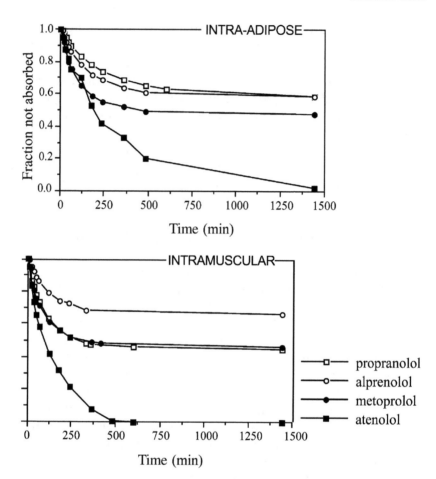

Figure 2 Typical example of the fraction of the injected dose remaining at the injection site after intra-adipose (s.c.) and intramuscular administration of a series β-blocking agents. In order of increasing lipophilicity: atenolol (black squares); metoprolol (black dots); alprenolol (open circles); propranolol (open squares). (From Ref. 15.)

anticipated based on its lipophilicity since the s.c. fat layer or adipose layer is less sensitive to irritating compounds and is less well perfused compared to the i.m. site (15).

Molecular weight appears inversely related to the absorption rate. It has been shown that relatively small molecules are absorbed primarily via the blood capillaries, while

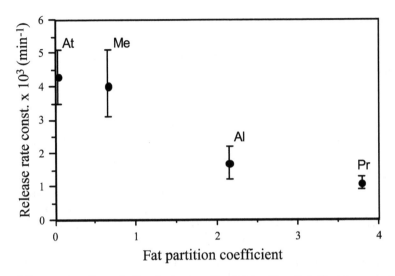

Figure 3 Correlation between the fat-buffer distribution constants and release rates on intra-adipose administration of atenolol (At), metoprolol (Me), alprenolol (Al), and propranolol (Pr). (From Ref. 14.)

compounds with molecular weights larger than approximately 16 kDa and particulate matter such as drug carriers appear to be absorbed mainly by the lymphatics, which results in a lower rate of absorption (6,16). (The reader is referred to chapter 6 in this book for a discussion of the biopharmaceutical principles with respect to injectable suspensions.)

2.2. Drug Absorption from Oily Vehicles

Drug solutions in oil and even suspensions in oil are often thought to be sustained release preparations. However, rapid absorption is often observed. Most likely, slow release is not a property of the oily vehicle but it is the result of relatively high lipophilicity of the drug or interactions between the drug and the vehicle.

Oily vehicles are absorbed slowly and remain present at the injection site for several months. As long as the oily vehicle is present at the site of injection and contains drug in solution, the drug will be released and absorbed from the injection site. The same applies to drug suspensions in oil. For example, the relatively hydrophilic anti-malaria drug artemisinine is relatively rapidly absorbed when injected as a suspension in oil (17). Due to drug absorption, drug concentration in the oily vehicle decreases and consequently suspended drug will dissolve and be absorbed from the site of injection into the blood circulation. Additionally, drugs in suspension may migrate in particulate form to the interface of the oily vehicle and the aqueous tissue fluid. At the interface, hydrophilic drugs may rapidly dissolve and become absorbed (Fig. 4A). In contrast, absorption of artemisinine from a suspension in an aqueous vehicle is much slower, which is most likely the result of rapid absorption of the aqueous vehicle and consequently low dissolution of the drug at the site of injection. As discussed in the previous section, precipitated drug is absorbed slowly and erratically (Fig. 4B).

Drug release and absorption from an oily vehicle into the blood circulation depends mainly on the lipophilicity of the drug. At the oily vehicle/tissue fluid interface, the transition of drugs from oily vehicles into the aqueous phase is controlled by the oil/water partition coefficient. More lipophilic drugs will transit slowly into the aqueous tissue fluid and consequently will be released and absorbed slowly. Hence, lipophilic derivatization can be used as a tool to optimize the sustained release characteristics of the formulation.

The importance of the partition coefficient is illustrated by experiments in which the influence of the lipophilicity of the vehicle on the in vivo release of testosterone-decanoate was studied (18). Disappearance of the drug from the injection site was determined to be proportionally related to the in vitro partition coefficient (Table 2). The slowest absorption rate occurred when ethyloleate, to which the drug has the highest affinity, was used as the vehicle. The mean absorption half-life value increased from 3.2 hr in light paraffin to 10.3 hr in ethyloleate. Thus through formulation in a vehicle with a high lipophilicity sustained release of this drug was achieved (18). These observations illustrate and endorse the importance of the choice of vehicle and the affinity of the drug for the vehicle on drug absorption following i.m. injection.

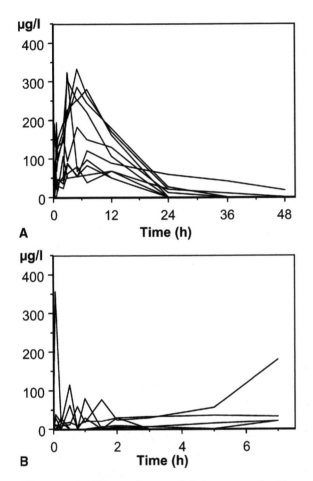

Figure 4 Plots of artemisinin concentrations in serum vs. time after an i.m. administered dose of 400 mg artemisinin to human volunteers. A) Suspension in oil; B) Suspension in an aqueous vehicle. Each line represents the data of an individual volunteer (*n* = 10). (From Ref. 17.)

3. DRUG ABSORPTION FROM DRUG CARRIER SYSTEMS

Drugs that are rapidly eliminated from the blood yield quickly declining blood drug levels and result in a short duration of

Table 2 Testosterondecanoate Fat/Phosphate Buffer Partition
and Absorption Rate Expressed as Half-Life in the Muscle

Solvent	Part coeff. $\times 10^{-3}$	$t_{1/2}$ in the muscle (h)
Ethyloleate	6.3	10.3
Octanol	5.3	9.7
Isopropylmyristate	4.3	7.8
Light liquid paraffin	1.3	3.2

Source: From Ref. 18.

therapeutic response. For a number of drugs, it is desirable to
maintain the concentration of the drug in blood within the
therapeutic range for a long period of time and avoid toxic
levels. In these cases, the use of drug delivery systems that
provide slow release of the drug over an extended period of
time is useful. Several drug carrier systems, e.g., liposomes
and microspheres, are currently being developed as sustained
systems following s.c. or i.m. administration (19).

Absorption of drugs encapsulated in carrier systems
after s.c. and i.m. administration is more complicated than
absorption from conventional formulations. Not only absorption of the free drug but also absorption of the carrier and
release of the drug from the carrier are important issues to
be considered. After release, the drug will behave similarly
to drug administered in conventional formulations and general biopharmaceutical principles will be applicable. However, if the drug is not released from the carrier, and the
carrier is absorbed from the injection site as an intact entity,
the drug will follow the kinetics and biodistribution of the
carrier, which is generally very different from the kinetics
of the free drug. Moreover, slow release of drug from the
circulating carrier will also affect drug concentration in the
blood circulation.

This section will deal with the absorption of s.c. and i.m.
administered drug carriers. First absorption of the carrier as
an intact entity from the local site of injection will be
discussed. Then attention will be paid to release mechanisms
of drugs from several carriers at the local injection site.

3.1. Absorption of Drug Carrier Systems

Following local administration, large molecules and particulate matter do not have direct access to the bloodstream as the permeability of blood capillaries in the interstitium is restricted to water and small molecules. Instead, large molecules and particulate matter may be taken up by lymphatic capillaries (16). Figure 5 presents a schematic illustration of drug absorption following s.c. injection of liposome-encapsulated drugs. Similar mechanisms hold for the i.m. route of administration. Lymphatic absorption is described for several carrier systems. Here absorption of liposomes will be discussed as a model for other drug carriers. Generally, similar phenomena occur for the other carrier systems.

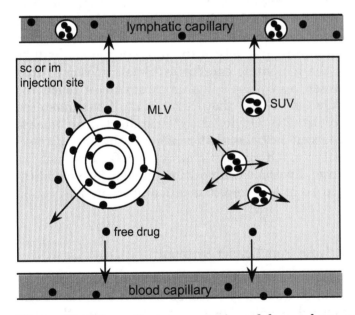

Figure 5 Schematic representation of drug release and absorption of injectable dispersed systems from the site of injection after s.c. or i.m. injection. Small molecules can enter the blood circulation either by entering blood capillaries or via lymphatic capillaries, whereas larger molecules and drug encapsulated in small particles can enter the blood circulation only via lymphatic capillaries. (From Ref. 19.)

Absorption of particles from the injection site after local parenteral administration depends mainly on one important carrier-related factor, particle size. For liposomes, several reports refer to a cut-off value of 0.1 µm, above which liposomes fail to appear in the blood to any substantial extent. Larger liposomes will remain at the s.c. injection site for a long period of time (20–23). This size-dependent retention at the injection site is likely to be related to the process of particle transport through the interstitium. The structural organization of the interstitium dictates that larger particles will have more difficulty to pass through the interstitium and will remain at the site of injection to a large, almost complete extent. Gradual release of the encapsulated drug from liposomes remaining at the injection site results in very low but prolonged drug levels in the blood. Therapeutic drug levels have been reported to last for several days (24–27). When smaller liposomes are administered, they will migrate through the aqueous channels in the interstitium and will be taken up by the lymphatic capillaries. Small liposomes that have been taken up by the lymphatic capillaries reach the general circulation where they behave as if administered by the i.v. route (23). Obviously, if sustained drug release is intended, larger particles that remain at the site of injection are preferred. Other liposome-related factors such as liposome charge, liposome composition, and surface modification appear to be of less importance for liposome absorption (23,28).

3.2. Drug Release from Carrier Systems

Drugs encapsulated in carrier systems that remain at the site of injection will be gradually released from the carrier. Drug release rates are determined by both the carrier and the drug characteristics.

The release mechanism of drugs from microspheres is dependent on the polymer and formulation technique used. From polyester microspheres, the drug is generally released by diffusion through aqueous channels or pores in the polymer matrix and by diffusion across the polymer barrier following erosion of the polymer. Surface erosion or bulk erosion occurs parallel with the hydrolytic degradation of the polymer, which influences the release pattern and stability of incorporated drugs after injection. Anderson and Shive summarized the factors affecting the hydrolytic degradation behavior of biodegradable polyesters and described their biocompatibility (29). These factors are shown in Table 3.

Hydrogel particles are considered to be interesting systems for peptide and protein delivery because of their good tissue compatibility and possibilities to manipulate the permeability for solutes. The release rate is dependent on the molecular size of the drug, the degree of cross-linking of the gel, and the water content (30).

Liposomes and microspheres often show a burst effect combined with sustained release. Several mechanisms of drug release from liposomes following local administration have been suggested. Liposomes remaining at the site of injection might gradually erode and eventually disintegrate completely (e.g., as a result of attack by enzymes, destruction by neutrophils). During this process, the entrapped drug is released. After release, free drug enters the blood circulation by either direct absorption into the blood capillaries or via lymphatic capillaries. As release of the

Table 3 Factors Influencing Hydrolytic Behavior of Biodegradable Polyesters[a]

Water permeability and solubility (hydrophilicity/hydrophobicity)
Chemical composition
Mechanism of hydrolysis (noncatalytic, autocatalytic, enzymatic)
Additives (acidic, basic, monomers, solvents, drugs)
Morphology (crystalline, amorphous)
Device dimensions (size, shape, surface-to-volume ratio)
Porosity
Glass transition temperature (glassy, rubbery)
Molecular weight and molecular weight distribution
Physico-chemical factors (ion exchange, ionic strength, pH)
Sterilization
Site of implantation

[a]From Ref. 29.

encapsulated drug may occur during an extended period of time, concentrations in the blood will be prolonged. The rate and duration of release from liposomes depends on plasma factors and liposome stability. Serum proteins, enzymes, phagocytosing cells, shear stress, and liposome aggregation at the injection site play a role in the destabilization or degradation of liposomes at the injection site and subsequent leakage of liposomal contents (31–37).

Stable liposomes composed of saturated lipids are known to release their content slower than liposomes with higher membrane fluidity. Schreier et al. reported the same relation between liposome stability and drug release after i.m. administration of liposomes (38). The rate of release of encapsulated drug as well as the erosion of liposomes at the injection site were found to be a function of the fluidity of the lipid membranes. The release rate of gentamicin from egg phosphatidylcholine liposomes was about seven times slower than from soy phosphatidylcholine liposomes with more fluid (unsaturated) bilayers when injected i.m. In line with these observations, Koppenhagen reported that after intratumoral injection of [111]Indium-labeled desferal encapsulated in "solid" liposomes, the amount of label remaining at the injection site was about 10-fold higher than when encapsulated in "fluid" liposomes, 6 days post-injection (39). From these observations, it may be concluded that drug absorption rates may be controlled (within certain limits) by using lipids with different degrees of bilayer fluidity.

Other liposome-related factors seem to be of less importance for drug release from the injection site. Several papers studied retention of differently charged liposomes at the injection site after i.m. injection. Results suggest that release from negatively charged liposomes at the i.m. injection site is somewhat less than the release of neutral and positively charged liposomes (27,40,41). Recently, the influence of liposome charge on the fate of methotrexate encapsulated in neutral, positively, and negatively charged liposomes was studied after i.m. injection (42,43). Plasma concentrations of methotrexate were not substantially influenced by liposome charge.

4. FACTORS INFLUENCING DRUG ABSORPTION

4.1. Influence of Injection Volume and Drug Concentration

4.1.1. Conventional Drug Formulations

Hydrophilic compounds such as atropine, sodium chloride, sugars, and polyols such as mannitol and sorbitol, are reported to be absorbed rapidly when the compounds are administered in relatively small injection volumes (44–46). The faster absorption rate of hydrophilic drugs from smaller injection volumes can be explained by higher concentrations of the drug in the aqueous vehicle which result in a higher diffusion rate into tissue fluid.

In contrast to the volume-dependent absorption rates of hydrophilic drugs in solution, hydrophilic drugs in suspension are found to be absorbed slowly from small injection volumes. In suspensions, the drug concentration does not depend on the volume and consequently the drug diffusion rate into tissue fluid is constant. Hirano et al.(47–50) studied the kinetic behavior of suspensions after i.m. and s.c. injection. They found that increasing concentrations and decreasing volumes lead to increased sustained release properties of their suspended model compounds and thus slower absorption. This phenomenon is explained by the aggregation of the separate particles by physicochemical forces and tissue tension. In 1958, Ober et al. stated that aggregation in concentrated particulate systems often gives rise to enlarged viscosity, specific rheological features and as a result a diminished dissolution rate after injection (51). In addition, it is known that viscosity and specific rheological features such as (pseudo) plastic behavior in suspensions and emulsions increase with increasing concentration and with decreasing particle size. Increased viscosity hampers diffusion of the drug out of an aggregated clot.

The influence of injection volume on the rate and extent of bioavailability of solutions of relatively lipophilic drugs in aqueous vehicles has been studied in rats (52). The study was performed to find an explanation for the incomplete absorption of the more lipophilic β-blocking agents as described in the former section of this chapter. Propranolol was used as a model compound. The rate and extent of absorption 8 hr after injection appeared to increase with increasing injection volume. This was explained by the observation that with increasing vehicle volume, the residence time of the vehicle at the injection site increased. Consequently, the drug was dissolved over a longer time-period resulting in a faster release rate compared to when the drug is present in the solid state (Fig. 6). Moreover a larger volume may increase the vehicle flow, including the drug in solution, from the depot. This effect explains the higher absorption rate during the initial phase.

The influence of injection volume on the absorption of drugs in solution was studied by determining the absorption of testosterone and some other lipophilic model drugs from different volumes of several oily vehicles (53). Lipophilic compounds are faster absorbed when they are dosed in smaller

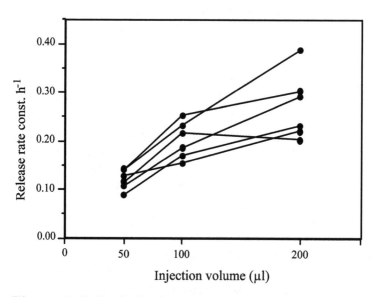

Figure 6 Individual release rate constants after i.m. injection of 3 mg propranolol HCl in 50, 100, and 200 µl aqueous solution in rats. Lines connect the individual values of the same rat ($n = 6$). (From Ref. 52.)

volumes of the oily vehicle. This effect is the same as occurs for solutions of hydrophilic drugs in aqueous vehicles, where the diffusion potential is dominating the absorption rate.

4.1.2. Drug Carrier Systems

In line with findings on drug absorption from hydrophilic suspensions, drug release rates from drug carrier formulations, are found to be slower when smaller volumes are injected and when higher dosages are used (54–59). The most likely explanation is that the observed effect is related to the formation of a pressure-induced aggregate at the injection site as discussed above. Smaller injection volumes result in a smaller and more compact depot from which drug release will be slower than from a larger depot. Studies with i.m. and s.c. injections of liposomal chloroquine in mice have confirmed that an increased volume at a fixed concentration of liposomes results in an increased absorption rate, whereas an increased dose or liposome concentration results in a decreased absorption rate. Hence, the absorption rate constants showed a positive correlation with injection volume after both routes of administration (60).

The finding that smaller volumes and higher concentrations in dispersed systems formulations may decrease the rate and extent of drug absorption has important clinical implications. Increasing the injected dose is often thought to lead to a proportional drug concentration increase in the blood circulation. However, dose increase in dispersed systems such as suspensions or liposomal dispersions often leads to a decrease in the absorption rate due to aggregation. This may cause even lower drug plasma concentrations and longer residence times at the site of injection. In general, in the case of dispersed injections, higher plasma concentrations can only be reached with multiple injections.

4.2. Injection Depth

Injection depth appears to be a major factor determining the absorption profile of lipophilic drugs. Generally, for i.m. injection a needle length larger than the s.c. fat layer is needed.

However, one should realize that the thickness of the s.c. fat layer varies greatly. For example, in the gluteal region, the s.c. layer shows a large interindividual variation. Moreover, large differences are found between males and females. Cockshott et al. investigated the thickness of the gluteal s.c. fat layer and found a large difference in mean skin to muscle distance between the investigated 63 males and 60 females (61). The skin to muscle distance was about 3–9 cm within the group of females whereas 1 cm to about 7 cm was measured in males. In line with these observations, large differences in absorption rates were found between males and females after injection of cefradine in the gluteal region (62). Differences in absorption rate between deep i.m. injections and shallow s.c. or intra-adipose injections are shown in Fig. 3. The absorption rate of the hydrophilic atenolol after i.m. injection is considerably faster than after s.c. administration. Atenolol was already completely absorbed within 8 hr after i.m. injection whereas 24 hr after s.c. or intra-adipose injection absorption still occurred (15).

Generally, absorption from the i.m. injection site is in most cases much faster than from the adipose s.c. fat layer which can be explained by structural and physiological differences in the tissue. The fatty connective tissues and adipose layers at the s.c. injection site are more lipomatous and much less perfused than muscular tissues (2). Lipophilic drugs will be more easily absorbed from hydrophilic tissue than from lipophilic tissue. Absorption of hydrophilic drugs from aqueous solutions is less dependent on injection depth.

These observations have important clinical implications. In 1974, a letter appeared in the Lancet (63) in which diazepam concentrations in plasma, 90 min after i.m. injection (administered by either 3 or 4 cm long needles) in the gluteal region were measured in females. With the 3-cm needle injections, plasma diazepam concentrations appeared very low and were therapeutically inadequate. In contrast, injections given with 4 cm needles resulted in much, therapeutically relevant diazepam concentrations. This indicates that the 3-cm injection was too shallow and that the injection was very likely placed in the s.c. fat layer instead of the intended gluteus maximus. Additionally, a too shallow injection might have great clinical impact in disease prophylaxis by vaccination. For example, despite timely post-exposure treatment with rabies vaccine in the gluteal region, a patient died from rabies encephalitis (64), probably due to too shallow injection.

In contrast to fast absorption from the i.m. injection site, s.c. injection generally results in slow absorption. Therefore, sustained release injections are better placed in the s.c. or lipomatous tissues. This has been demonstrated by Modderman et al. (65) and Pieters et al. (66). They found marked differences in dapsone absorption profile between males and females. Males showed high peak concentrations in the first week and a shorter mean residence time. These differences were not found if injections were given at two-third of the individually measured skin to muscle distance, which is a guarantee for s.c. injection. Additionally, the sustained release characteristics were much better after intra-adipose (s.c. in the gluteal region) than after i.m. injection. Release times up to a month were obtained with dapsone and up to 3 months with its more lipophilic derivative monoacetyldapsone (67).

4.3. Anatomical Site of Injection

The anatomical site of injection may be another important factor determining the absorption rate of drug carriers after s.c. administration. This was clearly demonstrated when small (0.1 μm) liposomes were administered at different sites of the body of rats. After s.c. injection into the flank of rats, disappearance from the site of injection was much lower than after injection in the footpad or into the dorsal side of the foot. In fact, after s.c. injection into the flank of rats, the injected liposomes remain to a large, almost complete extent, at the site of injection, whereas about 60% of the injected dose reaches the blood circulation after injection into the foot (Fig. 7). The observed site-dependent disposition is attributed to differences in the structural organization of the s.c. tissue at the different sites of injection (68). These results demonstrate that the anatomical site of s.c. injection should be considered carefully when designing injectable dispersions for local administration.

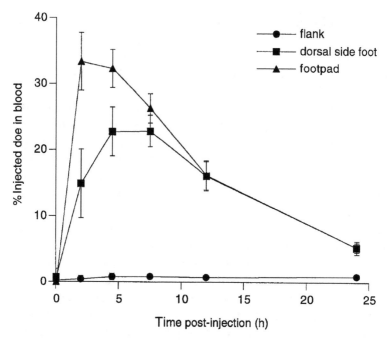

Figure 7   Recovery of liposomal label from blood after s.c. administration of liposomes at three different sites of injection. A single dose of radiolabeled liposomes (EPC:EPG:Chol, molar ratio 10:1:4; mean diameter, 0.1 μm, 2 μmol total lipid) was injected s.c. into the flank, into the dorsal side of the foot and into the footpad of rats. Values represent the mean percentage of injected dose circulating in the total blood volume ± SD of 4 animals. (From Ref. 68.)

5.   CARRIER KINETICS AND TARGETING

The i.v. route is the preferable route of administration to study carrier kinetics. Here, we will discuss mainly the biodistribution and pharmacokinetics of liposomes after i.v. administration as an example. Other particulate systems such as microspheres behave similarly to large liposomes because size and surface characteristics are the major determinants.

Studies on the in vivo fate or elimination of i.v. administered liposomes also provide information on other routes of injection, since liposomes, which are injected by other routes than the i.v. route, usually reach the circulation to some extent (23). Upon i.v. injection or intact absorption from local injection sites, liposomes and other particulate systems are mainly eliminated by accumulation in organs rich in cells belonging to the mononuclear phagocyte system (MPS). Due to rich blood supplies and the abundance of phagocytic MPS cells, the major sites of accumulation are the liver and spleen. A dummy dose of a particulate system, or other phagocytotic depressant, can decrease the uptake of a dispersed system by the MPS through presaturation of the system. Dose-dependent kinetics follow logically from these observations and have been demonstrated for liposomes (69–72).

A number of studies have illustrated that when liposomes are exposed to serum or plasma, they rapidly acquire a coating of proteinaceous molecules (73–76). Protein binding can be demonstrated by separating liposomes from serum or plasma incubations and then analyzing the liposome-protein complexes by sodium dodecylsulfate–polyacrylamide gel electrophoresis. Protein binding differs considerably in amount and pattern depending on the dose, size, lipid composition, bilayer rigidity, and surface characteristics (such as charge and hydrophilicity) of the vesicles. These liposome–blood protein interactions have a number of important consequences for the subsequent pharmacokinetic behavior of the vesicles in vivo and for their use as drug carrier systems. Generally, small liposomes are cleared more slowly than large liposomes, partly due to less mechanical obstruction but also resulting from a lower affinity of serum proteins (opsonines) which are involved with the liver uptake by phagocytic cells (77). Opsonic activity also seems to depend on the lipophilicity of the particle surface, and on the presence of divalent cations which probably play a role in conformational changes in the opsonins (78). In addition, the net surface charge influences the opsonization process. Positively charged and neutral liposomes of similar size appear to circulate longer than negatively charged liposomes (79). Increasing the negative charge results in a dramatic acceleration of liposome clearance (80). Additionally, bilayer fluidity, which can be influenced by the use of phospholipids with long and saturated acyl chains or by the incorporation of cholesterol in the liposomal bilayer, determines stability in plasma and thus blood circulation times of liposomes.

Liposomes are relatively rapidly cleared from the blood circulation by the MPS. To avoid this rapid clearance, surface-modified liposomes which escape from MPS uptake and achieve prolonged circulation half-times in vivo have been extensively investigated during the last decade. Steric hindering of negative charge by surface modification, hydrophilization, leads to increased circulation times and half-life by preventing interaction with opsonins. Increase of carrier half-life can be realized by the addition of carbohydrate moieties or the introduction of poly(ethyleneglycol) (PEG) derivatives absorbed or covalently bound to particulate materials or membrane lipids of liposomes. The most popular means to obtain sterically stabilized liposomes is to incorporate PEG conjugated to distearoyl-phosphatidyl-ethanolamine (DSPE-PEG) into the liposomal bilayer (28,81–86).

Prolonged circulation times of sterically stabilized liposomes result in increased amounts of liposomes extravasating in areas where the permeability of the endothelial barrier is increased, specifically infected tissue and tumor tissue. Therefore, sterically stabilized liposomes can be used to target drugs to these tissues (passive targeting) (87–89). An approach to site specific delivery of liposomes is to conjugate them with homing devices such as monoclonal or polyclonal antibodies (active targeting). These immunoliposomes demonstrate high selectivity of drug delivery in vitro (90–92). However, coupling of antibodies to liposomes increases MPS uptake. As a result, circulation times of immunoliposomes in blood are generally shorter than circulation times of conventional liposomes. Therefore, sterically stabilized immunoliposomes are currently under investigation (93–98). Results demonstrate that these liposomes combine the advantages of both systems, long circulation times, and site, specific drug delivery. To date, only a few reports on the therapeutic applications of long circulating immunoliposomes have appeared (99–102).

6. TISSUE PROTECTIVE EFFECT OF DISPERSED SYSTEMS

S.c. and i.m. administration of drugs may cause discomfort and pain at the site of injection. Injury by injection can be caused by disruption of the tissue by the needle or by specific toxic effects of the injected material. Several constituents such as drug and excipients of the pharmaceutical formulation may cause irritation and/or damage to the surrounding tissue. The use of carriers such as liposomes is an interesting tool to prevent or reduce tissue damage. Liposomes have been reported to protect tissue against irritating drugs after both i.v. and local injection in a number of papers (103–106).

The ability of liposomes to prevent tissue irritation by the non-steroid anti-inflammatory drug novaminsulfone and the anti-malarial drug chloroquine has been investigated in animal studies. These drugs are known to be irritating, causing hemorrhage, cell necrosis, inflammatory reactions, and eventually fibrosis (107). I.m. injection of free novaminsulfone resulted in severe tissue damage, whereas the same dose of novaminsulfone entrapped in liposomes was not irritating. Differences were less pronounced when tissue irritation was compared after i.m. injection of chloroquine and liposome encapsulated chloroquine. The difference between novaminsulfone and chloroquine was explained by the fact that novaminsulfone is known to possess stronger irritating properties (108). In line with these findings, Al-Suwayeh et al. showed that liposomal encapsulation of loxapine (i.m. injection) significantly reduced myotoxicity compared to the same dose of loxapine in a commercially available formulation containing propylene glycol (70% v/v) and polysorbate 80 (5% w/v) after i.m. injection (109).

Liposomal formulation factors such as bilayer rigidity have a marked influence on the liposomal protective effect against s.c. administered model drugs mitoxantrone and doxorubicin. Both model drugs are, in the chosen dosage, progressively vesicant and cause similar patterns of tissue damage. Liposome rigidity was an important factor in tissue protection. Gel-state liposomes with more rigid bilayers protected surrounding tissue more efficiently than fluid-state liposomes, which are more susceptible to leakage of the drug. The presence of a hydrophilic PEG-coating did not affect the protective effect of liposomes (110).

Additionally, in animal studies, local tissue damage was found to be strongly dependent on the route of administration. Initially, surrounding tissue was protected efficiently against the toxic effects of the encapsulated mitoxantrone, both after s.c. and i.m. injection. However, 7 days post-injection toxic effects after i.m. administration were reduced more than after s.c. administration (110). The differences between the protective effect after s.c. and i.m. administration were ascribed to higher clearance from the injection site of injected material after i.m. injection than after s.c. administration. Differences were attributed to a richer supply of blood and the presence of abundant lymphatic capillaries in muscle tissue and incre-ased blood and lymph flow through muscles during body movement of the animal. As surrounding tissue will only be exposed to very low concentrations of the free drug, tissue damage will be negligible. In contrast, the prolonged inflammatory response observed following s.c. administration of mitoxantrone- and doxorubicin-containing liposomes may be the result of a large proportion of the injected liposomal dose remaining at the injection site for a longer period of time.

Differences in damaging properties between novaminsulfone and chloroquine, mitoxantrone, and doxorubicin and in the potential of liposomal protection show that drugs differ in intrinsic irritating properties. Some drugs cause dose-related injury whereas others such as anthracyclines cause cumulative and/or irreversible damage. In the latter case, the impact of protection by encapsulation of the drug in liposomes may be of limited value (110). In general, liposomes might offer good protection against tissue damage, but each particular combination of drug and liposomes should be tested carefully and extensively for its clinical value.

7. SUMMARY

Dispersed systems for injection are developed in cases that solution dosage forms are impossible or to achieve a specific objective (e.g., sustained drug release, drug targeting, or reduction of toxic effects to sensitive tissues).

The rate of action of s.c. and i.m. injected drugs depends greatly on drug release from the formulation and absorption from the site of injection. Variables that are known to affect drug release and absorption after s.c. or i.m. injection include pharmaceutical aspects of the formulation, e.g., the initial drug concentration, the properties of the vehicle in which the drug is formulated, and the physicochemical properties of the drug. Biopharmaceutical aspects, e.g., the route of administration, injection site, injection technique, and injection depth, are also known to affect drug absorption. Moreover, physiologic aspects such as blood supply and temperature at the injection site and body movement may influence drug absorption after local administration.

Much attention has been paid to the development of particulate drug carrier systems such as liposomes and microspheres. Absorption of drugs encapsulated in drug carriers depends on carrier characteristics. Absorption of the carrier occurs only if the carrier is smaller than about 0.1 μm. Particles larger than about 0.1 μm remain at the injection site and disintegrate or release their contents gradually resulting in low but prolonged plasma concentrations. The rate and extent of release of the encapsulated drug may be controlled by manipulating physical and chemical characteristics of both drug and formulation.

Findings on the influence of formulation characteristics and biopharmaceutical aspects on the absorption of s.c. and i.m. injected drugs or drug carriers such as described in this chapter have important clinical implications and should be taken into consideration when designing formulations intended for s.c. and i.m. administration.

REFERENCES

1. Gibaldi M. Biopharmaceutics and Clinical Pharmacokinetics. Philadelphia: Lea & Febiger, 1977.

2. Zuidema J, Pieters FAJM, Duchateau GSMJE. Release and absorption rate aspects of intramuscularly injected pharmaceuticals. Int J Pharm 1988; 47:1–12.

3. Zuidema J, Kadir F, Titulaer HAC, Oussoren C. Release and absorption rates of intramuscularly and subcutaneously injected pharmaceuticals (II). Int J Pharm 1994; 105:189–207.

4. Titulaer HAC, Eling WMC, Zuidema J. Pharmacokinetic and pharmacodynamic aspects of artelinic acid in rodents. J Pharm Pharmacol 1993; 45:830–835.

5. Groothuis DG, Werdler MEB, van Miert ASJPAM, van Duin CTM. Factors influencing the absorption of ampicillin administered intramuscularly in dwarf goats. Res Vet Sci 1980; 29:116–117.

6. Ballard BE. Biopharmaceutical considerations in subcutaneous and intramuscular drug administration. J Pharm Sci 1968; 57:357–378.

7. Raeder JC, Nilsen OG. Prolonged elimination of midazolam after i.m. administration. Acta Anaesthesiol Scand 1988; 32:464–466.

8. Raeder JC, Breivik H. Premedication with midazolam in outpatient anaesthesia. A comparison with morphine-scopolamine and placebo. Acta Anaesthesiol Scand 1987; 31:509–514.

9. Tse FLS, Welling PG. Bioavailability of parenteral drugs. I. Intravenous and intramuscular doses. J Parent Drug Ass 1980; 34:409–421.

10. Kostenbauder HB, Rapp RP, McGroven JP et al. Bioavailability and single-dose pharmacokinetics of intramuscular phenytoin. Clin Pharmacol Ther 1976; 18:449–456.

11. Viswanathan CT, Booker HE, Welling PG. Bioavailability of oral and intramuscular phenobarbital. J Clin Pharmacol 1978; 18:100–105.

12. Cheng-Der Yu, Kent JS. Effect of propylene glycol on subcutaneous absorption of a benzidamole hydrochloride. J Pharm Sci 1982; 71:476–478.

13. Hung OR, Dyck B, Varvel JR, Shafer SL, Stansky DR. Intramuscular absorption kinetics of midazolam and diazepam in healthy volunteers. Clin Pharmacol Ther 1992; 51:130.

14. Kadir F, Zuidema J, Pijpers A, Melendez R, Vulto A, Verheijden JHM. Pharmacokinetics of intra-adiposely and intramuscularly injected carazolol in pigs. J Vet Pharmacol Ther 1990; 13:350–355.

15. Kadir F, Zuidema J, Pijpers A, Vulto A, Verheijden JHM. Drug lipophilicity and release pattern of some β-blocking agents after intra-adipose and intramuscular injection in pigs. Int J Pharm 1990; 64:171–180.

16. Charman WN, Stella VJ, eds. Lymphatic Transport of Drugs. Boca Raton, FL: CRC Press, 1992.

17. Titulaer HAC, Zuidema J, Kager PA, Wetsteyn JCFM, Lugt ChB, Merkus FWHM. The pharmacokinetics of artemisinin after oral, intramuscular and rectal administration to volunteers. J Pharm Pharmacol 1990; 42:810–813.

18. Al-Hindawi MK, James KC, Nicholls PJ. Influence of solvent on the availability of testosterone propionate from oily, intramuscular injections in rat. J Pharm Pharmacol 1986; 39: 90–95.

19. Oussoren C, Storm G, Crommelin DJA, Senior JH. Liposomes for local sustained drug release. In: Senior JH, Radomsky ML, Sustained Release Injectable Products. Denver, Colorado: Interpharm Press, 1999:137–180.

20. Allen TM, Hansen CB, Guo LSS. Subcutaneous administration of liposomes: a comparison with the intravenous and intraperitoneal routes of injection. BBA 1993; 1150(1):9–16.

21. Patel HM. Fate of liposomes in the lymphatics. In: Gregoriadis G, ed. Liposomes as Drug Carriers. John Wiley & Sons Ltd. 1988: 51–61.

22. Tümer A, Kirby C, Senior J, Gregoriadis G. Fate of cholesterol-rich liposomes after subcutaneous injection into rats. BBA 1983; 760:119–125.

23. Oussoren C, Zuidema J, Crommelin DJAC, Storm G. Lymphatic uptake and biodistribution of liposomes after subcutaneous injection. II. Influence of liposomal size, lipid composition and lipid dose. Biochem Biophys Acta 1997; 1328:261–272.

24. Kim S, Howell SB. Multivesicular liposomes containing cytarabine entrapped in the presence of hydrochloric acid for intracavitary chemotherapy. Cancer Treatment Rep 1987; 71(7/8):705–711.

25. Kim T, Kim J, Kim S. Extended-release formulation of morphine for subcutaneous administration. Cancer Chemother Pharmacol 1993; 333:187–190.

26. Stevenson RW, Patel HM, Parson JA, Ryman BE. Prolonged hypoglycemic effect in diabetic dogs due to subcutaneous administration of insulin in liposomes. Diabetes 1982; 31: 506–511.

27. Eppstein DA, van der Pas MA, Gloff CA, Soike KF. Liposomal interferon-β: sustained release treatment of simian varicella virus infection in monkeys. J Infect Dis 1989; 159:616–620.

28. Oussoren C, Storm G. Lymphatic uptake and biodistribution of liposomes after subcutaneous injection. III. Influence of surface modification with poly(ethyleneglycol). Pharm Res 1997; 14:1479–1484.

29. Anderson JM, Shive MS. Biodegradation and biocompatibility of PLA and PLGA microspheres. Adv Drug Deliv Rev 1997; 28:5–24.

30. Hennink WE, Talsma H, Borchert CJH, De Smedt SC, Demeester J. Controlled release of proteins from dextran hydrogels. J Control Release 1996; 39:47–55.

31. Hunt CA. Liposome disposition in vivo. V. Liposome stability in plasma and implications for drug carrier function. Biochim Biophys Acta 1982; 719:450–463.

32. Allen TM, Cleland LG. Serum induced leakage of liposome contents. Biochim Biophys Acta 1980; 597:418–426.

33. Damen J, Dijkstra J, Regts J, Scherphof G. Effect of lipoprotein-free plasma on the interaction of human plasma

high density lipoprotein with egg yolk phosphatidylcholine liposomes. Biochim Biophys Acta 1980; 620:90–99.

34. Hernandez-Caselles T, Villalain J, Gomez-Fernandez JC. Influence of liposome charge and composition on their inter-action with human blood serum proteins. Mol Cell Biochem 1993; 120:119–126.

35. Finkelstein MC, Weissmann G. Enzyme replacement via lipo-somes. Variations in lipid composition determine liposomal integrity in biological fluids. Biochim Biophys Acta 1979; 587:202–216.

36. Guo LSS, Hamilton RL, Goerke J, Weinstein JN, Havel RJ. Interaction of unilamellar liposomes with serum lipoproteins and apolipoproteins. J Lip Res 1980; 21:993–1003.

37. Mercadal M, Domingo JC, Bermudez M, Mora M, Africa de Madariaga M. N-palmitoylphosphatidylethanolamine stabi-lizes liposomes in the presence of human serum: effect of lipi-dic composition and system characterization. Biochim Biophys Acta 1995; 1235:281–288.

38. Schreier H, Levy M, Mihalko P. Sustained release of liposome-encapsulated gentamicin and the fate of phospholi-pid following intramuscular injection in mice. J Control Release 1987; 5:187–192.

39. Koppenhagen FJ. Liposomes as Delivery Systems for Recom-binant Interleukin-2 in Anticancer Immunotherapy. PhD Thesis, Utrecht University, Utrecht, 1997.

40. Kim CK, Han JH. Lymphatic delivery and pharmacokinetics of methotrexate after intramuscular injection of differently charged liposome-entrapped methotrexate to rats. J Microen-capsul 1995; 12(4):437–446.

41. Stevenson RW, Patel HM, Parson JA, Ryman BE. Prolonged hypoglycemic effect in diabetic dogs due to subcutaneous administration of insulin in liposomes. Diabetes 1982; 31:506–511.

42. Kim MM, Lee SH, Lee MG, Hwang SJ, Kim CK. Pharmacoki-netics of methotrexate after intravenous and intramuscular injection of methotrexate-bearing positively charged liposomes to rats. Biopharm Drug Dispos 1995; 16:279–293.

43. Jeong YN, Lee SH, Lee MG, Hwang SJ, Kim CK. Pharmaco-kinetics of methotrexate after intravenous and intramuscular injection of methotrexate-bearing negatively charged liposomes to rats. Int J Pharm 1994; 102:35–46.

44. Schriftman H, Kondritzer AA. Absorption of atropine from muscle. Am J Physiol 1957; 191:591–594.

45. Warner GF, Dobson EL, Pace N, Johnston ME, Finney CR. Studies of peripheral bloodflow: the effect of injection volume on the intramuscular radiosodium clearance rate. Circulation 1953; 8:732–734.

46. Sund RB, Schou J. The determination of absorption rates from rat muscles: an experimental approach to kinetic descriptions. Acta Pharmacol Toxicol 1964; 21:313–325.

47. Hirano K, Ichihashi T, Yamada H. Studies on the absorption of practically water-insoluble drugs following injection. II: Intramuscular absorption from aqueous suspensions in rats. Chem Pharm Bull 1981; 29:817–827.

48. Hirano K, Yamada H. Studies on the absorption of practically water-insoluble drugs following injection. VI: Subcutaneous absorption from aqueous suspensions in rats. J Pharm Sci 1982; 71:500–505.

49. Hirano K, Yamada H. Studies on the absorption of practically water-insoluble drugs following injection. VII: Plasma concentration after different subcutaneous doses of a drug in aqueous suspensions in rats. J Pharm Sci 1983; 72:602–607.

50. Hirano K, Yamada H. Studies on the absorption of practically water-insoluble drugs following injection. VIII: Comparison of subcutaneous absorption from aqueous suspensions in mouse, rat and rabbit. J Pharm Sci 1983; 72:609–612.

51. Ober SS, Vincent HC, Simon DE, Frederick KJ. A rheological study of procain penicillin G depot preparations. J Am Pharm Assoc 1958; 47:667–676.

52. Kadir F, Seijsener CBJ, Zuidema J. Influence of the injection volume on the release pattern of intramuscularly administered propranolol to rats. Int J Pharm 1992; 81:193–198.

53. Tanaka T, Kobayashi H, Oumura K, Muranashi S, Sezaki H. Intramuscular absorption of drugs from oily solutions. Chem Pharm Bull 1974; 22:1275–1284.

54. Hirano K, Ichihashi T, Yamada H. Studies on the absorption of practically water-insoluble drugs following injection. II: Intramuscular absorption from aqueous suspensions in rats. Chem Pharm Bull 1981; 29:817–827.

55. Hirano K, Yamada H. Studies on the absorption of practically water-insoluble drugs following injection. VI: Subcutaneous absorption from aqueous suspensions in rats. J Pharm Sci 1982; 71:500–505.

56. Hirano K, Yamada H. Studies on the absorption of practically water-insoluble drugs following injection. VII: Plasma concentration after different subcutaneous doses of a drug in aqueous suspensions in rats. J Pharm Sci 1983; 72: 602–607.

57. Hirano K, Yamada H. Studies on the absorption of practically water-insoluble drugs following injection. VIII: Comparison of subcutaneous absorption from aqueous suspensions in mouse, rat and rabbit. J Pharm Sci 1983; 72:609–612.

58. Kadir F, Eling WMC, Crommelin DJA, Zuidema J. Influence on release kinetics of liposomal chloroquine after s.c. and i.m. injection into mice. J Control Release 1991; 17:277–284.

59. Kadir F, Eling WMC, Crommelin DJA, Zuidema J. Kinetics and prophylactic efficacy of increasing dosages of liposome encapsulated chloroquine after intramuscular injection in mice. J Control Release 1992; 20:47–54.

60. Kadir F, Eling WMC, Crommeling DJA, Zuidema J. Influence of injection volume on the release kinetics of liposomal chloroquine administered intramuscularly or subcutaneously to mice. J Control Release 1991; 17:277–284.

61. Cockshott WP, Thompson GT, Howlett KL, Seely ET. Intramuscular or intralipomatous injections? N Engl J Med 1982; 307:356–358.

62. Vukovich RA, Brannick LJ, Sugerman AA, Neiss ES. Sex differences in the intramuscular absorption and bioavailability of cephradine. Clin Pharmacol Ther 1976; 18:215–220.

63. Dundee JW, Gamble JAS, Assaf RAE. Plasma diazepam levels following intramuscular injection by nurses and doctors. Lancet 1974; II:1461.

64. Shill M, Baynes RD, Miller SD. Fatal rabies encephalitis despite appropriate post-exposure prophylaxis. N Engl J Med 1987; 316:1257–1258.

65. Modderman ESM, Merkus FWHM, Zuidema J, Hilbers HW, Warndorff T. Sex differences in the absorption of dapsone after intramuscular injection. Int J Lepr 1983; 51:359–365.

66. Pieters FAJM, Zuidema J, Merkus FWHM. Sustained release properties of an intraadiposely administered dapsone depot injection. Int J Lepr 1986; 54:383–388.

67. Pieters FAJM, Woonink F, Zuidema J. A field trial among leprosy patients with depot injections of dapsone and monoacetyldapsone. Int J Leprosy 1988; 56:10–20.

68. Oussoren C, Zuidema J, Crommelin DJA, Storm G. Lymphatic uptake and biodistribution of liposomes after subcutaneous injection. I Influence of anatomical site of injection. J Liposome Res 1997; 7:85–99.

69. Allen TM, Smuckler EA. Liver pathology accompanying chronic liposome administration in mouse. Res Commun Chem Pathol Pharmacol 1985; 50(2):281–290.

70. Allen TM, Hansen C. Pharmacokinetics of stealth versus conventional liposomes: effect of dose. Biochem Biophys Acta 1991; 1068:133–141.

71. Allen TM. Toxicity of drug carriers to the mononuclear phagocyte system. Adv Drug Deliv Rev 1988; 2:55–67.

72. Harashima H, Kume Y, Yamane C, Kiwada H. Kinetic modeling of liposome degradation in blood circulation. Biopharm Drug Dispos 1993; 14:265–270.

73. Devine DV, Wong K, Serrano K, Chonn A, Cullis PR. Liposome-complement interactions in rat serum: implications for liposome survival studies. Biochim Bhiophys Acta 1994; 1191:43–51.

74. Patel HM. Serum opsonins and liposomes: their interaction and opsonophagocytosis. Crit Rev Ther Drug Carrier Syst 1992; 9:39–90.

75. Juliano RL. Factors controlling the kinetics and tissue distribution of liposomes, microspheres, and emulsions. Adv Drug Deliv Rev 1988; 2:31–54.

76. Moghimi SM, Patel HM. Serum mediated recognition of liposomes by phagocytic cells of the reticuloendothelial system—the concept of tissue specificity. Adv Drug Deliv Rev 1998; 32: 45–60.

77. Harashima H, Ohnishi Y, Kiwada H. In vivo evaluation of the effect of the size and opsonisation on the hepatic extraction of liposomes in rats: an application of Oldendorf method. Biopharm Drug Dispos 1992; 13:549–553.

78. Moghimi SM, Patel HM. Calcium as a possible modulator of Kupffer cell phagocytic function by regulation liver-specific opsonic activity. Biochim Biophys Acta 1990; 1028:304–308.

79. Senior JH. Fate and behaviour of liposomes in vivo: a review of controlling factors. Crit Rev Ther Drug Carrier Syst 1987; 3(2):123–193.

80. Gabizon A, Papahadjopoulos D. The role of surface charge and hydrophilic groups on liposome clearance in vivo. Biochim Biophys Acta 1992; 1103:94–100.

81. Storm G, Belliot SO, Daeman T, Lasic DD. Surface modification of nanoparticles to oppose uptake by the mononuclear phagocyte system. Adv Drug Deliv Rev 1995; 17:31–48.

82. Klibanov AL, Huang L. Long-circulating liposomes: development and perspectives. J Liposome Res 1992; 2(3):321–334.

83. Allen TA, Hansen CB, Lopes de Menezes DE. Pharmacokinetics of long-circulating liposomes. Adv Drug Deliv Rev 1995; 16:267–284.

84. Woodle M, Lasic DD. Sterically stabilized liposomes. Biochim Biophys Acta 1992; 1113:171–199.

85. Bakker-Woudenberg IAJM, Storm G, Woodle MC. Liposomes in the treatment of infections. J Drug Targeting 1994; 2: 363–371.

86. Torchilin VP, Trubetskoy VS. Which polymers can make nanoparticulate drug carriers long-circulating? Adv Drug Deliv Rev 1995; 16:141–155.

87. Gabizon A, Shiota R, Papahadjopoulos D. Pharmacokinetics and tissue distribution of doxorubicin encapsulated in stable liposomes with long circulating times. J Natl Cancer Inst 1989; 81(19):1484–1878.

88. Bakker-Woudenberg IAJM, Lokerse AF, ten Kate MT, Storm G. Enhanced localization of liposomes with prolonged circulating time in infected lung tissue. Biochim Biophys Acta 1992; 1138:318–326.

89. Bakker-Woudenberg IAJM, Lokerse AF, ten Kate MT, Mouton JW, Woodle MC, Storm G. Liposomes with prolonged blood circulation and selective localization in Klebsiella-pneumoniae-infected lung tissue. J Infect Dis 1993; 168: 164–171.

90. Peeters PAM, Storm G, Crommelin DJA. Immunoliposomes in vivo: state of the art. Adv Drug Deliv Rev 1987; 1:249–266.

91. Nässander UK, Steerenberg PA, Poppe H, Storm G, Poels LG, de Jong WH, Crommelin DJA. In vivo targeting of OV-TL 3 immunoliposomes to ascitic ovarian carcinoma cells (OVCAR-3) in athymic nude mice. Cancer Res 1992; 52:646–653.

92. Vingerhoeds MH, Storm G, Crommelin DJA. Immunoliposomes in vivo. Immunomethods 1994; 4:259–272.

93. Blume G, Cevc G, Crommelin DJA, Bakker-Woudenberg IAJM, Kluft C, Storm G. Specific targeting with poly(ethylene glycol)-modified liposomes: coupling of homing devices to the ends of polymeric chains combines effective target binding with long circulation times. BBA 1993; 1149:180–184.

94. Allen TM. Long-circulating (sterically stabilized) liposomes for targeted drug delivery. TIPS 1994; 15:215–220.

95. Allen TA, Agrawal AK, Ahmad I, Hansen CB, Zalipsky S. Anti-body mediated targeting of long-circulating (stealth) liposomes. J Liposome Res 1994; 4:1–25.

96. Bendas G, Krause A, Bakowsky U, Vogel J, Rothe U. Targetability of movel immunoliposomes prepared by a new antibody conjugation technique. Int J Pharm 1999; 181:79–93.

97. Maruyama K. PEG-immunoliposomes. Biosci Rep 2002; 22:252–266.

98. Koning G, Morselt HWM, Gorter A, Allen TM, Zalipsky S, Kamps JAAM, Scherphof GL. Pharmacokinetics of differently designed immunoliposome formulation in rats with or without hepatic colon cancer metastases. Pharm Res 2001; 18:1291–1298.

99. Bakker-Woudenberg IAJM, Lokerse AF, ten Kate MT, Mouton JW, Woodle MC, Storm G. Liposomes with prolonged blood circulation and selective localization in Klebsiella-pneumoniae-infected lung tissue. J Infect Dis 1993; 168: 164–171.

100. Park JW, Kirpiotin DB, Hong K, Shalaby R, Shao Y, Nielsen UB, Marks JD, Papahadjopoulos D, Benz CC. Tumor targeting using anti-her2 immunoliposomes. Journal of controlled release 2001; 74:95–113.

101. Lopes de Menezes DE, Pilarsky LM, Belch AR, Allen TM. Selective targeting of immunoliposomal doxorubicin against human multiple myeloma in vitro and ex vivo. BBA Biomembranes 2000; 1466:205–220.

102. Gabizon A, Goren D, Horowitz AT, Tzemach D, Lossos A, Siegal T. Long-circulating liposomes for drug delivery in cancer therapy: a review of biodistribution studies in tumor-bearing animals. Adv Drug Deliv Rev 1997; 24:337–344.

103. Forssen EA, Tökes ZA. Attenuation of dermal toxicity of doxorubicin by liposome encapsulation. Cancer Treatment Rep 1983; 67(5):481–484.

104. Axelsson B. Liposomes as carriers for anti-inflammatory agents. Adv Drug Deliv Rev 1989; 3:391–404.

105. Litterst CL, Sieber SM, Copley M, Parker RJ. Toxicity of free and liposome-encapsulated adriamycin following large volume, short-term intraperitoneal exposure in the rat. Toxicol Appl Pharmacol 1982; 64:517–528.

106. Kadir F, Oussoren C, Crommelin DJA. Liposomal formulations to reduce irritation of intramuscularly and subcutaneously administered drugs. In: Gupta PK, Brazeau GA, eds. Injectable Drug Development. Techniques to Reduce Pain and Irritation. Denver, Colorado: Interpharm Press, 1999:337–354.

107. Svendsen F, Rasmussen F, Nielsen P, Steiness E. The loss of creatine phosphokinase (CK) from intramuscular injection sites in rabbits. A predictive tool for local toxicity. Acta Pharmacol Toxicol 1979; 44:324–328.

108. Kadir F, Eling WMC, Abrahams D, Zuidema J, Crommelin DJA. Tissue reaction after intramuscular injection of liposomes in mice. Int J Clin Pharmacol Ther Toxicol 1992; 30:374–382.

109. Al-Suwayeh AA, Tebbett IR, Wielbo D, Brazeau GA. In vitro–in vivo myotoxicity of intramuscular liposomal formulations. Pharm Res 1996; 13(9):1384–1388.

110. Oussoren C, Eling WMC, Crommelin DJA, Storm G, Zuidema J. The influence of the route of administration and liposome composition on the potential of liposomes to protect tissue against local toxicity of two antitumor drugs. Biochem Biophys Acta 1998; 1369:159–172.

3

Characterization and Analysis of Dispersed Systems

JIM JIAO

Pharmaceutical Research and
Development, Pfizer Global
Research and Development, Pfizer Inc.,
Groton, Connecticut, U.S.A.

DIANE J. BURGESS

Department of Pharmaceutical
Sciences, School of Pharmacy,
University of Connecticut,
Storrs, Connecticut, U.S.A.

1. INTRODUCTION

The successful development and manufacture of pharmaceutical dispersions for parenteral administration relies on accurate analysis and characterization of these products. Applications of injectable emulsions, liposomes, and suspensions share one common problem: a few oversize particles in any of these submicron colloidal systems could cost the loss of hundreds of thousands of dollars and, even more importantly, more loss of life in the case of intravenous (i.v) delivery (1). It is imperative that high-quality injectable dispersed systems be developed and maintained, and that adequate, consistent, and comprehensive tests are available to ensure product quality and safety.

Appropriate assessment of an injectable dispersed system requires characterization of both chemical and physical stabilities. However, since the amount of active pharmaceutical ingredients dissolved in the continuous phase is limited by the drug solubility, the drug concentration available for chemical reactions is usually low and approximately constant over the product shelf life and drug degradation follows zero-order kinetics (2). As a result, chemical stability is relatively less of an issue compared to physical stability, which is often considered the determining factor for shelf life of dispersions such as emulsions and suspensions, as well as conventional liposome products. Physical properties are also very important with respect to the performance of dispersed systems. For example, the targeting of liposome systems may be dependent on particle size and/or charge (refer to chapter on Liposomes in this book). Drug release rates from microspheres are often dependent on particle size, as well as other factors such as polymer type and molecular weight (refer to chapter on Microspheres in this book).

The physical stability of dispersed systems can be characterized by evaluation of the optical, kinetic, and electric properties of the dispersed phase. The optical properties of a particle reflect its light scattering behavior and provide a basis for particle size and size distribution measurement by light microscopy and sizing techniques of light scattering, diffraction, and blocking. Kinetic properties resulting from Brownian motion and particle diffusion determine the rate of sedimentation or creaming of suspensions and emulsions that may eventually lead to caking and phase separation. Electrical properties resulting from the presence of a charge on the surface of a particle determine particle–particle interactions and affect flocculation and agglomeration. Electrical properties of particles also affect their interaction with cells in the body and can enable cellular uptake.

Commonly used characterization techniques for determination of particle size, zeta potential, and rheology are reviewed here.

2. PARTICLE SIZE MEASUREMENT

Particle size distribution is one of the most important character-istics of injectable dispersed systems. For example, sedimenta-tion or creaming tendencies of a dispersed system can be minimized by changing the particle size of the system. The sta-bility of injectable dispersed systems can also be conveniently monitored by measuring changes in particle size and size distri-bution. The biofate of some dispersed system dosage forms is dependent on their particle size distribution (3,4). Tomazic-Jezic et al. (5) reported an intensive phagocytosis of small- and med-ium-sized polystyrene particles (1.2 and 5.2 μm) after peritoneal injection of the polystyrene particles into mice but no engulf-ment of large polystyrene particles (12.5 μm) by the macro-phages was observed. The intravenous injection of emulsion droplets above 5 μm is clinically unacceptable because they cause the formation of pulmonary emboli (6,7). The size require-ment for suspensions administered intravenously is even smal-ler since unlike emulsion droplets solid particles lack flexibility that would enable passage through small capillaries.

A wide range of particle sizes is typical of dispersed sys-tems as evidenced by intravenous fat emulsions that contain droplets in the range of 10 nm–1 μm and by emulsions used as contrast media in computerized tomography that are of 1–5 μm size (8,9). Droplets larger than 5 μm are sometimes present because of inefficient homogenization or as a result of emulsion instability. Clearly, it is difficult to use a single method for determination of all particle sizes in such a wide range. For most colloidal dispersions the mean particle size as well as size distribution can be accurately determined by photon correlation spectroscopy (PCS). However, this method is not capable of measuring sizes greater than 3 μm in dia-meter. Laser diffraction can be used for measurement of par-ticles greater than 3 μm and is therefore useful for detecting larger particles. Photomicroscopy can also be used for sizing large particles. Table 1 shows various particle size analysis methods and approximate size ranges for each. Some of these methods including microscopy, electrical/optical sensing zone, and light scattering are reviewed in detail.

Table 1 Approximate Size Range of Methods for Particle Size Analysis

Particle sizing method	Size range (μm)	Comments
Optical microscopy (transmitted, reflected, polarized light, fluorescence, and confocal)	~0.5–600	Important tool for assessing particle size, shape, flocculation, aggregation, and coalescence, etc. Results are subjective and affected by sampling technique
Electron microscopy (transmission and scanning)	~0.01–10	High magnification and direct observation of particle size and shape. Samples need to be dry, coated, or frozen which can affect stability and size. Instrument is relatively expensive and difficult to operate
Electrical sensing zone (Coulter counter)	~0.5–500	Accurate but requires samples containing electrolyte to conduct current
Sedimentation	~1–500	Based on Stokes' law and applied to particles that settle in the dispersion medium by gravity without causing turbulence
Ultracentrifugation	~0.01–5	Based on Stokes' law and applied to particles that can be separated in the dispersion medium under centrifugal force
Sieving	~50–5000	No practical significance for colloidal dispersed systems

Method	Range	Description
Dynamic light scattering (Malvern, Nicomp, Brookhaven)	~0.01–3	Commonly used particle size method for injectable dispersed systems. Upper size limit 3 μm
Static light scattering	~0.02–2000	Commonly used particle size method for injectable dispersed systems
X-ray and neutron scattering	~0.005–10	Similar to light scattering techniques but with better resolution
Size exclusive chromatography or field-flow fractionation	~0.05–20	Particles are separated according to size by interacting with the stationary phase or field force applied to samples (electric, magnetic, or thermal). Requires sample be stable under separating conditions
Optical sensing zone (HIAC, AccuSizer)	~0.5–500	Based on the principle of light blockage and is the commonly used method for coarse-dispersed systems

2.1. Microscopic Methods

Microscopy is viewed by many as the ultimate tool for sizing particles since the direct observation of a sample is simple to interpret (10). Microscopy is also the most direct technique of assessing stability of dispersed systems because changes in particle size/shape and particle–particle interactions such as flocculation and aggregation can be visually observed. In general, microscopy is categorized into two groups depending on the radiation sources used: optical and electron. The optical microscope uses visible light as its radiation source and is only able to distinguish two particles separated by about 300 nm. The radiation source of the electron microscope is a beam of high-energy electrons having wavelengths in the region of 0.1 Å, which enables it to distinguish two particles separated by approximately 10 Å.

2.1.1. Optical Microscopy

Optical microscopy involves the use of transmitted light, reflected light, polarized light, fluorescence, and more recently techniques such as confocal microscopy. Each of these methods has particular strengths and applicabilities. Transmitted light microscopy requires a sample sufficiently thin to allow light to pass through it. This requirement is often accomplished by simple smearing of the sample on a slide. Most samples are transparent to white light or polarized light that is used as the light source to observe the samples. To enhance observation of dispersed systems, for example, emulsions, the fluorescence behavior of the organic phase is used. This approach involves incident light of violet or ultraviolet wavelength and observation of the fluorescent light in the visible region. The incident reflected beam is filtered out, and the returning light is due to the fluorescent behavior of the oil phase.

In cases where the sample cannot be made thin enough, a reflected light technique can be used. The reflecting microscope differs substantially from the petrographic (transmitted light) microscope. In particular, light must be directed onto the smooth surface of an opaque sample in such a manner that it can reflect up the microscope tube, rather than being transmitted through the material being investigated as is the case with transparent materials. For this reason, an illuminating system is located part way up the microscope tube, light is directed horizontally through an assemblage of diaphragms, lenses, and a polarizer such that polarized light is incident on a reflecting mechanism (mirror in the microscope tube) that directs some of the rays downward. The rays pass through an objective, hit the surface of interest, and reflect upward, partly passing through the reflector up the tube, through an ocular lens system, then reaching the observer.

Regular light microscopy has a limited depth of field or a narrow focal plane. Some samples, such as multiple emulsions containing large multiple droplets, have difficulty in being accommodated in the focal plane and signals above and below this plane are acquired as out-of-focus blurs that distort and degrade the contrast and sharpness of the final image. The confocal microscope solves these problems and adds a new dimension to optical microscopic analysis (11). Confocal microscopy uses a single-point source of light brightly illuminating a small volume inside thick samples. The area surrounding this point is weakly illuminated. A pinhole mask placed before the image detector allows only the single brightly illuminated spot to go through and be detected. The remaining area is blocked out so that a clear single-point image is formed. The point light source or sample station can be moved to map out the whole sample point by point. With the aid of fluorescent labeling, confocal microscopy is able to provide three-dimensional images of samples with 30–60 sec temporal and $\sim$0.2 μm spatial resolution which is often sufficient for the structural details of dispersion samples to be observed.

Using optical microscopy dispersed systems can be examined as is without sample treatment so that the network structure of particles such as flocculation and agglomeration may be directly observed. Video microscopy allows continuous monitoring of samples and therefore provides a better understanding of the destabilization process. Video microscopy has been used to determine multiple emulsion formation and droplet instability (12,13). However, one should be aware that samples placed on glass slides for light microscopic examination have a different environment than that of the bulk phase and these environmental changes may introduce different instability pathways from those that normally occur. Also microscopic examination requires the use of a coverslip placed over the sample (14). The coverslip can create enough pressure to deform, break or coalesce particles, and consequently alter destabilization pathways (Jiao et al., 2000).

2.1.2. Electron Microscopy

Particles smaller than about 0.5 µm begin to approach the resolution limit of the optical microscope which is a function of the visible light wavelength (typically 500–600 nm) and the numerical aperture of the objective (maximum at 1.3–1.4 with 100× oil immersion lens and 1.6 refraction index immersion oil). For observation of such small particles, electron microscopy has to be used. There are two types of electron microscopy: scanning and transmission. These are analogous to conventional optical microscopy but offer significant advantages in terms of depth of field and resolution. A transmission electron microscope (TEM) works much like a slide projector and shines a beam of electrons through the specimen. The electron beam passes through it and is affected by the structures of specimens and results in transmission of only a certain part of the electron beam. The transmitted beam is then projected onto the phosphor screen forming an enlarged image of the sample. In a scanning electron microscope (SEM), an electron gun emits a beam of high-energy electrons that travels downward through a series of magnetic lenses designed to focus the electrons to a fine point. Near the bottom, a set of scanning coils moves the focused beam back and forth across the specimen row by row. As the electron beam hits each spot on the sample, secondary electrons are knocked loose from its surface. A detector counts these electrons and sends the signals to an amplifier. The final image is built up from the number of electrons emitted from each spot on the sample. SEM shows detailed three-dimensional images at much higher magnifications than is possible with a light microscope.

There are, however, disadvantages associated with electron microscopic methods. First, the samples have to withstand the vacuum inside the SEM and need to be dry and coated with a thin layer of gold to conduct electricity, resulting in possible loss of the structure that exists when particles remain in the continuous phase. For example, drug crystals containing hydrates in suspension may lose associated water upon drying and become pseudomorphs (Grant, 2001). Emulsion droplets cannot be directly measured unless the sample is frozen and a freeze-etching electron microscopy technique is applied (15). Second, the instruments are more expensive and the observation process is time consuming compared to optical microscopy, thus making them inapplicable for routine size analysis.

2.2. Electrical Sensing Zone Method

In this, instrument particles are diluted in an electrolyte solution and passed through a fine capillary that connects two larger chambers containing immersed electrodes. A potential difference is applied between the electrodes. The resistance change that occurs when a particle passes through the orifice between the chambers is proportional to particle size (volume). The most common instrument for this technique is the Coulter counter, which is widely used for biological or industrial applications in quality control or research. This technique has high accuracy and resolution. The Coulter counter uses an aperture instead of a fine capillary to separate the two electrodes that are immersed in a weak electrolyte solution between which an electric current flows and suspended particles are drawn through. As each particle passes through the aperture (or "sensing zone"), it momentarily displaces its own volume of conducting liquid thus increasing the impedance of the aperture. This change in impedance produces a tiny but proportional current flow into an amplifier that converts the current fluctuation into a voltage pulse large enough for accurate measurement. Scaling these pulse heights in volume units enables a size distribution to be acquired and displayed. In addition, if a metering device is used to draw a known volume of the particle suspension through the aperture, a count of the number of pulses will yield the concentration of particles in the sample.

A wide range of aperture sizes is available to cover particle sizes between 0.5 and about 500 µm. However, the concentration of suspended particles cannot be too high because the appearance of two or more particles in the sensing zone at one time would be measured as a single larger particle. The size range analyzed in a single aperture is also limited because particles of the same or larger size than the aperture diameter lead to blockage, and particles smaller than about 2% of the aperture diameter do not produce a signal above the background noise. Furthermore, electrolytes may change the properties of dispersed system and therefore affect the results. Some Coulter models have mercury gauges or switches in them. The mercury may be in a pressure gauge, on–off switch, timing count gauge, vacuum gauge, and possibly other gauges, depending on the model and year. There is no evidence that the mercury in such equipment is spilling, leaking, or causing other problems. However, operation, repair, and maintenance service of such instruments requires extra precautions.

2.3. Optical Sensing Zone Method

The principle of the optical sensing zone method is analogous to that of the electrozone method. The optical sensing instrument consists of an autodiluter, an optical sensor, and a capillary placed on the path of a beam of light. When a solution containing particles flows through the capillary, the particles interrupt the incident light beam and decrease the amount of light that reaches the optical sensor. The decrease in light transmission produces a voltage pulse proportional to the projected area of each particle. For this reason, it is also referred to as light extinction or blockage method. This approach is in contrast to "ensemble" methods, such as Fraunhofer diffraction and sedimentation, which must process information produced by many particles simultaneously. The method uses simple statistics to convert differential and cumulative population data (counts vs. size) to the respective number, area, and volume distributions. It has advantages such as high resolution because of the single-particle counting mechanism, high reproducibility due to auto dilution, quick measurement, and the absence of electrolyte in the counting medium so that particles can be suspended in any medium including organic solvents.

However, the low concentration limits of the sensors require extensive dilutions of concentrated dispersions, making the system more suitable for particle contamination monitoring rather than particle sizing. Typically clean room conditions are required for optimal sensitivity.

2.4. Dynamic Light Scattering Method (Photon Correlation Spectroscopy)

Of the many techniques available for particle sizing light scattering offers many advantages: speed, versatility, small sample size, non-destructive measuring, and measurement times independent of particle density. The oscillating electromagnetic field of light induces oscillations of the electrons in a particle. These oscillatory changes form the source of the scattered light. Many features of the scattered light have been used to determine particle size. These include: (1) changes in the average intensity as a function of angle; (2) changes in the polarization; 3) changes in the wavelength; and (4) fluctuations about the average intensity (Brookhaven, 1999).

The phenomenon of fluctuation about the average intensity is the basis for quasi-elastic light scattering (QELS), the technique utilized in the photon correlation spectroscopy (PCS) method. In the PCS instrument, the distance between the detector monitoring scattered light from each particle and the scattering volume containing a large number of particles is fixed. Since the small particles are moving around randomly in the liquid, undergoing diffusive Brownian motion, the distance that the scattered waves travel to the detector varies as a function of time. The scattered wave can interfere constructively or destructively depending on the distances traveled to the detector just like water and sound waves. The decay times of the fluctuations are related to the diffusion constants and, therefore, the particle size. Small particles moving rapidly cause faster decaying fluctuations than large particles moving slowly. The decay times of these fluctuations may be determined either in the frequency domain (using a spectrum analyzer) or in the time domain (using a correlator). The correlator generally offers the most efficient means for this type of measurement. In QELS, the total time over which a measurement is made is divided into small time intervals called decay times. These intervals are selected to be small compared with the time it takes for a typical fluctuation to relax back to the average. The scattered light intensity in each of these intervals, as represented by the number of electrical pulses, fluctuates about a mean value. The intensity autocorrelation function (ACF) is formed by averaging the products of the intensities in these small time intervals as a function of the time between the intervals.

2.4.1. Basic Equations

The random motion of small particles in a liquid gives rise to fluctuations in the time intensity of the scattered light. The fluctuating signals are processed by forming the ACF, $C(t)$, t being the time delay. For short times, the correlation is high. As t increases the correlation is lost, and the function approaches the constant background term B. In between these two limits the function decays exponentially for a monodisperse suspension of rigid, spherical particles and is given by

$$C(t) = Ae^{-2\Gamma t} + B \tag{1}$$

where A is an optical constant determined by the instrument design, and Γ is related to the relaxation of the fluctuations by

$$\Gamma = Dq^2 \text{ (rad/sec)} \tag{2}$$

The value of q is calculated from the scattering angle θ (e.g., 90°), the wavelength of the laser light λ_0 (e.g., 0.635 μm), and the refraction index n (e.g., 1.33) of the suspending liquid.

The equation relating these quantities is

$$q = \frac{2\pi n}{\lambda_0} 2 \sin\left(\frac{\theta}{2}\right) \tag{3}$$

The translational diffusion coefficient, D, is the principal quantity measured by QELS. It is an inherently important property of particles and macromolecules. QELS has become the preferred technique for measuring diffusion coefficients of submicron particles. Notice that no assumption about particle shape has to be made in order to obtain D other than assuming a general globular shape. Not until needle shaped particles have aspect ratios of greater than about 5, does a rotational diffusion term appear in Eq. (1).

Particle size is related to D for simple common shapes like a sphere, ellipsoid, cylinder, and random coil. Of these, the spherical assumption is useful in most cases. For a sphere,

$$D = \frac{k_B T}{3\pi\eta(t)d} \ (\text{cm}^2/\text{sec}) \tag{4}$$

where k_B is Boltzmann's constant, T is the temperature in K, $\eta(t)$ (in centipoises) is the viscosity of the liquid in which the particle is moving; and d is the particle diameter. This equation assumes that the particles are moving independently of one another. Generally the determination of particle size consists of four steps: (1) measurement of the ACF; (2) fitting the measured function to Eq. (1) to determine Γ; (3) calculating D from Eq. (2) given n, θ, and Γ; and (4) calculating particle diameter d from Eq. (4) given T and η.

2.4.2. Data Interpretation

The type of diameter obtained with PCS is the hydrodynamic diameter (the particle diameter plus the double layer thickness) that is the diameter a sphere would have in order to diffuse at the same rate as the particle being measured. When a distribution of sizes is present, the effective diameter measured is an average diameter weighted by the intensity of light scattered by each particle. This intensity weighting, which is discussed below, is not the same as the population or number weighting used in a single-particle counter such as in electron microscopy. However, for narrowly dispersed samples, the average diameters obtained are usually in good agreement with those obtained by single-particle techniques. The measured ACF determined using the above equations provides accurate particle size results for mono- or narrowly dispersed rigid spheres. When a broad distribution of spheres is present in the sample, Eq. (1) must be modified before data analysis of the measured ACF can proceed. Each size contributes its own exponential, hence the mathematical problem becomes

$$g(t) = \int G(\Gamma) e^{-\Gamma t} d\Gamma \tag{5}$$

here the baseline B of Eq. (1) has been subtracted and the square root of the remaining part has been taken eliminating the factor of 2 in the exponent.

The left-hand side of Eq. (5) is the measured data, and the desired distribution information, $G(\Gamma)$, is contained under the integral sign on the right-hand side. The solution to this Laplace transform equation is non-trivial. The insufficient conditioning of this transform, combined with measurement noise and baseline drifts, makes the function particularly difficult to solve. It is therefore imperative to acquire the best possible statistics.

2.4.3. Fit to a Known Distribution

Several schemes have been proposed for obtaining information from Eq. (5). These may be grouped as follows: (1) fit to a known distribution; (2) cumulant analysis; (3) inverse Laplace transform. The fit to a known distribution method requires the assumption of a particular distribution. For example, if the distribution is assumed to consist of only two peaks, each of which is very narrow, then Eq. (5) reduces to the sum of two exponentials. The measured ACF is then fit to this sum of exponentials. The ratio of the fitted, pre-exponential factors contains information on the ratio of scattered intensities for each of the two peaks, which in turn can be related to their mass ratio. Similarly the fitted Γ contains information on the diameters of the two peaks.

The assumption of a given form (here bimodal) for the distribution is problematic. One might think that different models (unimodal, bimodal, trimodal, etc.) could be compared using a statistical criteria such as a goodness-of-fit parameter like chi-squared. Unfortunately, it is well known that this approach often leads to ambiguous results with equations such as Eq. (5). An alternative approach is to assume an analytical equation for the distribution and fit the results to the measurements. However, this approach again suffers from the assumption of a particular form of the distribution and has generally not proven successful.

2.4.4. Cumulant Analysis

The method of cumulants is more general compared to the above method (Brown et al., 1975). An understanding of this method provides the researcher with an insight into the "intensity"-weighted character of the results from PCS. Cumulant results are very often reported in the scientific literature. They are often the starting point for further analysis. During a measurement, the cumulant results (effective diameter and polydispersity index) are calculated and displayed on an ongoing basis. No assumption has to be made about the form of the distribution function in the method of cumulants. The exponential in Eq. (5) is expanded in a Taylor series about the mean value. The resulting series is integrated to give a very general result. This result shows that the logarithm of the ACF can be expressed as a polynomial in the delay time, t. The coefficients of the powers of t are called the cumulants of the distribution. In practice, only the first couple of cumulants are obtained reliably, and these are identical to the moments of the distribution. In general, the first moment of any distribution is the average and the second is the variance.

The first two moments of the distribution $G(\Gamma)$ are as follows:

$$\Gamma = Dq^2 \tag{6}$$

$$\mu_2 = \left(D^2 - D^{*2}\right)q^4 \tag{7}$$

where q is the scattering vector given previously and D^* is the average diffusion coefficient. Equation (7) shows that μ_2 is proportional to the variance of the intensity-weighted diffusion coefficient distribution. As such, it carries information on the width of the size distribution. The magnitude and units of μ_2 are not immediately useful for characterizing a size distribution. Furthermore, distribution with the same relative width (same shape) may have very different means and variances. For these reasons, a relative width (reduced second moment) is conveniently defined as follows:

$$\text{Polydispersity} = \frac{\mu_2}{\Gamma^2} \tag{8}$$

Polydispersity has no units. It is close to zero (0–0.2) for monodisperse or nearly monodisperse samples, small (0.2–0.8) for narrow distributions, and larger for broader distributions.

2.4.5. Intensity Weighting of the Averaging Process

The intensity of light scattering by a suspension of particles with diameter d is proportional to the number of particles N, the square of the particle mass M, and the particle-form factor $P(q,d)$ which depends on size, scattering angle, index of refraction, and wavelength. Using "intensity" weighting, the following equations may be derived for the moments of the diffusion coefficient distribution:

$$D = \frac{\sum NM^2 P(q,d)D}{\sum NM^2 P(q,d)} \tag{9}$$

$$D^2 = \frac{\sum NM^2 P(q,d)D^2}{\sum NM^2 P(q,d)} \tag{10}$$

here the sums are carried out over all the particles. For particles much smaller than the wavelength of light (say, 60 nm and smaller), $P(q,d)$ equals 1. Likewise, it equals 1 if measurements are extrapolated to zero angle. For narrow distributions, an average value of $P(q,d)$ can be used. In this case, it cancels out of both equations. In other cases, suitable approximations exist for $P(q,d)$ (Allcock and Frederick, 1981). To further understand the intensity weighting process and how it is related to other techniques, consider the simple case where $P(q,d) = 1$. Then

$$D = D_z = \frac{\sum NM^2 D}{\sum NM^2} \qquad (11)$$

This is called the z-average. More familiar are the number, area, and weight or volume averages. Furthermore, since the diffusion coefficient is inversely proportional to the diameter for a sphere, the "average" obtained in this type of light scattering experiment is the inverse z-average given by

$$\frac{1}{d_z} = \frac{\sum NM^2(1/d)}{\sum NM^2} \qquad (12)$$

since M is proportional to d^3, and since an average rather than an inverse average is preferred, the effective diameter, in this case, is defined by the following equation:

$$d_{psc} = \left(\frac{1}{d_z}\right)^{-1} = \frac{\sum Nd^6}{\sum Nd^5} \qquad (13)$$

Comparing this to the number average ($d_n = \sum Nd/\sum N$), the area average ($d_a = \sum Nd^3/\sum Nd^2$), and the weight average diameter ($d_w = \sum Nd^4/\sum Nd^3$), all four averages are equally descriptive of the same sample. One may be preferred for a particular application: d_n when numbers of particles are important; d_a when surface area is important; d_w when mass or volume is important; and d_{pcs} when light scattering-dependent properties are important.

Examination of the above equations shows that, quite generally, $d_n \leq d_a \leq d_w \leq d_{pcs}$. The equal sign occurs only for monodisperse samples. For this reason, the weight average diameter is always less than or equal to the effective diameter. For narrow distribution, they are nearly equal, and for broader distributions they will differ considerably. In order to transform the cumulant results to weight average results, it is necessary to assume a form for the size distribution. (Please note that this is quite different from assuming a form prior to determining the cumulants as discussed before.) The lognormal distribution by weight has been widely used in applications of particle sizing. The lognormal distribution by weight is given by

$$
dW = \left\{ \frac{1}{\ln \sigma_g \sqrt{2\pi}} \exp \left[-\left(\frac{\ln d(-\ln \mathrm{MMD})^2}{\ln \sigma_g \sqrt{2}} \right)^2 \right] \right\} d(\ln d)
$$

$$(14)$$

where MMD is the mass median diameter (50% by mass, weight, or volume is above and 50% is below this value) and σ_g is the geometric standard deviation. With these two parameters, all other distributions, averages, tables, and graphs can be produced.

For narrow, unimodal size distributions any other two parameter equation is equally adequate. For broad and unimodal distributions, the lognormal is a better assumption than a symmetric form like a Gaussian (neither grinding, accretion, nor aggregation lead to symmetrical size distributions). Furthermore, manipulation of the lognormal equation is particularly convenient. The lognormal parameters are described in any number of standard tests on size distributions (Hinds, 1982). To monitor relative changes in size distribution either the cumulant results or the lognormal fit will usually suffice. However, when one requires more detailed information (for example, is it bimodal or just broad and skewed?) then the cumulant/lognormal analysis is no longer adequate.

2.4.6. Non-spherical Particles

Describing the "size" distribution of non-spherical particles presents difficulties for any technique including PCS. If shape information is needed, then an image analysis is necessary. Even then problems arise with sample preparation, statistical relevance of the small number of particles sized, and finally, choosing the right statistical description of "size". Other techniques have their pitfalls. The settling velocity of non-spherical particles depends, in a non-trivial fashion, on shape. Thus, results from sedimentation techniques are shape dependent. Highly porous, rough, or multifaceted particles present much more surface area than smooth spheres of equivalent volume, therefore, size distributions based on surface area measurements may be highly biased toward large sizes. Non-spherical particles may interact quite differently with the flow than spherical ones.

Angular light scattering measurements have the capability, in some circumstance, of distinguishing between simple shapes like spheres, ellipsoids, rods, and random coils (polymers). However, this is not possible with only one scattering angle unless extra information is supplied. For example, if it were known that the particles were all prolate ellipsoids (cigar shaped), with the same aspect ratio (major/minor ratio), then the diffusion coefficient results obtained by PCS could be used in conjunction with an equation relating the translational diffusion coefficient of an ellipsoid to its aspect ratio to obtain the other dimension (Cummins, 1964).

2.4.7. Advanced Data Interpretation

As explained before, the general solution of Eq. (5) is surprisingly difficult. Several approximate solutions have been proposed over the more than 20 years since the initial QELS measurements were performed. Most are of very limited utility. The reason for the many failures has been addressed in a paper by Pike (16). Grabowski and Morrison (17) formulated an approach for the solution of this problem. This approach called the non-negatively constrained least squares (NNLS) algorithm is used at the heart of PCS. The prime assumptions with NNLS are: (1) only positive contributions to the intensity-weighted distribution are allowed; (2) the ratio between any two successive diameters is constant; (3) a least squares criterion for judging each iteration is used. In order to calculate the weight (or mass or volume) and number fractions from the intensity fractions, scattering factors have to be calculated. These are called spherical Mie factors. In order to calculate these corrections, the complex refractive index of the particle (in addition to the refractive index of the liquid) must be known. It is worth repeating that the most important pieces of information obtained in PCS using any multimodal size distribution analysis are the positions of the peaks and the ratio of the peak areas. The widths of the peaks, however, are not particularly reliable. They will narrow, typically, with increasing experiment duration, until a limit is reached.

2.4.8. Particle Size Characterization of Injectable Dispersed Systems

Table 2 shows selected papers on particle size characterization of experimental injectable dispersions. As an example, Bakan et al. (18) reported a study on a lipid emulsion for hepatocyte-selective delivery of polyiodinated triglycerides as a contrast agent for computed tomography. They successfully incorporated the lipophilic compound into the delivery vehicle to form a stable chylomicron-remnant-like emulsion capable of localizing material to the liver following intravenous injection. Particle size was determined using laser PCS and transmission electron microscopy. PCS was conducted using Nicomp 370 submicron particle analyzer (Particle Sizing Systems, Santa Barbara, CA, USA). Samples were diluted approximately 1000-fold with deionized water prior to analysis. The Nicomp 370 was calibrated against NIST-traceable latex standards (Duke Scientific, Palo Alto, CA, USA) to within 10% or less of the certified diameter. Transmission electron microscopy was performed using both freeze-fracture and negative staining techniques. For freeze-fracture studies, a 2–4 µL sample of concentrated emulsion was frozen in liquid nitrogen, loaded in a double-replica breaking device, and placed on the specimen stage of a Balzers Model BAF 301 freeze-fracture/etch unit. The specimen stage was evacuated and the temperature was raised from -170 to $-115°C$. Samples were fractured, etched for 60 s, and coated with 1–2 nm of platinum which was then dried at an angle of 45° while the sample was rapidly rotating. The platinum was stabilized with 10–20 nm of vapor carbon applied at a 90° angle. Replicas were floated off the sample in double-distilled water and submerged in methanol. The methanol was replaced slowly with 50% chlorine bleach and held overnight. Samples were rinsed three times with double-distilled water and placed on 400 mesh copper grids for visualization. For negative staining, emulsion samples were diluted in 1% ammonium molybdate and freshly glow-discharged carbon-coated 400-mesh copper grids were floated on 25 mL of the diluted sample for 1 min. Grids were removed and air-dried prior to examination. All images were collected using a Phillips Electronics Instrument Model CM-10 transmission electron microscope using a 60 kV accelerating voltage and 30 μm objective aperture.

2.4.9. Limitations

Table 1 provides a general guidance on which sizing instruments should be used for what particle size ranges. There are overlaps where more than one sizing instrument can measure a particular size range. It is common that different sizing instruments provide different particle sizes on the same sample due to different instrument operation principles, e.g., microscopy vs. PCS. Even instruments with the same fundamental basis made by different vendors can produce different results because of different detector designs and proprietary algorithms, not to mention the effect that sampling technique has on the results. One has to be aware of such instrumentation-related differences for stability assessment and product development, especially when comparing particle size data from different sources or collaborating multiple laboratory efforts where each laboratory has its own particle sizing capability. Inconsistencies in particle size and size distribution often cause difficulties or delay in formulation development and technology transfer from laboratory to production scale.

3. ZETA POTENTIAL

Almost all solids acquire a surface charge when placed in polar liquids. The charge can arise in a number of ways including ionization of surface groups such as carboxyl, amino

Table 2 Selected Papers Reporting Particle Size Characterization of Dispersed Systems for Parenteral Administration

Dispersed system	Application	Particle size range	Sizing method	Injection route	Literature source
Liposomes	Intrahepatic distribution of phosphatidylglycerol and phosphatidylserine	200–400 nm	Dynamic laser light scattering	i.v. in rats	Daemen et al. (49)
Lipid emulsion	Target delivery of highly lipophilic antitumor agent 13-O-palmitoylrhizoxin	94–474 nm	Dynamic laser light scattering	i.v. in rats	Kurihara et al (50)
Polymethyl methacrylate and polystyrene suspensions	Tissue distribution and phagocytosis of particulates from wear of injected biomaterials	1.2–12.5 μm	Laser light diffraction	i.p. in mice	Tomazic-Jezic et al. (5)
Liposomes	Target delivery of heme oxygenase inhibitor tin mesoporphyrin	< 200 nm	Nicomp 270 submicron particle analyzer	i.v. in rats	Cannon et al. (51)
Liposomes	Formulation development for Atovaquone-containing liposomes	~260 nm	Photon correlation spectroscopy	in vitro	Cauchetier et al. (52)

Polyethylene suspension	Bone resorption activity of macrophages on polyethylene wear debris of artificial hip joint	0.24–7.62 μm	Laser light diffraction	in vitro	Green et al. (53)
Lipid emulsion	Hepatocyte-selective imaging agent for computed tomography	~200 nm	Nicomp 370 submicron particle analyzer and TEM	i.v. in rats	Bakan et al. (18)
Microsphere	Development of cross-linked dextran microspheres for parenterals	10–20 μm	Laser light diffraction	s.c. in rats	Cadée et al. (54)
Super paramagnetic iron oxide	Enhancement of liver and spleen uptake of iron oxide for blood pool MR-angiography	65 nm	SEM	i.v. in rabbits	Christoph et al.

groups, adsorption of ions such as surfactants, multivalent ions, polyelectrolytes, and unequal dissolution of the ions comprising the surface molecules such as metal oxides and silver halides. Surface charge affects the physical interactions of particles and plays an active role in governing dispersed system stability (19,20).

For emulsions, liposomes, and nanoparticles, their potential as drug carriers is closely related to their in vivo distribution. After intravenous administration, these carriers are rapidly removed from the circulation as a result of phagocytosis by mononuclear cells. Much research has been devoted to developing carrier systems which can avoid phagocytosis and thus circulate longer. It is known that when the surface of particles is covered with hydrophilic chains (such as polyethylene glycol and polyoxyethylene) then opsonization of the particles is reduced. This technique has been successfully applied to liposomes (21), nanoparticles (22,23), and emulsions (24–27). The surface properties of colloidal systems are critical in determining the drug carrier potential, which will control interactions with plasma proteins (28). Zeta potential measurements provide information on particle surface charge and on how this is affected by changes in the environment (e.g., pH, presence of counter ions, adsorption of proteins). Charge shielding by PEG or other hydrophilic groups can be used to predict the effectiveness of the barrier function against opsonization in vivo. Zeta potential can also be used to determine the type of interaction between the active substance and the carrier; i.e., whether the drug is encapsulated inside the particle or simply adsorbed on the particle surface. This is important because adsorbed drug may not be protected from enzymatic degradation, or may be released rapidly after administration.

Attractive forces exist even for completely non-polar particles. Such forces have their origin in the permanent and the momentarily induced dipole effects that arise when non-polar particles approach each other at random. For simple molecules, the potential energy of the attractive forces varies as r^{-6}, where r is the intermolecular distance. However, for colloidal particles, the potential energy of the attractive forces varies between D^{-1} and D^{-2} as a result of the sum of the attractive forces over all individual molecular forces, where D is the shortest distance between particles. Consequently, colloidal particles attract each other over much longer ranges than individual molecules. This leads to aggregation unless a repulsive force is present.

Repulsive forces between particles, mainly electrostatic repulsion, depend on particle surface charge. The surface charge varies with solution conditions such as pH, ionic strength, and concentrations of surfactants or other reagents. In general, a charged surface tends to gather ions of the opposite charge (counter ions) close to it. The closest counter ions may remain permanently at the particle surface and be carried along with the particle in motion. Those ions that are further away from the particle surface will be replaced by other ions as the particle moves. The distribution of ions in the region around the particle is a difficult theoretical problem to solve. Close to the particle surface, the ion distribution may be dominated by effects related to the shape and size of the surface and by ions in the solution. Whereas far from the surface, the ion distribution is primarily, dominated by electrostatic considerations. This gives rise to the formation of a double layer, the inner and diffuse regions being characterized by different ionic behaviors. The overall effect of the ionic atmosphere is to shield the surface charge. The higher the ionic strength of the medium the more compact the diffuse region becomes due to the strong inter-ionic attraction. The more compact the diffuse layer, the greater the shielding effect. At some distance from the surface within the double layer the ions are no longer dragged along with a moving or diffusing particle, but remain in the bulk solution. This is referred to as the plane of shear where electrokinetic and stationary ions are separated. The potential at the plane of shear is, by definition, the zeta (ζ) potential and can be measured using eletrophoretic techniques (29).

3.1. Electrophoresis

When an electric field is applied to a liquid, the dispersed particles in the liquid having a surface charge will move towards either the positive or the negative pole. The direction of particle movement is a clear indication of the sign of their charge. The velocity with which they travel is proportional to the magnitude of the charge. The technique that measures both the direction and the velocity of the particles under the influence of a known electrical field is termed electrophoresis. Classically, this technique involved measuring the velocity of individual particles in a suspension or emulsion, viewed in a microscope fitted with a reticule, and the transit times across the reticule being recorded using some more or less sophisticated timer. This technique is clearly limited to particles large enough to be visible (in practice $\sim 0.4\,\mu m$ or above). It has less obvious limitations in that Brownian motion can blur particle positions, particles of high potential may move so fast across the reticule that they do not remain in focus long enough to time accurately. In addition, many particles must be measured to get a high accuracy, which is a time consuming and tiring process. These considerations were sufficiently weighty to severely limit the use of this technique.

3.2. Laser Doppler Electrophoresis

It is well known that light scattered from a moving particle experiences a frequency shift called the Doppler shift. Modern zeta potential instruments measure zeta potential using the Doppler shift of laser light scattered by particles moving in an electric field. In such instruments, a laser beam passes through the sample in a cell carrying two electrodes to provide the electrical field. Light scattered at certain angles (typically 15° angle) is detected for the Doppler shift which is proportional to the velocity of the moving particles. Normally the Doppler shift is of the order of a 10^2 Hz, a negligible value compared to the frequency of light 10^{14} Hz. To measure such small changes in the frequency of scattered light an optical mixing or interferometric technique has to be used. This is done in practice using a pair of mutually coherent laser beams derived from a single source and following similar path lengths. Usually a portion of the beam is split off (the reference beam or the local oscillator) and then recombined with the scattered beam after it is modulated at 250 Hz. What this means is that in the absence of an electrical field, a power spectrum of the signal from the detector would have a sharp peak at 250 Hz. When an electrical field is applied, any resultant Doppler shift would occur from this frequency, and it is easy to detect a shift of 100 Hz from 250 Hz. This arrangement is referred to as heterodyne measurement and is much more efficient than the traditional technique. The frequency is converted successively to velocity, electrophoretic mobility, and finally zeta potential. Also, this automatically allows for detection of charge sign. The electronics is arranged such that if the resultant shift is <250 Hz, the sign of the zeta potential is negative and vice versa.

3.3. Zeta Potential Calculation

Since the measured velocity is proportional to the applied field, the electrophoretic mobility can be defined as follows:

$$\nu = \mu_E E \tag{15}$$

where ν is the measured velocity, E the applied electric field, and μ_E is the electrophoretic mobility. The mobility is clearly the velocity in unit electric field. It is this quantity that can be measured directly. The situation is analogous to the measurement of particle size by PCS, in which a diffusion coefficient is measured and a particle size is derived by applying Stokes–Einstein law. In the case of zeta potential, a modification of the Stokes–Einstein law applies in which the driving force is the electric field rather than random diffusion,

$$\frac{\nu}{E} = \mu_E = \frac{Q}{6\pi a \eta} \tag{16}$$

where Q is the effective charge on the particle, η the viscosity, and a is the particle radius. The effective charge arises from both the actual surface charge and the charge in the double layer. The thickness of the double layer is quantified by κ, a parameter with the dimensions of inverse length. Its dimensionless number κa effectively measures the ratio of particle radius to double layer thickness. Q can be estimated using some approximations. Providing that the value of charge is low (zeta potential $< 30\,\text{mV}$ or so) the Henry equation can be applied (30):

$$\mu_E = \frac{2\varepsilon\zeta}{3\eta}\left(1 + \frac{1}{16}(\kappa a)^2 - \frac{5}{48}(\kappa a)^3 - \frac{1}{96}(\kappa a)^5 \right.$$

$$\left. - \left[\frac{1}{8}(\kappa a)^4 - \frac{1}{96}(\kappa a)^6\right]\exp(\kappa a)\int_{\infty}^{\kappa a}\frac{e^{-t}\mathrm{d}t}{t}\right) = \frac{2\varepsilon\zeta}{3\eta}f(\kappa a)$$

$$(17)$$

For small κa, the Henry function $f(\kappa a)$ approaches 1, for large κa it approaches 1.5, these corresponding to limiting cases where the particle is either much smaller or much larger than the double layer thickness. These are also known as the Huckel and Smoluchowski limits, respectively, and are the usual relationships used in the zeta potential calculations. The Huckel limit applies when the ionic concentration approaches zero, as in virtually non-conductive media. The Smoluchowski limit applies when the salt concentration is high enough to significantly compress the double layer. For typical colloidal dispersions and salt concentrations, it is much easier to arrange for $\kappa a > 1$ than it is for $\kappa a < 1$. Therefore, the Smoluchowski equation is much simpler to apply and used almost exclusively to convert mobility to zeta potential. In the Smoluchowski limit, a unit of mobility corresponds to 12.85 mV zeta potential in an aqueous media at 25°C.

3.4. Contribution of Zeta Potential Measurements

Zeta potential measurements help characterize the surface of amphiphilic β-cyclodextrin nanospheres and predict their behavior in different environments (31). Cyclodextrins are used in the pharmaceutical industry to solubilize lipophilic drugs. It was shown possible to prepare nanospheres of about 100 nm diameter from amphiphilic β-cyclodextrins (acylated at C2 and C3) by a nanoprecipitation method. In such a system, drug could be incorporated either within the cyclodextrin molecules or between them in the matrix, or be adsorbed on the surface, thus allowing a high payload. In the study, it was shown that particles of identical size could be formed in the presence or absence of a non-ionic surfactant, Pluronic F68. At pH 7.4 in 1 mm KCl, the zeta potential of the two preparations, measured using a Malvern Zetasizer equipped with a tubular cell of 2.6 mm diameter, was similar (about −25 mV). However, nanoparticles prepared without the surfactant had a significant change in zeta potential as pH moved away from neutral pH, indicating that they would aggregate in acidic or basic medium. On the other hand, particles prepared with the surfactant had a constant zeta potential irrespective of pH, probably because the surfactant prevented binding of counter-ions within the cavity of the cyclodextrin. Such particles would be expected to be stable.

Zeta potential and size measurement were used to confirm the adsorption of poloxamer onto the liposomes. In the study conducted by Bochot et al. (32), liposomes dispersed in a gel that solidifies at physiological temperature were investigated as a possible solution to overcome the challenge of designing ocular dosage forms to retain the drug in the precorneal area for sufficient absorption. The zeta potentials were measured after liposomes of different compositions were mixed with the gel consisting of 27% poloxamer 407 (Table 3). The results indicated that poloxamer 407 interacted strongly with liposomes that were either negatively (phosphatidylglycerol-containing) or positively (stearylamine-containing) charged. This was confirmed by the size increases of the charged liposome vesicles but no size increase in the neutral ones in these samples.

Zeta potential was also used for preparation and characterization of positively charged submicron emulsions that were reported to be stable in the presence of physiological cations and can interact in vivo with negatively charged biological membranes, resulting in an enhanced drug uptake and site-specific targeting (24). The positive charge of the submicron emulsions was introduced by adding cationic

Table 3 Effect of Poloxamer 407 on Zeta Potential of Liposomes

	Zeta potential (mV ± SD)	
Liposome composition	Without poloxamer	With poloxamer
PC/CHOL	−9.5 ± 2.1	−12.4 ± 0.8
PC/CHOL/PG	−25.0 ± 0.5	−15.2 ± 0.1
PC/CHOL/SA	+14.6 ± 0.4	−8.2 ± 1.7
PC/CHOL/PEG-DSPE	−8.6 ± 1.9	−12.9 ± 0.4

PC, Phosphatidylcholine; CHOL, cholesterol; PG, phosphatidylglycerol; SA, stearyla-mine; PEG-DSPE, distearoylphosphatidylethanolamine coupled to PEG.

lipids, polymers, and surfactants, such as stearylamine (33,34), chitosan (35,36), and cetyltrimethylammonium bromide (37) to the emulsions. Elbaz et al. (33) reported that the zeta potential of the submicron emulsion used in their study changed from −14.6 mV in the absence of stearylamine to a positive value (up to +21.0 mV) in the presence of 0.3% w/w of stearylamine. Similar behavior was observed by Jumma and Muller (36) that addition of positively charged chitosan to lipid emulsions led to a change of surface charge of oil droplets from negative (−11 mV) to positive values (+23 mV).

3.5. Limitations

Several parameters can significantly affect zeta potential measurement. Electrical double layer is directly dependent on the ionic strength of the sample to be measured. Sampling techniques such as dilution may substantially change zeta potential due to change in ionic strength upon dilution. This change can be magnified if a viscosity difference is introduced after dilution. Electro-osmosis effect, which is not discussed in this chapter, should be taken into consideration [there are many good references on this topic such as Hunter (38)]. In addition, zeta potential data may be inaccurate if samples being measured have low mobilities due to there being close to the isoelectric point, contain high salt concentration ($> \sim 20$ mM), or if they consist of oils or organic solvents.

4. RHEOLOGY

Many injectable dispersed formulations, whether intended for local drug delivery (for example, to heal bone injury) or as intramuscular implants to prolong drug release (39), require formulations possessing adequate rheological properties to achieve these desired therapeutic effects. Appropriate rheological properties are also needed to prevent phase separation of emulsions due to creaming or caking of suspensions due to sedimentation on storage. Rheology plays an important role in the formulation, mixing, handling, processing, transporting, storing, and performance of such systems.

Rosenblatt et al. (40) reported a rheological study on a concentrated dispersion of phase-separated collagen fibers in aqueous solution used to correct dermal contour defects through intradermal injection. The effect of electrostatic forces on the rheology of injectable collagen was studied using oscillatory rheological measurements on dispersions of varying ionic strengths (0.06–0.30). The associated relaxation time spectra, interpreted using the theory of Kamphuis et al. for concentrated dispersions, shows that collagen fibers become more flexible as ionic strength increases. This result was analyzed at the molecular level from the perspective that collagen fibers are a liquid-crystalline phase of rigid rod collagen molecules which have phase-separated from solution. Electrostatic forces affect the volume fraction of water present in the collagen fibers which in turn alters the rigidity of the fibers. Flexible collagen fiber dispersions displayed emulsion-like flow properties whereas more rigid collagen fiber dispersions displayed suspension-like flow properties. Changes in fiber rigidity significantly altered the injectability of collagen dispersions which is critical in clinical performance.

4.1. Bulk Rheology

Bulk rheology measures flow properties such as viscosity, elasticity, yield stress, shear thinning, and thixotropy. There are typically three test procedures that can be used to determine these properties.

4.1.1. Flow Tests

Flow tests measure non-Newtonian (shear-thinning) behavior when the viscosity is not constant but decreases with increasing stress. Many injectable dispersed systems are shear-thinning, which can be an extremely useful property. They possess a high viscosity under low stress to prevent sedimentation or creaming but a low viscosity at high stress for ease of injection, or a relatively low-viscosity fluid that is a liquid formulation prior to injection but undergoes a rapid change in physical form to a semi-solid or solid depot once injected in the body (41). The reasons for shear thinning are often complex and vary between materials. In polymer gels, such as pectin and gelatin, this occurs due to rupture of junction zones of attraction between adjacent polymer molecules. In flocculated colloidal dispersions such as yoghurts, the flocs are broken down into smaller units, and eventually into primary particles.

Not only can the viscosity of materials depend on the magnitude of the applied stress, it may also depend on the length of time for which the stress is applied. The viscosity of many dispersions will decrease with time upon stressing, and will take time to recover following removal of the stress. This effect is known as thixotropy. Like pseudoplasticity, a reasonable degree of thixotropy is often useful, for example it enables surface coatings to flow out after application, removing irregularities, before they stiffen. The complete viscometric characterization of a liquid would therefore require the shear rate to be monitored as a function of shear stress, time, and temperature, but in practice such completeness is seldom necessary. It is usually enough to identify the conditions and parameters of interest and make the corresponding measurements.

A large body of viscosity data has been published for many materials, and it is remarkable that the curves of viscosity, plotted against shear rate, of almost all of these are very similar in shape. The main features of such curves are most easily seen if the data are plotted on a logarithmic axes (i.e., with 0.1, 1, 10, 100, etc., equally spaced). At low shear rates, a Newtonian region exists, followed at higher rates by shear thinning a region which often takes the form of a power law (straight line on logarithmic axes). At still higher rates, a second Newtonian region is observed. In some cases, the material will eventually shear thicken.

If the shear thickening region is ignored, then it is possible to describe a curve of the form by four parameters. Two of these are the low shear viscosity, η_0, and the high shear viscosity, η_∞. The shear-thinning part of the curve usually approximates to a straight line when plotted on logarithmic axes, and can therefore be described by a two-parameter power law relationship. Combining gives the equation:

$$\eta = \frac{\eta_0 - \eta_\infty}{1 + (K_\gamma)^m} + \eta_\infty \qquad (18)$$

where K is known as the characteristic time of the material. The greater the value of K, the further to the left the curve lies, and the greater the value of the index m, the greater the degree of shear thinning.

4.1.2. Creep Tests

Many important processes are driven by very low stresses, such as those produced by gravity or surface tension. These include settling and creaming in dispersions. Moreover the handling properties of some materials are affected by elasticity, the ability to recover in some part their original shape after forced deformation.

These properties can be investigated using a creep test, in which a very low shearing stress is applied to the sample, and the resulting strain (displacement) is monitored. It is usual to plot the compliance, J, defined as the strain divided by the stress, against time. If a low stress is placed on a solid sample, it will respond by deforming almost instantaneously to a new position, and then stopping. When the stress is removed, the sample will immediately recover its original dimensions. The compliance will depend on the material: the stiffer the material, the lower the compliance. Typical values are about 10^{-3}/Pa for a 2% gelatin solution in water, 5×10^{-5}/Pa for foam rubber, 3×10^{-7}/Pa for natural rubber, and 2×10^{-9}/Pa for nylon.

If the stress is placed on a liquid material, it will deform continuously, the rate of deformation being inversely proportional to the viscosity of the liquid. When the stress is removed, the sample will cease to deform, and will not show any recovery. Typical values of viscosity are 1 mPa s for water, 1 Pa s glycerin, about 1000 Pa s for polymer melts. Many dispersed systems are neither completely solid nor completely liquid. They show properties that are some way between these limiting forms of behavior, and are called viscoelastic. When a stress is placed on a viscoelastic material, the deformation may be retarded. When the stress is removed, the recovery may also be retarded. In some cases, a sample will show both retarded deformation and continuous flow (liquid-like properties), in which case recovery after removal of the stress will be incomplete.

The elastic properties of the material can then be read from the retarded deformation and the recovery parts of the curve, while the continuous flow part of the curve provides the viscosity at the applied stress. This is identical to the low shear viscosity obtained from a flow curve. For dilute dispersions of monodisperse hard spheres, the settling rate can be predicted from Stokes' law combined with Archimedes' principle shown by the following equation:

$$V = \frac{2r^2 g(\rho_p - \rho_m)}{9\eta} \tag{19}$$

where V is the sedimentation velocity, r is the particle radius, g is the acceleration due to gravity, ρ_p and ρ_m are the density of the particles and medium, respectively, and η is the viscosity.

Most industrial dispersions do not fulfill the criteria for this expression, but the viscosity will nonetheless give a qualitative indication of the relative stability to sedimentation of comparable dispersions. In some cases, the stability against particle aggregation or emulsion coalescence can also be predicted from low shear viscosity. Food "stabilizers", for example, do not impart stability in the thermodynamic sense, but reduce the rate of particle motion by increasing the viscosity of the medium. It is important that industrial and biological materials should have the correct degree of elasticity. An IM depot formulation, for example, will fail to pass through a syringe if it is too elastic, but will give poor depot performance if it is insufficiently so (42).

4.1.3. Oscillatory Tests

Many dispersion materials show behavior which is neither completely liquid nor completely solid, but is somewhere between. Such materials are termed viscoelastic. It is viscoelasticity which is responsible, at least in part, for the handling properties of these materials. There are several ways of examining the viscoelastic properties of materials. But the most common way is to use oscillatory rheology.

If a sinusoidal stress (s) (force acting over an area) is placed on a solid sample, a sinusoidal displacement (strain, g) will result which is in phase with the applied stress. The modulus, or stiffness, of the material can be obtained by dividing the amplitude of the stress, s_0, by the amplitude of the strain, g_0. If a sinusoidal stress is applied to a liquid sample, the stress is in phase with the rate of change of strain, and a phase lag of 90° is therefore introduced between the stress and the strain. For viscoelastic materials, the phase angle, d, will be somewhere between 0° and 90°. The ratio of the stress to the strain amplitude gives the stiffness of the material, and the phase angle describes its viscoelastic nature.

The degree to which a material behaves as a solid or liquid depends on the timescale of the observation. Water is usually described as a liquid, of course, but if examined over timescales of less than about a nanosecond, would appear to be a solid. Ice behaves as a liquid under very high stresses, over periods of years, hence glacier flow. It happens that the materials which are listed above as being viscoelastic show a transition from liquid to solid behavior over typical laboratory timescales. To examine more precisely the transition time, the frequency (w) of the applied stress can be varied.

The usual method of performing an oscillation experiment is to apply a sinusoidal stress to a sample, over a range of frequencies, and to monitor the strain and phase angle. The stress is kept low so that it can be assumed that the unperturbed properties of the sample are determined. Rather than reporting s_0/g_0 and phase angle directly, it is more usual to report the storage modulus, G', and loss modulus, G''. These are defined as $G' = s_0 \cos(d/g_0)$ and $G'' = s_0 \sin(d/g_0)$. The advantage of this is that G' represents the "solid" component of the material, and G'' the "liquid" component. The viscosity of a liquid with no solid component would actually be G''. Just as polymers show broad transitions in melting point, sometimes over many decades of temperature, they also show rheological property changes over broad frequency ranges. In general, the higher the polymer molecular weight, the broader the range. If examined over short timescales, they may appear to be solid, while over longer timescales, they may flow like a liquid.

4.1.4. Applications

Suspension

Ramstack et al. (43) reported that increased viscosity of an injection vehicle containing the fluid phase of a suspension significantly reduces in vivo injection failures. Injectable compositions for microspheres can be made by mixing dry microparticles with an aqueous injection vehicle to form a suspension, and then mixing the suspension with a viscosity-enhancing agent to increase the viscosity of the fluid phase of the suspension to the desired level for improved injectability.

Emulsion Examples

Viscosity can be monitored by standard rheological techniques. The rheological properties of emulsions, reviewed by Sherman (1983), can be complex and depend on the identity of surfactants and oil used, the ratio of the disperse and continuous phases, particle size, as well as other factors. Flocculation will generally increase viscosity, thus, monitoring viscosity on storage is important for assessing shelf-life stability.

Viscosity can be used to assess multiple emulsion stability. This method is based on change in viscosity of the external aqueous phase as water is lost from the internal to the external aqueous phase of W/O/W emulsions due to rupture of the oil layer. As the overall viscosity of the emulsion system is dependent on the continuous phase viscosity, Kita et al. (44) attempted to estimate the stability of W/O/W emulsions which had relatively low volume fractions of internal aqueous phase (< 0.2) by measuring viscosity as a function of time. Viscosity was related to the volume fraction of the internal aqueous phase using a modified Mooney's equation:

$$\ln \eta_{rel} = \frac{a(\phi_{wi} + \phi_o)}{1 - \lambda(\phi_{wi} + \phi_o)} \tag{20}$$

where η_{rel} is the relative viscosity, a is the shape factor, λ is the crowding factor, $(\phi_{wi} + \phi_o)$ represents the dispersed phase volume fraction and where ϕ_{wi} and ϕ_o are the volume fractions of the internal aqueous and oil phases, respectively. The equation can be written as a function of the volume fraction of the internal aqueous phase ϕ_{wi} as follows:

$$\phi_{wi} = \frac{a[\log \eta_{rel}\{(2.303/a) - (2.303\lambda\phi_o/a)\} - \phi_o]}{a + (2.303)\lambda \log \eta_{rel}} \tag{21}$$

ϕ_o remains constant, however, ϕ_w decreases with increasing rupture of the oil layer as the internal aqueous droplets are mixed with external aqueous phase. Kita et al. (44) used the viscometric method to estimate stability of W/O/W emulsions with relatively low volume fractions of internal aqueous phase (< 0.2). Emulsions with higher volume fractions do not exhibit Newtonian flow at low shearing rates and therefore cannot be assessed using this method. Ingredients such as glucose, bovine serum albumin, and electrolytes did not allow an accurate estimation of multiple emulsion stability as these molecules alter the dispersion state of the droplets.

Matsumoto and Kohda (45) calculated the rate of swelling and shrinkage of the internal aqueous phase of W/O/W emulsions under the influence of an osmotic gradient from the rate change in viscosity. The rate of change of viscosity was determined to be proportional to the osmotic pressure difference across the oil phase. The authors estimated the flux of water across the oil layer from viscosity changes in the initial stages of aging using modified Mooney's equation (18). This enabled the authors to estimate a water permeation coefficient (P_0) for the oil layer. P_0 was determined to be in the range of 10^{-4} to 10^{-5} cm/s at 25°C. The rate of swelling or shrinkage was calculated by subtracting the quantity of water taken up into the oil layer by solubilizing micelles from the total water flux. When glucose or sodium chloride was present in the internal aqueous phase, the viscosity of the emulsions increased initially and then decreased. This was explained by the migration of water from the outer to the internal aqueous phase to satisfy the osmotic gradient caused by glucose or sodium chloride in the internal phase leading to swelling of the internal droplets with an increase in viscosity. Further swelling of the internal droplets was considered to result in rupture of the oil layer causing release of the internal water with consequent decrease in the external phase viscosity.

4.1.5. Interfacial Rheology

Surfactants added to dispersed systems adsorb at interfaces reducing interfacial tension and forming an interfacial film which resists coalescence or agglomeration following particle collision. It has been shown that the stronger this film the more stable the dispersions and that the interfacial film can play a more crucial role than the reduction of interfacial tension in maintaining long-term emulsion stability to coalescence for certain emulsion systems (46). The strength of the interfacial film, which can be a monolayer, a multilayer or a collection of small particles adsorbed at the interface, depends on the structure and conformation of surfactant molecules at the interface (47). The structure and conformation can be affected by formulation variables including surfactant type and concentration, other additives and their concentrations, storage temperature, ionic strength, and pH. For the film to be an efficient barrier, it should remain intact when sandwiched between two particles. If broken, the film should have the capacity to reform rapidly. This requires that the film possess a certain degree of surface elasticity. It has been shown that interfacial elasticity correlates well with interfacial film strength and can be used to predict emulsion stability (Opawale and Burgess 1997).

An oscillating shear interfacial rheometer consists of four interconnecting systems: a moving-coil galvanometer; a Du Nouy ring attached to the galvanometer; an amplitude controller for motion of the ring; and a data processor. The equation of motion for the instrument and the associated theory has been explained by Sheriff and Warburton (48). A normalized resonance mode is used where the frequency of phase resonance was $> 2\,Hz$. At phase resonance, the input stress leads the strain by $90°$. The outputs are the strain amplitude and/or the frequency of phase resonance. The amplitude of motion of the ring is measured via a proximity probe transducer and automatic analysis of the signal generated provides the dynamic interfacial rigidity modulus (interfacial elasticity, G'_s (mN/m). G'_s is defined as

$$G'_s = g_f I_o 4\pi^2 (f^2 - F_0^2) \tag{22}$$

where I_0 is the moment of inertia of the ring, f and f_0 are the sample and reference interfacial resonance frequencies, respectively, and g_f is the geometric factor. The g_f is defined as

$$g_f = \frac{4\pi(R_1^2\,R_2^2)}{(R_1 + R_2)(R_2 - R_1)} \tag{23}$$

where R_1 is the radius of the ring and R_2 is the radius of the sample cell.

4.1.6. Limitations

Accurate measurement of the mechanical and rheological properties of injectable dispersed systems relies on accuracy of force applied to the samples since rheology describes the interrelation between force, deformation, and time. For non-Newtonian systems, measuring viscosity at low shear stress (or yield) may be significantly influenced by precision of controlled stress force, history of sample to be measured, and timescale for the measurement. One of the main limitations of most commercial rheometers lies in the frequency range. Frequencies above 100 Hz are often hard to achieve, and frequencies below 0.01 Hz require significant time investment to collect data.

5. CONCLUSIONS

Particle size, zeta potential, and rheological properties are important and useful indicators of injectable dispersed system stability. However, there are several pitfalls that one has to be aware of when characterizing and analyzing injectable dispersed systems using these parameters. To obtain accurate and reproducible results for particle size, surface charge, and rheological properties requires knowledge of the injectable dispersed systems under development, understanding instrumentation operation basis, and careful experimental planning.

REFERENCES

1. R&D Feature Stories. Particle Sizing Gets Serious. http://www.rdmag.com/features.

2. Florence AT, Attwood D. Physicochemical Principles of Pharmacy. 3rd ed. Hampshire: Macmillan Press, 1998.

3. Singh M, Ravin LJ. Parenteral emulsions as drug carrier systems. J Parenter Sci Technol 1986; 40(1):34–41.

4. Grimes G, Vermess M, Gallelli JF, Griton M, Chatterji DC. Formulation and evaluation of an ethiodized oil emulsion for intravenous hepatography. J Pharm Sci 1979; 68:52.

5. Tomazic-Jezic VJ, Merritt K, Umbreit TH. Significance of the type and the size of biomaterial particles on phagocytosis

and tissue distribution. J Biomed Mater Res 2001; 55(4): 523–529.

6. Burnham WR, Hansrani PK, Knott CE, Cook JA, Davis SS. Stability of a fat emulsion based intravenous feeding mixture. Int J Pharm 1983; 13:9.

7. Floyd AG. Top ten considerations in the development of parenteral emulsions. PSTT 1999; 2:134–143.

8. Collins-Gold LC, Lyons RT, Bartholow LC. Parenteral emulsions for drug delivery. Adv Drug Deliv Rev 1990; 5:189–208.

9. Laval-Jeantet AM, Laval-Jeantet M, Bergot C. Effect of particle size on the tissue distribution of iodized emulsified fat following intravenous administration. Invest Radiol 1982; 17: 617.

10. Mikula RJ. Emulsion characterization. In: Emulsion in the Petroleum Industry. American Chemical Society, 1992.

11. Conn PW. Confocal microscopy. In: Methods in Enzymology. Vol. 307. San Diego, CA: Academic Press, 1999.

12. Hou W, Papadopoulos KD. $W_1/O/W_2$ and $O_1/W/O_2$ globules stabilized with Span 80 and Tween 80. Colloids Surf A 1997; 125:181–187.

13. Ficheux MF, Bonakder L, Leal-Calderon F, Bibette J. Some stability criteria for double emulsions. Langmuir 1998; 14: 2702–2706.

14. McCrone WC, McCrone LB, Delly JG. Polarized Light Microscopy. McCrone Research Institute, 1997.

15. Davis SS, Burbage AS. Electron micrography of water-in-oil-in-water emulsions. J Colloid Interface Sci 1977; 62(2): 361–363.

16. Pike ER. The analysis of polydisperse scattering data. In: Chen SH, Chu B, Nossal R, eds. Scattering Techniques Applied to Supermolecular and Nonequilibrium Systems. New York: Plenum Press, 1981.

17. Grabowski E, Morrison I. Measurements of suspended particles by quasi-elastic light scattering. In: Dahneke B, ed. Particle Size Distributions from Analysis of Quasi-Elastic Light Scattering Data. New York: Wiley-Interscience, 1983.

18. Bakan DA, Longino MA, Weichert JP, Counsell RE. Physico-chemical characterization of a synthetic lipid emulsion for hepatocyte-selective delivery of lipophilic compounds: application to polyiodinated triglycerides as contrast agents for computed tomography. J Pharm Sci 1996; 85(9):908–914.

19. Haines BA, Martin A. Interfacial properties of powdered material: caking in liquid dispersions II. J Pharm Sci 1961; 50:753–756.

20. Matthews BA, Rhodes CT. Some studies of flocculation phenomena in pharmaceutical suspensions. J Pharm Sci 1968; 57:569–573.

21. Lian T, Ho RJ. Design and characterization of a novel lipid–DNA complex that resists serum-induced destabilization. J Pharm Sci 2003; 92(12):2373–2385.

22. Gref R, Domb A, Quellec P, Blunk T, Mller RH, Verbavatz JM, Langer R. The controlled intravenous delivery of drugs using PEG-coated sterically stabilized nanospheres. Adv Drug Deliv Rev 1995; 16:215–233.

23. Chauvierre C, Labarre D, Couvreur P, Vauthier C. Novel spolysaccharide-decorated poly(isobutyl cyanoacrylate) nano-particles. Pharm Res 2003; 20(11):1786–1793.

24. Yang SC, Benita S. Enhanced absorption and drug targeting by positively charged submicron emulsions. Drug Dev Res 2000; 50:476–486.

25. Lee MJ, Lee MH, Shim CK. Inverse targeting of drugs to reti-culoendothelial system-rich organs by lipid microemulsion emulsified with poloxamer-338. Int J Pharm 1995; 113:175–187.

26. Liu F, Liu D. Long circulating emulsions (oil-in-water) as carries for lipophilic drugs. Pharm Res 1995; 12:1060–1064.

27. Song YK, Liu DX, Maruyama K, Takizawa T. Antibody mediated lung targeting of long-circulating emulsions. PDA J Pharm Sci Technol 1996; 50:372–377.

28. Allen TM, Moase EH. Therapeutic opportunities for targeted liposomal drug delivery. Adv Drug Deliv Rev 1996; 21:117–133.

29. Martin A. Physical Pharmacy. 4th ed. Baltimore, MD: Williams & Wilkins.

30. Hiemenz PC. Principles of Colloid and Surface Chemistry. 2nd ed. New York: Marcel Dekker, 1986.

31. Skiba M, Fessi H, Puisieux F, Duchene D, Wouessidjewe D. Development of new colloidal drug carrier from chemically modified cyclodextrins: nanospheres, and influence of physicochemical and technological factors on particle size. Int J Pharm 1996; 129:113–121.

32. Bochot A, Fattal E, Grossiord JL, Puisieux F, Couvreur P. Characterization of a new ocular delivery system based on a dispersion of liposomes in a thermosensitive gel. Int J Pharm 1998; 162(1/2):119–127.

33. Elbaz E, Zeevi A, Klang S, Benita S. Positively charged submicron emulsions: a new type of colloidal drug carrier. Int J Pharm 1993; 96:R1–R6.

34. Klang SH, Signos CS, Benita S, Frucht-Pery J. Evaluation of a positively charged submicron emulsion of piroxicam on the rabbit corneum healing process following alkali burn. J Control Release 1999; 57:19–27.

35. Calvo P, Remuna-Lopez C, Vila-Jato JL, Alonso MJ. Development of positively charged colloidal drug carrier: chitosan-coated polyester nanocapsules and submicron-emulsions. Colloid Polym Sci 1997; 275:46–53.

36. Jumma M, Muller BW. Physicochemical properties of chitosan lipid emulsions and their stability during the autoclaving process. Int J Pharm 1999; 183:175–184.

37. Samama JP, Lee KM, Biellmann JF. Enzymes and microemulsions: activity and kinetic properties of liver alcohol dehydrogenase in ionic water-in-oil microemulsions. Eur J Biochem 1987; 163:609–617.

38. Hunter RJ. Zeta Potential in Colloid Science, Principles and Applications. London: Academic Press, 1981.

39. Alessandro M, Sara L. Sustained release injectable products. Am Pharm Rev 2003; 3 Apr Fall.

40. Rosenblatt J, Devereux B, Wallace DG. Effect of electrostatic forces on the dynamic rheological properties of injectable collagen biomaterials. Biomaterials 1992; 12(12):878–886.

41. Tipton AJ, Dunn RL. In situ gelling systems—sustained release injectable products. In: Senior J, Radomsky M, eds. Interpharm Press, 2000:241–278.

42. Ansel HC, Allen LV, Popovich NG. Pharmaceutical Dosage Forms and Drug Delivery Systems. 7th ed. Philadelphia, PA: Lippincott Williams & Wilkins, 1999.

43. Ramstack JM, Riley MG, Zale SE, Hotz JM, Johnson OL. Preparation of injectable suspensions having improved injectability. US Patent 57787520000525, 2000.

44. Kita Y, Matsumoto S, Yonezawa D. Viscometric method for estimating the stability of W/O/W type multiple-phase emulsions. J Colloid Interface Sci 1977; 62:87–94.

45. Matsumoto S, Kohda M. The viscosity of W/O/W emulsions: an attempt to estimate the water permeation coefficient of the oil layer from the viscosity changes in diluted system on aging under osmotic pressure gradient. J Colloid Interface Sci 1980; 73(1):13–20.

46. Burgess DJ. In: Pezzuto JM, Johnson ME, Manasse HR, eds. Biotechnology and Pharmacy. New York: Chapman and Hall, 1993:116.

47. Swarbrick J. Coarse dispersions.   In: Gennaro AR, ed. Remington: The Science and Practice of Pharmacy, 1997.

48. Sheriff M, Warburton B. Measurement of dynamic rheological properties using the principle of externally shifted and restored resonance. Polymer 1974; 15:253–254.

49. Daemen T, Velinova M, Regts J, de Jager M, Kalicharan R, Donga J, van der Want JJL, Scherphof GL. Different intrahepatic distribution of phosphatidylglycerol and phosphatidylserine liposomes in the rat. Hepatology 2003; 26(2):416–423.

50. Kurihara A, Shibayama Y, Yasuno A, Ikeda M. Lipid emulsions of palmitoylrhizoxin: effects of particle size on blood disposition of emulsion lipid and incorporated compound in rats. Biopharm Drug Dispos 1996; 17(4):343–353.

51. Cannon JB, Martin C, Drummond GS, Kappas A. Targeted delivery of a heme oxygenase inhibitor with a lyophilized liposomal tin mesoporphyrin formulation. Pharm Res 1993; 10(5):715–721.

52. Cauchetier E, Fessi F, Boulard Y, Deniau M, Astier A, Paul M. Preparation and physicochemical characterization of atrovaquone-containing liposomes. Drug Dev Res 1999; 47(4): 155–161.

53. Green TR, Fisher J, Matthews JB, Stone MH, Ingham E. Effect of size and dose on bone resorption activity of macrophages by in vitro clinically relevant ultra high molecular weight polyethylene particles. J Biomed Mater Res 2000; 53(5):490–497.

54. Cadée JA, Brouwer LA, den Otter W, Hennink WE, van Luyn MJA. A comparative biocompatibility study of microspheres based on crosslinked dextran or poly(lactic-co-glycolic)acid after subcutaneous injection in rats. J Biomed Mater Res 2001; 56(4):600–609.

55. Bremer C, Allkemper T, Baermig J, Reimer P. RES-specific imaging of the liver and spleen with iron oxide particles designed for blood pool MR-angiography. J Magn Reson Imaging 1999; 10(3):461–467.

56. Brotherton CM, Davis RH. Electroosmotic flow in channels with step changes in zeta potential and cross-section. J Colloid Interface Sci 2004; 270(1):242–246.

57. Burgess DJ, Yoon JK. Interfacial tension studies on surfactant systems at the aqueous/perfluorocarbon interface. Colloid Surf B 1993; 1:283–293.

58. Burgess DJ, Yoon JK. Influence of interfacial properties on perfluorocarbon/aqueous emulsion stability. Colloid Surf 1995; 4:297–308.

59. Burgess DJ, Sahin NO. Interfacial rheology of β-casein solution. In: Herb CA, Prud'homme RK, eds. Structure and Flow in Surfactant Solutions. ACS Symposium Series, 1994: 380–393.

60. Burgess DJ, Sahin NO. Influence of protein emulsifier interfacial properties on oil-in-water emulsion stability. Pharm Dev Technol 1998; 3(1):21–29.

61. Chavany C, Saison-Behmoaras T, Le Doan T, Puisieux F, Couvreur P, Helene C. Adsorption of oligonucleotides onto polyisohexylcyanoacrylate nanoparticles protects them

against nucleases and increases their cellular uptake. Pharm Res 1994; 11:1370–1378.

62. Couvreur P, Dubernet C, Puisieux F. Controlled drug delivery with nanoparticles: current possibilities and future trends. Eur J Pharm Biopharm 1995; 41:2–13.

63. De Chasteigner S, Fessi H, Devissaguet J-P, Puisieux F. Comparative study of the association of itraconazole with colloidal drug carriers. Drug Dev Res 1996; 38:125–133.

64. Jiao J, Burgess DJ. Ostwald ripening of water-in-hydrocarbon emulsions stabilized by non-ionic surfactant Span 83. J Colloid Interface Sci 2003; 264(2):509–516.

65. Jiao J, Burgess DJ. Rheology and stability of W/O/W multiple emulsions containing Span 83 and Tween 80. AAPS Pharm Sci 2003; 5(1), article 7 (http://www.aapspharmsci.org).

66. Jiao J, Rhodes DG, Burgess DJ. Multiple emulsion stability: pressure balance and interfacial film strength. J Colloid Interface Sci 2002; 250:444–450.

67. Lundberg B. Preparation of drug-carrier emulsions stabilized with phosphatidylcholine-surfactant mixtures. J Pharm Sci 1994; 83:72–75.

68. Lundberg B. A submicron lipid emulsion coated with amphipathic polyethylene glycol for parenteral administration of paclitaxel (Taxol). J Pharm Pharmacol 1997; 49:16–21.

69. Passirani C, Barratt G, Devissaguet J-P, Labarre D. Interactions of nanoparticles bearing heparin or dextran covalently bound to poly(methyl methacrylate) with the complement system. Life Sci 1998; 62:775–785.

70. Passirani C, Barratt G, Devissaguet J-P, Labarre D. Long-circulating nanoparticles bearing heparin or dextran covalently bound to poly(methyl methacrylate). Pharm Res 1998; 15(7):1046–1050.

71. Prankerd RJ, Stella VJ. The use of oil-in-water emulsions as a vehicle for parenteral drug administration. J Parenter Sci Technol 1990; 44:139–149.

72. Puisieux F, Barratt G, Benita S, Couarraze G, Couvreur P, Devissaguet J-Ph, Dubernet C, Fattal E, Fessi H, Vauthier C.

Polymeric micro- and nanoparticles as drug carriers. In: Dumitru S, Szycher M, eds. Polymeric Materials for Biomedical Applications. New York: Marcel Dekker Inc., 1993:749–794.

73. Robert C, Malvy C, Couvreur P. Inhibition of the Friend retrovirus by antisense oligonucleotides encapsulated in liposomes: mechanism of action. Pharm Res 1993; 10:1427–1433.

74. Sheriff M, Warburton B. The theory of a universal oscillatory rheometer for the study of linear viscoelastic materials using the principle of normalized resonance. In: Theoretical Rheology. London: Applied Science Publishers, 1975:299–316.

75. Strickley RG, Anderson BD. Solubilization and stabilization of an anti-HIV thiocarbamate, NSC 629243, for parenteral delivery using extemporaneous emulsions. PDA J Pharm Sci Technol 1993; 10:1076–1082.

76. Tarr BD, Sambandan TG, Yalkowsky SH. A new parenteral emulsion for the administration of taxol. Pharm Res 1987; 4:162–165.

77. Vippagunta SR, Brittain HG, Grant DJW. Crystalline solids. Adv Drug Deliv Rev 2001; 48(1):3–26.

78. Wiersma PH, Loeb AL, Overbeek JTG. J Colloid Interface Sci 1966; 22:78.

79. Zhu H, Grant DJW. Dehydration behavior of nedocromil magnesium pentahydrate. Int J Pharm 2001; 215(1/2):251–262.

4

In Vitro/In Vivo Release from Injectable Dispersed Systems

**BRIAN C. CLARK and
PAUL A. DICKINSON**
Pharmaceutical and Analytical R&D,
AstraZeneca, Macclesfield, U.K.

IAN T. PYRAH
Safety Assessment, AstraZeneca,
Macclesfield, U.K.

1. IN VITRO RELEASE

1.1. Basis of Dissolution Testing

Dissolution testing is an in vitro procedure designed to discriminate important differences in components, composition, and/or method of manufacture between dosage forms (1). A dissolution test for solid oral dosage forms, utilizing a rotating basket apparatus, was first included in the United States Pharmacopeia (USP) 18 in 1970. The current USP (2) includes general methods for disintegration, dissolution, and drug release. The disintegration and dissolution tests are intended primarily for immediate release solid oral dosage forms, the former controlling the time taken for a tablet or capsule to break down and the latter controlling release of the active ingredient(s). The drug release test is intended for application to modified release articles including delayed and extended release tablets; seven apparatus are described, the choice being based on knowledge of the formulation design and actual dosage form performance in the in vitro test system. Specific guidance is given with regard to the utility of Apparatus 1 (basket) or 2 (paddle) at higher rotation frequencies, Apparatus 3 (reciprocating cylinder) for bead-type delivery systems, Apparatus 4 (flow cell) for modified release dosage forms containing active ingredients of very limited solubility, Apparatus 5 (paddle over disc) or Apparatus 6 (cylinder) for transdermal patches, and Apparatus 7 (reciprocating disc) for transdermal systems and non-disintegrating oral modified release dosage forms. The usage of the various apparatus across all modified release dosage forms described in the current USP is shown in Fig. 1.

The state of science is such that in vivo testing is necessary in the development and evaluation of dosage forms. It is

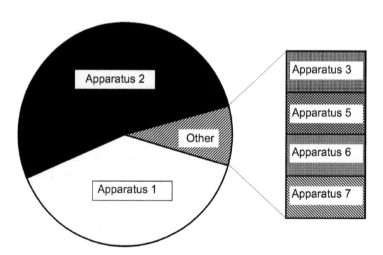

Figure 1 Modified release dosage forms—usage of USP apparatus for drug release, as indicated in USP27-NF22.

a goal of the pharmaceutical scientist to find a relationship between an in vitro characteristic of a dosage form and its in vivo performance (refer to the chapter on In Vitro–In Vivo Correlation for Modified Release Parenteral Drug Delivery Systems in this book).

1.2. General Considerations

Whatever dissolution apparatus is used, close control must be applied to several parameters, including geometry, dimensions, materials of construction, environment, temperature, and time, in order that reliable results may be obtained. Consideration should be given to the potential for extraction of interferants from the equipment, or adsorption of the active substance, and method validation should include an assessment of recovery for a completely dissolved dosage unit.

The dissolution test is carried out at constant temperature (normally 37°C, body temperature, for oral and parenteral dosage forms, or 32°C, skin temperature, for transdermal systems), although other temperatures may be used with justification. Temperature should be controlled with a tolerance of ± 0.5°C and should be measured and verified as being within limits over the duration of the test.

The dissolution test may be carried out using water, or a medium chosen to mimic physiological conditions, and may include a buffer system to maintain pH, additives such as surfactants or albumin (to mimic protein binding of lipophilic drugs when administered intravenously). A bacteriostatic agent may be incorporated to control microbiological growth, which can be a particular problem for real-time release testing of extended release formulations. Deaeration of dissolution media should be considered, as degassing resulting in the formation of air bubbles on the surface of the dosage form will significantly affect surface area and hence rate of release. The use of non-aqueous media is not normally recommended, as a meaningful in vitro–in vivo correlation is unlikely.

The test is normally carried out on a unit dose of the formulation. The dose may be dispersed in the dissolution medium, contained within a cell, or for solid dosage forms retained by a sinker, for example, a platinum wire, designed to minimally occlude the dosage form.

The composition and volume of dissolution medium should be chosen to ensure that, when all of the drug substance has dissolved, the concentration of the resulting solution will be less than one-third of that of a saturated solution; thus, the dissolution medium acts as a sink, in which the concentration of dissolved drug will be low enough not to inhibit ongoing release. The usual volume of dissolution medium is 500–1000 mL, but other volumes may be used with justification.

Under certain circumstances, usually to mimic the change in environment as an enteric-coated or extended release tablet or capsule moves through the gastrointestinal tract, the dissolution medium may be modified/changed at a predetermined intermediate time-point.

The dissolution medium should be stirred, or the sample compartment rotated or oscillated, to ensure homogeneity of solution. Other than this, the dissolution apparatus should not contribute and should be isolated from, any vibration or other motion which could affect the rate of release.

Test duration is normally 30–60 min for immediate release formulations, but may be much longer for extended release products. Evaporative losses during the test must be minimized or compensated for. For tests longer than 24 hr, measures (sanitization of equipment and/or inclusion of antimicrobial additive) must be taken to prevent microbiological proliferation.

The release profile may be characterized by determining the concentration of drug released at each of a minimum of three time-points—an early time-point to determine "dose dumping," a late time-point to evaluate completeness of release, and an intermediate time-point to define the in vitro release profile. Measurement may be continuous (e.g., by use of a flow cell or fiber-optic probe) or discrete, and if a sample of dissolution medium is withdrawn for analysis, it should be replaced (if the assay method is non-destructive), or an equal volume of fresh dissolution medium added and the amount of drug removed corrected for in subsequent calculations, or the test may be continued with diminished volume. If a sample-and-replace approach is used, care should be taken to minimize any perturbation to temperature, perhaps by pre-heating the replacement medium. For products containing two or more active ingredients, release should be measured for each active ingredient.

The analytical methodology used to determine drug concentration should be selective for the active ingredient. Where the formulation is dispersed in the dissolution medium, separation of dissolved from undissolved drug may be accomplished by filtration, centrifugation or by the use of an analytical technique sensitive only to dissolved drug. Care must be taken to ensure that the sampling and subsequent analysis does not influence the distribution of drug between undissolved and dissolved forms. Degradation of the drug substance under the conditions of the test should be evaluated during method development (3,4); if significant degradation is apparent, it may be appropriate to sum active and degradants, or to utilize a non-specific method, such that the reported results are indicative of release. It is normal practice to report results as cumulative release, as a percentage of the labeled content of drug (Q).

The drug release test is normally performed in replicate, initially using 6 units but with the scope for additional testing (up to a total of 24 units) if acceptance criteria are not met. Acceptance criteria should control mean release and the range of individual values for a batch of the formulation. The drug release test is normally considered to be stability indicating.

1.3. Applicability to Injectable Dispersed Systems

Current guidance is that no product where a solid phase exists, including suspensions and chewable tablets, should be developed without dissolution or drug release characterization. In the context of injectable dispersed systems, it is therefore appropriate to apply a drug release test to suspensions and microspheres. Drug release characterization is also relevant for emulsion and liposomal products where the formulation is designed to control the release of the active substance(s).

1.4. Mechanistic Studies

The development of in vitro drug release methodology should be underpinned by an understanding of the mechanism of drug release. This requires knowledge of the drug substance, the release-controlling excipients, and any interactions in the formulation. Depending on the characteristics of the dosage form and the route of administration, in vitro drug release may involve hydration, swelling, aggregation, disintegration, diffusion, hydrolysis, and/or erosion. In vivo release may be additionally complicated by enzymatic action, encapsulation by tissue, complexation, or partitioning into tissue.

1.4.1. Emulsions

Submicron emulsions are typically used for parenteral nutrition or the intravenous administration of a hydrophobic, lipophilic drug substance. Characterization studies should include investigation of particle size distribution (particles >5 µm are likely to cause pulmonary embolism), zeta (surface) potential, which is a key indicator of the physical stability of the emulsion, pH, which is a determinant factor for surface potential and which is liable to decrease on storage due to the formation of free fatty acids, and drug substance content. The drug substance will partition between the disperse (oil) phase, the continuous (aqueous) phase, and the oil–water interface where the drug may associate with the emulsifying agent(s). A quantitative assessment of drug distribution is required if the mechanism of release is to be understood, and this may be determined using a combination of ultrafiltration and ultracentrifugation techniques (5).

1.4.2. Liposomes

Liposomes may be used for drug delivery to confer sustained release, for tumor targeting, to increase bioavailability or expand the therapeutic window. The characterization techniques described above for emulsions may also be applied to liposomes, although the partitioning of the drug substance is complicated by the existence of internal and external aqueous phases.

1.4.3. Suspensions

Injectable suspensions may be used for drug delivery primarily for insoluble drug substances; for intravenous administration, a submicron particle size distribution is essential. Characterization studies should encompass particle size distribution, partitioning of the drug substance between solid and solution; and the potential for Ostwald ripening should be considered.

1.4.4. Microspheres

Microsphere drug delivery systems are usually based on biodegradable polymers (6) such as poly(lactic acid), poly(lactide-co-glycolide), polyanhydrides, cross-linked polysaccharides, gelatin or serum albumin, and are intended for subcutaneous or intramuscular administration. Drug loading is determined by potency, duration of release, and other factors, but is generally in the range 0.1–15% by weight. Characterization studies should include the particle size distribution, drug distribution within the formulation (solid solution, drug polymer salt, discrete domains of drug in the polymer matrix), and the surface and bulk morphology. Solid state imaging techniques are important in elucidating structural information.

1.5. Methodology

Experimental methodology for the determination of in vitro release from injectable disperse systems may be considered to fall into four categories (7): membrane diffusion, sample and separate, in situ, and continuous flow methods.

1.5.1. Membrane Diffusion Techniques

These techniques are characterized by their use of a dialysis membrane to partition the sample and test media, thereby facilitating the determination of concentration of released drug. The membrane is selected to have a molecular weight cut-off allowing permeation of the drug substance, and it is assumed that diffusion of drug through the membrane is not a rate-limiting step. The dialysis membrane must be conditioned by soaking in dissolution medium prior to use, in order to remove extractables which may interfere in the subsequent analysis.

The dialysis sac diffusion technique involves placing a suitably sized sample (unit dose if possible), along with a suitable carrier medium (continuous phase, suspending medium or dissolution buffer), into a dialysis sac or tube. This is sealed and placed in a large volume of dissolution buffer, which is stirred to ensure uniform mixing, and the concentration of drug arising from diffusion through the membrane is determined at an appropriate frequency. The dialysis sac diffusion technique has been used to measure in vitro release from liposomes (8), submicron emulsions (9,10), and microspheres (11).

In a variation of the method, release is determined by assay of microspheres remaining within a dialysis tube at each test time-point (12); this approach also allows measurement of mass loss, hydration, and polymer degradation. The technique is simple to apply, separates the sample from the dissolution medium simplifying subsequent assay, and is applicable to a wide range of formulation types, but suffers the significant disadvantage that the sample within the dialysis sac is largely undiluted and therefore sink conditions do not apply. In the example of an emulsion formulation of a lipophilic drug substance, release rate measured using this technique will be determined largely by the partition coefficient between disperse and continuous phases within the dialysis bag and will not be indicative of release in the blood stream, which can be considered a true sink due to binding of the lipophilic drug substance to blood proteins. This issue may be resolved by the inclusion of a solubilizing agent, in the form of a hydrophilic β-cyclodextrin derivative, in the dissolution medium to maintain sink conditions (13). Further applications may include the study of depot formulations administered by subcutaneous or intramuscular injection, where the depot may become encapsulated by tissue leading to membrane-mediated release.

A modification to the above approach, the bulk equilibrium reverse dialysis sac technique (5,10), avoids this problem by placing the sample directly into an appropriate volume of dissolution buffer in equilibrium with several dialysis sacs each containing 1 mL of the same dissolution buffer. At appropriate intervals, one dialysis sac and a 1 mL sample from the bulk dissolution buffer are removed and the drug contents of the dialysis sac and the bulk solution are assayed. In this approach, release may be studied under sink conditions. If the active substance is chemically stable under the conditions of the test, and if the sample is accurately dispersed, analysis of the bulk solution is unnecessary as the percentage release can be calculated from the assay of the dialysis sac alone. In this approach, the formulation is diluted in a large volume of dissolution medium and sink conditions may be considered to apply. The technique may therefore have utility in the study of intravenous emulsions and liposomes.

This approach has been further developed into a fully automated system, microdialysis sampling, initially applied to tablets (14,15), and subsequently to implants (16). A schematic illustration of such an apparatus is shown in Fig. 2.

The test sample is added to a suitable volume of continuously stirred dissolution medium and the microdialysis probe, consisting of narrow-bore dialysis tubing, is positioned below the surface. A perfusion medium is continuously pumped through the probe and collected for analysis by high-performance

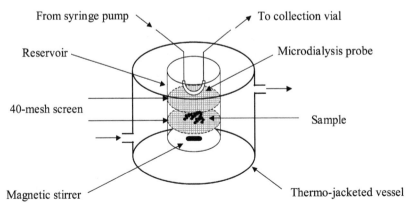

Figure 2 Microdialysis sampling.

liquid chromatography (HPLC). The perfusion medium may be buffered to ensure compatibility with the HPLC column, and the flow rate and surface area of the microdialysis probe may be manipulated to ensure that the drug concentration is within the range of the assay method.

As for the reverse dialysis sac technique, this approach allows sink conditions to be maintained, and therefore may be applicable to intravenous formulations.

The rotating dialysis cell is a further variation on the membrane diffusion theme. This approach was first used to assess in vitro release from parenteral oil depot formulations (17) and has also been used to assess drug salt release from suspensions (18). The apparatus consists of a small (10 mL) and a large (1000 mL) compartment separated by a dialysis membrane, as shown in Fig. 3.

In use, approximately 5 mL of sample is introduced into the dialysis cell which is placed in a large (typically 1000 mL) volume of dissolution medium. The dialysis cell is rotated at a constant speed, typically 50 rpm, and the concentration of drug arising through diffusion into the sink solution is measured at appropriate intervals.

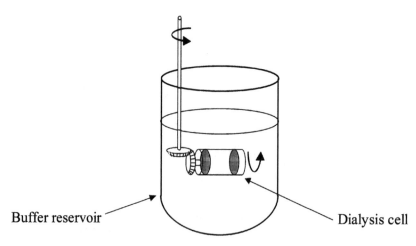

Buffer reservoir   Dialysis cell

Figure 3 Rotating dialysis cell.

It is considered that this approach, in which the apparatus acts as a two-compartment model, may mimic release in vivo where the route of administration is into a small compartment (e.g., intra-articular) or where release into the systemic circulation is mediated by passive diffusion through a membrane.

1.5.2. Sample and Separate Techniques

This category covers methods in which the sample is diluted with dissolution medium under sink conditions, a sample is withdrawn at appropriate intervals, and undissolved material removed leaving a solution containing dissolved drug.

This approach has been applied to assess drug release from PLGA microspheres, using USP Apparatus 2 (paddle method); samples were withdrawn, filtered, and the filtrate analyzed by HPLC (19). A variation involved shaking several tubes (one per test time-point) containing sample, taking one tube at each test time-point, and centrifuging to separate free drug in solution from undissolved material then determining dissolved drug concentration using HPLC (20). Tube-to-tube variability may be eliminated by replacing the supernatant removed for assay with an equal volume of fresh dissolution medium, vortexing to resuspend, then continuing the test with the same tube (21).

The centrifugal ultrafiltration technique developed by Millipore (22) in the form of the Ultrafree®-MC unit, illustrated in Fig. 4, utilizes an ultrafiltration membrane having a nominal molecular weight limit (NMWL) of 5000–100,000 Da. A maximum 400 µL sample of dissolution medium containing the suspended formulation is withdrawn from the dissolution vessel at appropriate intervals and transferred to a centrifugal filter unit with an NMWL value chosen to allow passage of the drug. The unit is placed in a microcentrifuge tube and centrifuged at up to 5000 g using a fixed angle microcentrifuge. The resulting ultrafiltrate is assayed to determine the free drug substance concentration.

The centrifugal ultrafiltration method has been applied to the determination of in vitro release from a submicron emulsion (23).

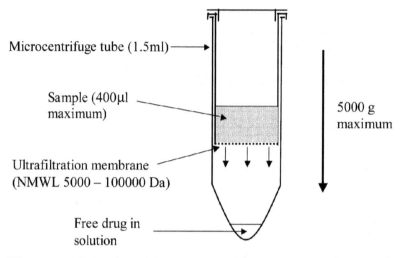

Microcentrifuge tube (1.5ml)

Sample (400µl maximum)

5000 g maximum

Ultrafiltration membrane (NMWL 5000 – 100000 Da)

Free drug in solution

Figure 4 Centrifugal ultrafiltration apparatus.

1.5.3. In Situ Techniques

In this approach, the sample is diluted in the dissolution medium and release is measured in situ, without separation of undissolved material, using a suitable analytical methodology specific to dissolved drug. This approach is little used in the determination of drug release from injectable dispersed systems, as correction for interference from undissolved drug may be problematic. Differential pulse polarography has been successfully used to determine the release of pyroxicam from polymeric nanoparticle dispersions (24).

1.5.4. Continuous Flow Methods

This category includes single-pass methods in which dissolution medium is pumped through a cell containing the sample and the eluant is analyzed continuously or fractions are collected for subsequent assay; and loop methods in which the dissolution medium is continuously recirculated.

This technique is mainly applicable to microspheres and other solid dosage forms which may be retained in the flow-through cell by use of an appropriate filter. The sample may be mixed with glass beads to minimize aggregation as well as to alter the flow pattern within the sample bed to help avoid channeling effects that would lead to inaccurate release patterns. The flow-through cell is placed vertically in a thermo-jacketed vessel and dissolution medium pumped from the reservoir, through a delay coil to allow temperature equilibration, through the flow-through cell from bottom to top, then through an in-line measurement device such as a UV spectrophotometer before being returned to the reservoir. The volume of dissolution medium remains constant throughout. A schematic illustration is shown in Fig. 5.

The flow-through apparatus has been used extensively to evaluate drug release from oral solid dosage forms and has been applied to injectable dispersed systems, mainly microspheres. Release from microwave-treated gelatin microspheres has been investigated under sink and non-sink conditions, using deionized water as dissolution medium (25). In a study of verapamil hydrochloride-loaded microspheres intended for oral administration, a surfactant was added to the dissolution medium to improve wetting, and the flow rate was controlled to maintain sink conditions in the flow-through cell; different dissolution media were evaluated, and in the "half-change" method, step changes in pH were introduced at predetermined

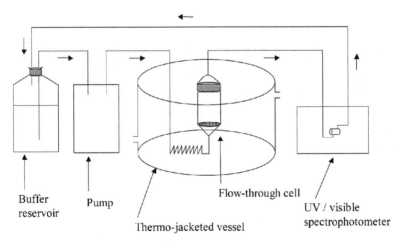

Buffer reservoir   Pump   Flow-through cell

                                                         UV / visible
                    Thermo-jacketed vessel   spectrophotometer

Figure 5   USP Apparatus 4 (flow-through cell).

time-points (26). A novel approach was used to investigate release of glial cell line-derived neurotrophic factor (GDNF) from PLGA microspheres; the apparatus utilized an unpacked HPLC column as the sample compartment; dissolution medium was passed through the column to a fraction collector, and protein release determined by gamma counting, ELISA, and/or bioassay methods. A study of the dissolution of a poorly soluble compound in unmicronized and micronized form concluded that homogeneous mixing of the sample with the glass beads in the flow-through cell was effective in achieving maximum dissolution with minimum variability for unmicronized powders, but for micronized powders poor wetting resulted in particles being carried into the filter, resulting in anomalously low release. Presuspending drug in dissolution medium modified to include a suspending medium (0.3% HPMC) and a surfactant (0.2% Tween 80), introducing the sample as a slug below the glass beads, and reducing flow rate, were shown to lead to release profiles in line with particle size (27).

1.6. Method Development

Preliminary method development should be based on a knowledge of the dosage form and route of administration, and the in vitro procedure should emulate in vivo conditions so far as is reasonably practical.

For extended release dosage forms, which may be designed to release drug over prolonged periods up to 12 months, the development of an accelerated in vitro release procedure may offer considerable benefits in reducing development time-lines and, for marketed products, in resource efficiency and enhanced responsiveness to manufacturing problems. In vitro release may be accelerated by the choice of appropriate conditions, in particular increased temperature and extreme pH, but the same requirements for biorelevance must be met.

When in vivo data from exploratory studies are available, method optimization should be carried out with the aim of achieving an in vivo–in vitro correlation for fast, intermediate, and slow-releasing batches. An experimental design approach should be utilized; the following are generally considered to be critical parameters for investigation:

- Volume of dissolution medium (sink conditions)
- Composition and pH of dissolution medium
- Temperature
- Agitation/flow.

Systematic variation of selected parameters to optimize discrimination, duration of release, and release profile for two or more batches or formulation variants known to behave differently in vivo should lead to the definition of a biorelevant release test.

For detailed information on the development of an in vitro-in vivo correlation, refer to the chapter on In Vitro-In Vivo Correlation for Modified Release Parenteral Drug Delivery Systems in this book.

2. DATA MANIPULATION

2.1. Calculation of Cumulative Release

For an analytical system in which volume is constant, as is the case for in situ methods and recirculatory continuous flow systems with in-line concentration measurement, cumulative release may be determined as follows:

$$R_n = 100\frac{C_n V}{D}$$

where R_n is the percentage cumulative release at time-point n, C_n the concentration at time-point n, and D is the drug content of the sample; for a regulatory test, it is normal practice that the test comprise, multiple determinations of a single unit dose, and for D to represent the labeled dose.

Where the release test involves the withdrawal of a sample of dissolution medium in which the formulation is homogeneously dispersed, and the test is continued with diminished volume, the formulation:dissolution medium ratio is unaffected by the sampling operation and hence the concentration of dissolved drug at subsequent time-points is the

same as would be the case in the constant volume procedure
described above. Cumulative release is given by

$$R_n = 100 \frac{C_n V_0}{D}$$

where R_n, C_n, and D are as previously described and V_0 is the
initial volume of dissolution medium.

Where the method involves in-situ filtration such that
supernatant medium containing dissolved drug is withdrawn
and not replaced, a correction factor must be applied as
follows:

$$R_n = 100 \left\{ \frac{C_n V_n + \sum_{i=1}^{n-1} C_i V_s}{D} \right\}$$

where R_n, C_n, and D are as previously described, and V_s is the
volume of supernatant medium withdrawn at each time-point.

2.2. Mathematical Description of Release Profile

In general, drug dissolution from solids can be described
using the Noyes–Whitney equation as modified by Nernst
and Brunner:

$$\frac{dM}{dt} = \frac{DS(C_s - C_t)}{h}$$

where M is the amount of drug dissolved in time t, D is the
diffusion coefficient of the solute in the dissolution medium,
S is the surface area of the expressed drug, h is the thickness
of the diffusion layer, C_s is the solubility of the solute and C_t is
the concentration of the solute in the medium at time t; the
equation may be simplified by assuming that, for dissolution
testing under sink conditions, C_t is zero.

This model assumes that a layer of saturated solution
forms instantly around a solid particle, and that the dissolution rate-controlling step is transport across this so-called
diffusion layer. Ficks law describes the diffusion process:

$$\frac{m}{t} = \frac{DA\Delta C}{L}$$

where m/t is the mass flow rate (mass m diffusing in time t), D is the diffusion constant, ΔC is the concentration difference, A is the cross-sectional area, and L is the diffusion path length.

The cube root law developed by Hixson and Crowell takes into account Fick's law and may be considered to describe the dissolution of a single spherical particle under sink conditions:

$$w^{1/3} = w_0^{1/3} - k_{1/3}t, \quad k_{1/3} = \left(\frac{4\pi\rho}{3}\right)^{1/3}\frac{DC_s}{\rho h}$$

where w is particle weight at time t, w_0 is the initial particle weight, $k_{1/3}$ is the composite rate constant, ρ is the density of the particle, and D, C_s, and h are as previously defined.

The expressions above may be used as the basis for mathematical models of drug release (28).

2.3. Comparison of Release Profiles

A comparison of dissolution profiles may be necessary to support changes in formulation, site, scale or method of manufacture. The comparison should be based on at least 12 units of reference (prechange) and test (postchange) product.

A common procedure is the model-independent approach (29,30), which involves calculation of a difference factor (f_1) and a difference factor (f_2) to compare profiles. This approach is suitable where the dissolution profile is based on three or more time-points, only one of which occurs after 85% dissolution; the time-points for the reference and test batches must be the same and no modification to the release test is permissible. The reference profile may be based on the mean dissolution values for the last prechange batch, or the last two or more consecutively manufactured prechange batches; for mean data to be meaningful, the RSD should be $< 20\%$ for the initial time-point and $< 10\%$ for subsequent time-points.

The difference factor (f_1) calculates the percentage difference between the two curves at each time-point and is a measurement of the relative error between the two curves:

$$f_1 = 100\frac{\sum_{t=1}^{n}|R_t - Tt|}{\sum_{t=1}^{n}R_t}$$

where n is the number of time-points, R_t is the dissolution value of the reference (prechange) batch at time t, and T_t is the dissolution value of the test (postchange) batch at time t. For curves to be considered similar, f_1 should be close to zero and within the range 0–15.

The similarity factor (f_2) is a logarithmic reciprocal square root transformation of the sum of squared error and is a measurement of the similarity between the two curves:

$$f_2 = 50 \log \left\{ \frac{100}{\sqrt{[1 + (1/n) \sum_{t=1}^{n} (R_t - T_t)^2]}} \right\}$$

where n, t, R_t, and T_t are as defined for the calculation of difference factor.

For curves to be considered similar, the similarity factor (f_2) should be close to 100 and within the range 50–100.

Where batch-to-batch variation within the reference and test batches is greater than 15% RSD, misleading results may arise and an alternative approach is preferable. A model-dependent method, involving the derivation of a mathematical function to describe the dissolution profile followed by determination of the statistical distance between the reference and test batches (31), may be used to compare the test and reference profiles taking into account variance and covariance of the data sets and allowing the use of different sampling schemes for the reference and test lots.

A comparison of ANOVA-based, model-dependent, and model-independent methodologies for immediate release tablets (32) concluded that the ANOVA-based and model-dependent methods have narrower limits and are more discriminatory than the similarity/difference factor methods.

3. IN VIVO RELEASE

3.1. Introduction

The following section discusses the preclinical in vivo evaluation of injectable dispersed systems particularly the evaluation of extended release systems intended for intramuscular or subcutaneous administration and should be read in conjunction with the chapters in this book by Oussoren et al. (Biopharm) and Young et al. (IVIVC).

Preclinical testing of parenteral modified release formulations is performed for two major reasons:

- To provide data that support the ethical dosing of new chemical entities and formulations, clinically with regard to safety and efficacy.
- To support the pharmaceutical development of formulations with the predicted desired clinical performance (which is usually assessed by pharmacokinetic performance).

This section will focus on the design of in vivo preclinical experiments aimed at supporting the pharmaceutical development of modified release particulate drug delivery systems with special emphasis on those in which the excipient modifying release is a poly-lactic acid or poly-lacticco-glycolic acid ester polymer (PLA/PLGA). When performing preclinical in vivo evaluations, a fundamental assumption is that the model/ species chosen is likely to be predictive of the clinical situation. There is, however, a lack of systematic investigations to establish which animal species are the most predictive of the clinical situation. Indeed an AAPS, FDA, and USP co-sponsored workshop on "Assuring Quality and Performance of Sustained and Controlled Release Parenterals" recommended the initiation of research in this area (33). Notwithstanding this lack of systematic research, from a knowledge of first principles and review of the literature, it is possible to draw conclusions about the relative merits of different animal models.

During the pharmaceutical development of PLGA-based dispersed systems, the primary aim of preclinical experiments, in common with in vitro dissolution testing, is to characterize the release of drug from the delivery system. Therefore, the primary requirement for the preclinical model is that the absorption/injection site is sufficiently similar to that in humans such that the release mechanism and release kinetics of drug from the PLA/PLGA system are qualitatively equivalent to that which will be experienced in the clinic. There is a substantial body of evidence supporting the hypothesis that the release of drug from PLA/PLGA-based systems is predominately controlled by the characteristics of the delivery system and dependent mainly on a combination of diffusion (early phase) and hydrolytic erosion (later phase) (34). For PLA/PLGA-dispersed systems, it is clear that this release is also the rate-determining step for the pharmacokinetics (35), otherwise there would be no pharmacokinetic driver to produce such a complicated system. It is therefore not unreasonable to assume that if the injection sites of preclinical species do not differ too markedly for human tissue in terms of biochemistry and tissue reaction then the release profiles are likely to be comparable across species. Exhaustive comparative analysis of the subcutaneous and intramuscular tissue interstitial fluid has apparently not been performed; however, we know that interstitial fluid is in equilibrium with serum/ plasma and the serum data for preclinical species and man are available (Table 1).

For larger molecular weight species (plasma proteins), the equilibrium point between plasma and interstitial fluid is dependent on the endothelial properties of each tissue (37) and, for both small and large molecular weight species, the metabolic fate within the tissue (38). Lymph to plasma (interstitial) concentration ratios are available for different species. While somewhat variable, and dependent upon measurement technique (39), they are broadly comparable across species although for albumin, dogs and rabbits may exhibit a lower ratio than seen in humans while rats show the most similar ratios. It should also be noted that tissue differences in concentration ratios also exist with interstitial albumin concentration being higher in skeletal muscle than subcutaneous tissue under some conditions (37,40). Thus it can be concluded that in terms of biochemistry interstitial fluid is likely to be broadly similar across species.

The histopathological reaction observed following injection of PLA/PLGA microspheres is typical of a response to an inert foreign body in which the aim of the tissue reaction is the removal of the material from the host without the generation of an antigen-specific immune response. The cells

Table 1 Mean Values of the Inorganic Components in the Serum of the Male of Each Species Listed

	Mice (albino)	Rat (albino)	Rabbit	Dog	Man
Sodium (mEq/L)	138 (128–145)	147 (143–156)	146 (138–155)	147 (139–153)	141 (135–155)
Potassium (mEq/L)	5.25 (4.85–5.85)	5.82 (5.4–7)	5.75 (3.7–6.8)	4.54 (3.6–5.2)	4.1 (3.6–5.5)
Chloride (mEq/L)	108 (105–110)	102 (100–110)	101 (92–112)	114 (103–121)	104 (98–109)
Bicarbonate (mEq/L)	26.2 (20–31.5)	24 (12.6–32)	24.2 (16.2–31.8)	21.8 (14.6–29.4)	27 (22–33)
Phosphorous (mg/dL)	5.6 (2.3–9.2)	7.56 (3.11–11.0)	4.82 (2.3–6.9)	4.4 (2.7–5.7)	3.5 (2.5–4.8)
Calcium (mg/dL)	5.6 (3.2–8.5)	12.2 (7.2–13.9)	10 (5.6–12.1)	10.2 (9.3–11.7)	9.8 (8.5–10.7)
Magnesium (mg/dL)	3.11 (0.8–3.9)	3.12 (1.6–4.44)	2.52 (2–5.4)	2.1 (1.5–2.8)	2.12 (1.8–2.9)

The bracketed values indicate the range in literature values.
(From Ref. 36.)

involved in this reaction are overwhelmingly those of the macrophage series, but the detailed form of the response is dependent upon the size of the microspheres injected.

Following injection of microspheres of less than approximately 10 μm in diameter, the response is characterized by the progressive invasion and phagocytosis of the mass of microspheres by single macrophages. The single macrophage, however, is incapable of phagocytosing larger microspheres, and if these are injected the host reaction includes large numbers of multinucleate giant cells that are formed from the fusion of individual macrophages. The macrophages or giant cells engulf and presumably digest the microspheres. In all instances, a two- to three-cell thick rim of fibrous tissue surrounds the invading phagocytic cells and small blood vessels invade alongside the phagocytes (Fig. 6).

This is a subcutaneous injection site and hair follicles are clearly visible (*). An area of the microsphere tissue reaction is illustrated. There is a thin fibrous capsule (arrow), under which there is the advancing wall of macrophages and giant cells (line) that is engulfing microspheres (arrowhead). To the center of the reaction site, the microspheres are lost as a tissue processing artifact.

The size of the lesion is directly related to the number of microspheres injected. It is the mass of invading phagocytic cells which are palpable at the injection site, which explains the delay between injection of the microspheres and the formation of a clinically obvious lump. These lesions are progressive, and resolution is complete to a point where there is no histopathological abnormality detected at the injection site.

This tissue reaction is identical in subcutaneous and intramuscular injection sites, and is very similar across species including rat, mouse, and primate.

For drugs which are non-irritant, the tissue reaction to drug laden microspheres is indistinguishable from that to control microspheres. If the drug is an irritant or has other proinflammatory properties, the cellular infiltrate contains large numbers of lymphocytes and fibrosis is prominent.

To summarize, it seems reasonable to conclude that the absorption site environment across species should be

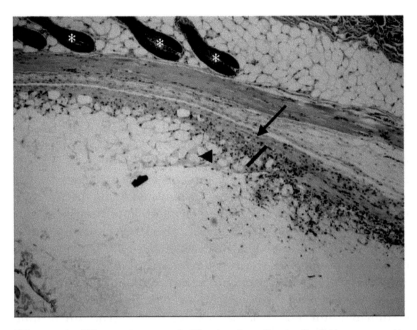

Figure 6 Photomicrograph illustrating the typical tissue reaction to PLGA microspheres, showing hair follicles (*), the fibrous capsule (↓), giant cells (|) and microspheres (▼).

sufficiently similar so as not to affect the release kinetics of drugs from PLA/PLGA microspheres. Review of the literature in which PLA/PLGA-based systems have been administered to more than one species or strain or have been administrated both subcutaneously and intramuscularly seem, to confirm this conclusion. It was shown during the development of a 1-month leuprolide acetate PLGA microsphere formulation (41,42) that the rate of release of leuprolide acetate from PLGA microspheres (as measured by loss from the injection site) was the same after administration to both subcutaneous and skeletal muscle tissue. Furthermore, the rat strain (Sprague–Dawley vs. Wistar) did not affect the performance of the microspheres as assessed by pharmacodynamic endpoints (41). This research group also demonstrated that the plasma concentration–time curves for leuprolide acetate in dogs and rats after intramuscular administration of leuprolide acetate-loaded PLGA microspheres had essentially the same pattern indicating non-species-specific release of leuprolide acetate from the microspheres (43). Furthermore, the same group also demonstrated similar performance for rat and dog after subcutaneous and skeletal muscle administration for a 3-month leuprolide acetate PLA microsphere formulation (44).

Although a case has been made for the similarity of all animal models including humans for the evaluation of the release (pharmacokinetic) behavior of preformed PLA/PLGA-based systems, a few cautionary points should be considered. If there is likely to be a specific immune interaction in a species that is not present in other species then this may make this species inappropriate for the pharmacokinetic evaluation of the PLGA system. This has been highlighted previously for liposomal systems (45). The foregoing discussion can only be considered to apply to "preformed" controlled release systems. For other formulations which are dependent on the formation of the rate-controlling structure in vivo (for instance precipitation), there may be sufficient difference between species for the structure forming step, which is likely to be rapid, to be sufficiently different to give different formulation behavior. For this type of formulation, the identification of the most appropriate preclinical species may need to be performed empirically as part of the development program. Finally, it should be remembered that rabbits, and possibly dogs, have a slightly higher body temperature (rabbit 38.5–39.5°C; dog 37.5–39.0°C) than other preclinical species (mouse 36.5–38.0°C; rat 35.9–37.5°C; non-human primate 37.0–39.0°C) which maybe an important consideration depending on the glass transition temperature of the formulation.

3.2. Choice of Animal Species

As all preclinical species are likely to be equally predictive for preformed PLA/PLGA delivery systems other selection criteria become important. These are discussed briefly below.

3.2.1. Ethical Considerations

Within the European Union and United Kingdom in particular, there is an ethical and legal obligation to use the species with the lowest neurophysiological sensitivity to meet the objectives of the experiment.

3.2.2. Dose Volume and Sample Volume Considerations

Both ethically and scientifically, it is desirable not to give dose volumes or take blood volumes that will unduly change the physiology of the animal and therefore cause unnecessary discomfort or invalidate the scientific integrity of the study. Therefore, there is a balance to be struck between delivering sufficient drug to give quantifiable systemic plasma drug concentrations (see Sec. 4) or volumes of complex formulations that can be practically administered (see Sec. 5) and what can be ethically and scientifically justified. Currently accepted European good practice guide on dosing and sampling volumes are reported (46). A strategy to allow experimental design to meet these limits is discussed in the case study.

3.2.3. Toxicological Species

Where possible, it would seem appropriate that the formulation development program is performed in species that are to be used in the safety assessment evaluation of the compound and formulation as this should reduce the number of studies that need to be performed (for instance avoidance of pharmacokinetic sighting studies prior to the start of a full safety assessment study). Choice of toxicological species is driven by the regulatory requirement to provide data in a rodent and non-rodent, demonstration of pharmacological activity in the chosen species, and the ethical consideration to use animals of the lowest neurophysiological sensitivity to meet the objectives of the experiment.

4. BIOANALYSIS

There is little value in taking such care to choose species and design the live phase of any preclinical evaluation if the drug blood/plasma concentration analysis is then lacking. Many compounds are considered for parenteral-controlled release due to their poor oral bioavailability and usually short elimination half-life (in many cases, these compounds are peptides and proteins). This presents the bioanalyst with considerable challenges as these compounds are usually unstable in blood/plasma and also difficult to resolve from endogenous material in plasma. The advent of quantitative HPLC–MS–MS has made this task easier (47,48); however, radioimunnoassay may still need to be considered as an analytical method. Development of the assay and assessment of assay performance should meet accepted criteria, for instance those proposed (49) and documented in FDA guidance documents (50). For biopharmaceuticals, special attention should be given to the stability of the drug in blood/plasma which is likely to be relatively poor and requires special steps to make the stability manageable, for example, inclusion of protease inhibitors; special care to avoid hemolysis on plasma collection; storage on ice and specialized collection procedures. At the planning stage of an in vivo study, it is particularly important to talk through these aspects of sample collection with the animal technicians who will perform the study as this is likely to be somewhat different to the procedures they usually follow.

5. INJECTABILITY

In vitro measurements of content and dose uniformity should be reviewed in the light of in vivo (clinical) behavior. It is important to note that under clinical conditions, this behavior may be substantially compromised. Injectability has been identified as an important performance parameter. Injection into tissue differs in two ways from that experienced when using standard in vitro techniques due to changes in fluid dynamics and the potential for "coring" of tissue within the needle bore (Fig. 7).

Both these factors increase the potential for needle blockage possibly preceded by filtering out of the microspheres. That is if the microsphere size, morphology, and suspending agent characteristics are not optimal then the suspending fluid is able to pass into the tissue while the microspheres

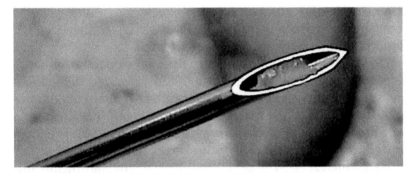

Figure 7 Example of tissue coring that can partially occlude the needle and lead to filtering of the microspheres and eventually blockage (example shown is a wide bore needle for clarity).

are trapped in the needle and syringe luer. The trapped microspheres eventually reach a critical mass and on further pressure to the syringe plunger compress to form a "plug" which blocks the needle/syringe. When there is a potential for sieving to occur, for instance with earlier development formulations, it is important to note that injection volume may not necessarily equate to microsphere dose and a prudent step maybe to assay for drug remaining in the syringe even if a correct volume has been injected. A useful model to qualitatively assess in vivo injectability is injection into meat. We have found that subcutaneous injection into chicken carcasses produced for the food industry mimics subcutaneous injectability in preclinical species.

6. CONCLUSIONS

In vivo and in vitro studies are essential components of the drug development process. The objective of such studies is to determine a relationship between an in vitro characteristic of a dosage form and its in vivo performance, such that the in vitro test may be used to predict in vivo performance. In vitro testing is used in early development to select batches for in vivo pharmacokinetic/pharmacodynamic studies, but the ultimate objective is a test capable of distinguishing, prior to medical use, clinically effective batches from those which would be ineffective and/or unsafe if used.

In vitro dissolution tests were first developed for immediate release solid oral dosage forms then extended to modified release formulations. In recent years, the application of dissolution testing has been extended to "special" dosage forms including injectable dispersed systems, and for such products administered by a non-oral route the term "drug release" or "in vitro release" test is preferred. Due to the significant differences in formulation and hence in physicochemical and release characteristics, it is not possible to specify a generally applicable apparatus or method, rather different techniques are employed on a case-by-case basis (51).

Preclinical in vivo studies are performed to provide safety, efficacy, and pharmacokinetic data to support formulation development and clinical use. Animal models are selected on the basis of their relevance to humans, with reference to the formulation and route of administration. For depot formulations in particular, it is important to study histopathological reactions at the injection site, as this may mediate drug release and is important in assessing the tissue compatability of both the formulation and the drug substance. Injectability may be an issue for injectable dispersed systems, and should be assessed prior to the commencement of in vivo studies. Ethical considerations govern the choice of animal species (that with the lowest neurophysiological sensitivity, other factors being equal), dosing, and sampling regimes; where possible the scale of in vivo testing should be minimized by, for example, use of the same species for toxicological and pharmacokinetic studies, and by the early development of a predictive in vitro release test.

A thorough understanding of the in vivo drug release mechanism of the dosage form, underpinned by comprehensive physicochemical characterization of the drug substance and delivery system, is a necessary foundation both for the development of a discriminatory in vitro release test and for the development of a high-quality product. Application of the principles outlined in this chapter should lead to a biorelevant release test and facilitate the development of a meaningful in vitro–in vivo correlation using techniques described elsewhere in this publication. This information should be considered an essential component of the Chemistry and Manufacturing Controls section of a New Drug/Marketing Authorization Application for Injectable Dispersed Systems.

REFERENCES

1. Pharmacopeial Forum 2004; 30(1):1092 The Dissolution Procedure: Development and Validation © 2004 The United States Pharmacopeia Convention, Inc.

2. United States Pharmacopeia. National Formulary (USP28-NF 22) 2005.

3. Kim HC, Burgess DJ. Effect of drug stability on analysis of release data from controlled release microspheres. J Microencapsul 2002; 19(5):631–640.

4. Kim HC, Burgess DJ. Isosbestic point—spectrophotometric method to measure in vitro release from controlled release PLGA microspheres. Presented by Burgess DJ, at the CRS Annual Meeting, Seoul, South Korea, July 2002.

5. Benita S, Levy MY. Submicron emulsions as colloidal drug carriers for intravenous administration: comprehensive physicochemical characterisation. J Pharm Sci 1993; 82:1069–1079.

6. Pitt CG. The controlled parenteral delivery of polypeptides and proteins. Int J Pharm 1990; 59:173–196.

7. Washington C. Drug release from microdisperse systems: a critical review. Int J Pharm 1990; 58:1–12.

8. Elorza B, Elorza MA, Frutos G, Chantres JR. Characterization of 5-fluorouracil loaded liposomes prepared by reverse-phase evaporation or freezing-thawing extrusion methods: study of drug release. Biochim Biophys Acta 1993; 1153:135–142.

9. Benita S, Friedman D, Weinstock M. Pharmacological evaluation of an injectable prolonged release emulsion of physostigmine in rabbits. J Pharm Pharmacol 1986; 38:653–658.

10. Chidambaram N, Burgess DJ. A novel method to characterize in vitro release from submicron emulsions. AAPS PharmSci

August 31, 1999; 1(3). Available online at http://aapspharma-ceutica.com/scientificjournals/pharmsci.

11. Kostanski JW, DeLuca PP. A novel in vitro release technique for peptide-containing biodegradable microspheres. AAPS PharmSciTech 2000; 1(1) article 4.

12. Woo BH, Kostanski JW, Gebrekidan S, Dani BA, Thanoo BC, DeLuca PP. Preparation, characterization and in vivo evaluation of 120-dat poly(D,L-lactide) leuprolide microspheres. J Control Release 2001; 75:307–315.

13. Saarinen-Savolainen P, Jarvinen T, Taipale H, Urtti A. Method for evaluating drug release from liposomes in sink conditions. Int J Pharm 1997; 159:27–33.

14. Shah KP, Chang M, Riley CM. Automated analytical systems for drug development studies. II—a system for dissolution testing. J Pharm Biomed Anal 1994; 12:1519–1527.

15. Shah KP, Chang M, Riley CM. Automated analytical systems for drug development studies. III—a system for dissolution testing. J Pharm Biomed Anal 1995; 13:1235–1245.

16. Dash AK, Haney PW, Garavalia MJ. Development of an in vitro dissolution method using microdialysis sampling technique for implantable drug delivery systems. J Pharm Sci 1999; 88:1036–1040.

17. Schultz K, Mollgaard B, Frikjaer S, Larsen C. Rotating dialysis cell as in vitro release method for oily parenteral depot solutions. Int J Pharm 1997; 157:163–169.

18. Parshad H, Frydenvang K, Liljefors T, Cornett C, Larsen C. Assessment of drug salt release from solutions, suspensions and in situ suspensions using a rotating dialysis cell. Eur J Pharm Sci 2003; 19:263–272.

19. Castelli F, Conti B, Maccarrone DE, Conte U, Puglisi G. Comparative study of 'in vitro' release of anti-inflammatory drugs from polylactide-co-glycolide microspheres. Int J Pharm 1998; 176:85–98.

20. Genta I, Perugini P, Pavanetto F, Maculotti K, Modena T, Casado B, Lupi A, Iadarola P, Conti B. Enzyme loaded biodegradable microspheres in vitro ex vivo evaluation. J Control Release 2001; 77:287–295.

21. Berkland C, Kipper MJ, Narasimhan B, Kim K, Pack DW. Microsphere size, precipitation kinetics and drug distribution control drug release from biodegradable polyanhydride microspheres. J Control Release 2004; 94:129–141.

22. Millipore on-line catalogue: http://www.millipore.com/catalogue.nsf/docs/c7553.

23. Santos-Magalhaes NS, Cave G, Seiller M, Benita S. The stability and in vitro release kinetics of a clofibride emulsion. Int J Pharm 1991; 76:225–237.

24. Charalampopoulos N, Avgoustakis K, Kontoyannis CG. Differential pulse polarography: a suitable technique for monitoring drug release from polymeric nanoparticle dispersions. Anal Chim Acta 2003; 491:57–62.

25. Vandelli MA, Romagnoli M, Monti A, Gozzi M, Guerra P, Rivasi F, Forni F. Microwave-treated gelatine microspheres as drug delivery system. J Control Release. 2004; 96:67–84..

26. Kilicarslan M, Baykara T. The effect of the drug/polymer ratio on the properties of the verapamil HCl loaded microspheres. Int J Pharm 2003; 252:99–109.

27. Bhattachar SN, Wesley JA, Fioritto A, Martin PJ, Babu SR. Dissolution testing of a poorly soluble compound using the flow-through cell dissolution apparatus. Int J Pharm 2002; 236:135–143.

28. Wang J, Flanagan DR. General solution for diffusion-controlled dissolution of spherical particles 1. Theory. J Pharm Sci 1999; 88(7):731–738.

29. Moore JW, Flanner HH. Mathematical comparison of dissolution profiles. Pharm Tech 1996; 20(6):64–74.

30. FDA Center for Drug Evaluation and Research. Guidance for Industry: Dissolution Testing of Immediate Release Solid Oral Dosage Forms. FDA Center for Drug Evaluation and Research, 1997, pp. 8–9.

31. Sathe M, Tsong Y, Shah VP. In vitro dissolution profile comparison: statistics and analysis, model dependent approach. Pharm Res 1996; 13(12):1799–1803.

32. Yuksel N, Kanik AE, Baykara T. Comparison of in vitro dissolution profiles by ANOVA-based, model-dependent and-independent methods. Int J Pharm 2000; 209:57–67.

33. Burgess DJ, Hussain AS, Ingallinera TS, Chen M. Assuring quality and performance of sustained and controlled release parenterals. AAPSPharmSci 2002;4(2):7(http://www.aapspha rmsci.org/scientificjournals/pharmsci/journal/040207.htm).

34. Hutchinson FG, Furr BJA. Biodegradable polymer systems for the sustained release of polypeptides. J Control Release 1990; 13:279–294.

35. Maulding HV. Prolonged delivery of peptides by microcapsules. J Control Release 1987; 6:167–176.

36. Mitruka BM, Rawnsley HM. Clinical Biochemical and Hematological Reference Values in Normal Experimental Animals. Tunbridge Wells: Abacus Press Ltd, 1977.

37. Taylor AE, Granger DN. Exchange of macromolecule across the microcirculation. In: Renkin EM, Michel CC, eds. Handbook of Physiology: The Cardiovascular System. Vol. IV. Microcirculation. Part 1. Bethesda, MD: American Physiological Society, 1984.

38. Kammermeier H. The immediate environment of cardiomyocytes: substantial concentration differences between the interstitial fluid and plasma water for substrates and transmitters. J Mol Cell Cardiol 1995; 27:195–200.

39. Fogh-Andersen N, Altura BM, Altura BT, Siggaard-Andersen O. Composition of interstitial fluid. Clin Chem 1995; 41: 1522–1525.

40. Ellmerer M, Schaupp L, Brunner GA, Sendlhoffer G, Wutte A, Wach P, Pieber TR. Measurement of interstitial albumin in human skeletal muscle and adipose tissue by open-flow microperfusion. Am J Physiol Endocrinol Metab 2000; 278: E352–E356.

41. Okada H, Heya T, Ogawa Y, Shimamoto T. One-month release injectable microcapsules of a luteinizing hormone-releasing hormone agonist (leuprolide acetate) for treating experimental endometriosis in rats. J Pharmacol Exp Ther 1988; 244: 744–750.

42. Okada H, Heya T, Igari Y, Ogawa Y, Toguchi H, Shimamoto T. One-month release injectable microcapsules of leuprolide acetate inhibit steroidogenesis and genital organ growth in rats. Int J Pharm 1989; 544:231–239.

43. Ogawa Y, Okada H, Heya T, Shimamoto T. Controlled release of LHRH agonist, leuprolide acetate, from microcapsules: serum drug level profiles and pharmacological effects in animals. J Pharm Pharmacol 1989; 41:439–444.

44. Okada H, Doken Y, Ogawa Y, Toguchi H. Sustained suppression of the pituitary-gonadal axis by leuprorelin three-month depot microspheres in rats and dogs. Pharm Res 1994; 11:1199–1203.

45. Moghimi SM, Patel HM. Serum mediated recognition of liposomes by phagocytic cells of the reticuloendothelial system—the concept of tissue specificity. Adv Drug Deliv Rev 1998; 32:45–60.

46. Diehl K-H, Hull R, Morton D, Pfister R, Rabemampianina Y, Smith D, Vidal J-M, van der Vorstenbosch C. A good practice guide to the administration of substance and removal of blood, including routes and volumes. J Appl Toxicol 2001; 21:15–23.

47. Lee MS, Kerns EH. LC/MS applications in drug development. Mass Spec Rev 1999; 18:187–279.

48. Hühmer AFR, Aced GI, Perkins MD, Gürsoy RN, Seetharama Jois DS, Larive C, Siahaan TJ, Schöneich C. Separation and analysis of peptides and proteins. Anal Chem 1997; 69: 29R–57R.

49. Shah VP, Midha KK, Findlay JWA, Hill HM, Hulse JD, McGilveray IJ, McKay G, Miller KJ, Patnaik RN, Powell ML, Tonelli A, Viswanathan CT, Yacobi A. Bioanalytical method validation—a revisit with a decade of progress. Pharm Res 2000; 17:1551–1557.

50. FDA (CDER). Bioanalytical Method Validation. FDA (CDER), May 2001.

51. Siewert M, Dressman J, Brown C, Shah V. FIP/AAPS guidelines for dissolution/in vitro release testing of novel/special dosage forms. Diss Tech 2003; 10(1):6–15.

5

In Vitro/In Vivo Correlation for Modified Release Injectable Drug Delivery Systems

DAVID YOUNG, COLM FARRELL, and THERESA SHEPARD

GloboMax Division of ICON plc,
Hanover, Maryland, U.S.A.

1. INTRODUCTION

A number of Food and Drug Administration (FDA) guidances discuss the development and role of in vitro–in vivo correlation (IVIVC) in oral solid dosage forms (1–4). One of these guidances, the FDA IVIVC Guidance (1), has defined IVIVC as

> a predictive mathematical model describing the relationship between an in vitro property (usually the rate or extent of drug dissolution or release)...and a relevant

in vivo response, e.g., plasma drug concentration or amount of drug absorbed.

Four types of IVIVC approaches (i.e., Level A, Level B, Level C, Multiple Level C) are defined within this guidance. A Level A correlation is "a predictive mathematical model for the relationship between the entire in vitro dissolution release time course and the entire in vivo response time course." A Level B correlation is "a predictive mathematical model for the relationship between summary parameters that characterize the in vitro and in vivo time courses, e.g., models that relate mean in vitro dissolution time to the mean in vivo dissolution time." A Level C correlation is "a predictive mathematical model for the relationship between the amount dissolved in vitro at a particular time (or the time required for in vitro dissolution of a fixed percent of the dose) and a summary parameter that characterizes the in vivo time course (e.g., C_{max} or AUC)." A Multiple Level C correlation is "a Level C correlation at several time points in the dissolution profile." Although each type of IVIVC may have its place in the product development process, the Level A correlation is accepted as the most informative IVIVC for oral drug delivery systems.

As modified release parenteral dosage forms have become more viable alternative drug delivery systems, scientists and regulatory agencies have begun to investigate the development and role of IVIVC for these dosage forms (5,6). Since the Level A correlation is the only type of IVIVC that encompasses the entire time course of the in vivo curve and is accepted as the most informative and valuable, this chapter will present the general principles of Level A IVIVC as well as some of the approaches that can be used to develop a Level A IVIVC for modified release parenteral dosage forms. Since it is not the intent of this chapter to provide examples of Level A IVIVC for every type of parenteral formulation (e.g., microspheres, liposomes, implants, oily suspensions), this chapter will use the microsphere delivery system to describe the IVIVC issues relevant to other modified release parenteral drug delivery systems.

2. A GENERAL APPROACH TO DEVELOPING A LEVEL A IVIVC

The Level A correlation can be developed using a two-stage deconvolution procedure, a one-stage convolution procedure, a compartmental modeling approach, or any modeling technique that relates in vitro dissolution to the in vivo curve (1). Independent of the procedure, the entire in vivo time course must be described from the in vitro data.

The most common approach used in the development of the Level A correlation and the only approach discussed in this chapter is the two-stage procedure. The first stage is to estimate the in vivo absorption or in vivo dissolution time course using deconvolution or a mass balance approach such as Wagner–Nelson. Equation (1) presents the convolution/deconvolution equation that can be used to perform the first stage of deconvolution:

$$c(t) = \int_0^t c_\delta(t - u)x'_{\text{vivo}}(u)\,\mathrm{d}u \tag{1}$$

where c is the plasma drug concentration of the formulation to correlate (e.g., the extended release formulation), x_{vivo} the cumulative amount absorbed or released in vivo of the formulation to correlate (e.g., the extended release formulation), x'_{vivo} the in vivo absorption or release rate (i.e., the first derivative of x_{vivo}), and c_δ is the unit impulse response (i.e., the plasma concentration time course resulting from the instantaneous in vivo absorption or release of a unit amount of drug).

In the second stage, a model is developed to describe the relationship between the in vitro release (IVR) and the in vivo absorption (or release) estimated in stage 1. Prior to the publication of the FDA IVIVC Guidance (1) and the first meeting solely dedicated to IVIVC (7), an IVIVC model was thought to be a linear "point-to-point" relationship between the cumulative amount released in vitro and the cumulative amount absorbed (or released) in vivo for one formulation. Since 1996, the view of the IVIVC model has changed. An IVIVC no longer exists when the in vitro–in vivo relationship is developed for a single formulation. The accepted criteria now require that the mathematical model describes the in vitro–in vivo relationship for two or more formulations (1,7). In addition, models more complex than linear correlation models are now accepted with nonlinear and/or time-variant models becoming very common (1,7).

Once a model is developed to describe the relationship between in vivo and in vitro response, the next task is to determine the validity of the model. Within the FDA IVIVC Guidance, the predictability of the IVIVC model is used to validate the IVIVC model. This predictability of an IVIVC model is a verification of the model's ability to describe the in vivo bioavailability from:

1. The data set that was used to develop the model (internal predictability) and/or
2. A data set not used to develop the model (external predictability).

The $C_{\max}$ and AUC predicted by the IVIVC model are compared to the observed $C_{\max}$ and AUC. Percent prediction errors (%PE) are estimated from the following equation

$$\%PE = \frac{\text{Observed value} - \text{Predicted value}}{\text{Observed value}} \times 100 \qquad (2)$$

All IVIVC models should be evaluated for their internal predictability. In order to evaluate the robustness beyond the internally used data, external predictability can be used.

For regulatory purposes, the FDA IVIVC Guidance sets an acceptable criterion for internal and external predictability. The internal predictability criteria for a regulatory acceptable IVIVC model is that the $C_{\max}$ and AUC %PE for each formulation is less than or equal to 15% and the average absolute %PE of $C_{\max}$ and AUC for all formulations is less than or equal to 10%.

If the internal predictability is greater than the acceptable criteria or the drug is a narrow therapeutic index drug, the FDA requires the more robust analysis of the model using external predictability. The external predictability criteria for an acceptable IVIVC model is that the C_{max} and AUC %PE for the external formulation is less than or equal to 10%. If the %PE is between 10% and 20%, the predictability is inconclusive and additional data sets and/or formulations should be evaluated. If the %PE is greater than 20%, this generally indicates that the IVIVC model does not adequately predict in vivo bioavailability parameters for regulatory use.

The importance of a predictable IVIVC model (based on the above criteria) cannot be overemphasized from a regulatory perspective (1). However, if the IVIVC is to be used for development purposes (e.g., improving a formulation), the criteria defined in the FDA IVIVC Guidance are not required (7). The only requirement is the belief that the model has enough robust validity to assist the formulator in the further development of the formulation.

2.1. Level A Model Development

In order to understand the basic two-stage approach to developing a Level A IVIVC, an example is presented for five modified release oral formulations with differing IVR profiles (Fig. 1a) and an immediate release solution. The plasma concentration data after administering each formulation to human normal volunteers were obtained (Fig. 1b). The mean in vivo and in vitro data were then used for the analysis. The plasma concentration profile for the solution is not presented. Deconvolution was performed using the plasma concentration data from the five modified release dosage forms and the unit impulse response from the solution. The cumulative amount absorbed over time is provided in Fig. 1c. The relationship between in vivo and in vitro is presented in Fig. 1d and follows a linear relationship.

2.2. Predictability of the Level A IVIVC

An example demonstrating how to evaluate the predictability of an IVIVC model is presented in Figs. 2 and 3. Figure 2 represents the screen shot from the program PDx-IVIVC (8,9). The in vitro and in vivo plasma data were placed in

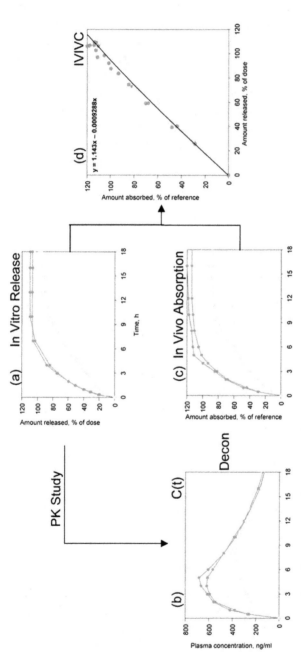

Figure 1 Diagram of the basic two-stage approach to IVIVC. (a) % Released in vitro vs. time for five formulations to be administered for the IVIVC study (top center), (b) plasma concentration vs. time for the five formulations with in vitro release profiles described in (a), (c) % released/absorbed in vivo vs. time for the five formulations after performing the deconvolution using a solution administered through the same route, and (d) the % released in vitro vs. % release/absorbed in vivo with the solid line representing the predicted relationship based on the IVIVC model.

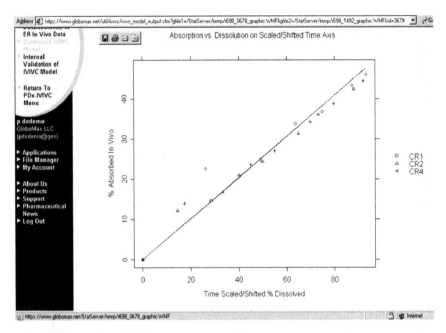

Figure 2 Output from the PDx-IVIVC software comparing % absorbed in vivo to time scaled/shifted % dissolved in vitro. The bullets represent the observed data and the solid line the predicted line from the IVIVC model.

the program for four formulations (three modified release and one solution). The program was used to perform the deconvolution and to develop an IVIVC model. The plot of % released/absorbed in vivo vs. % released in vitro is presented in Fig. 2, the bullets represent the raw data and the solid line represents the predicted line from the IVIVC model.

In order to demonstrate the validation process, Fig. 3 illustrates how the PDx-IVIVC program then compares the predicted and observed C_{max} and AUC for the modified release formulations. The results show that the %PE for each formulation and the mean %PE are all within the criteria for a predictable regulatory standard IVIVC. For external prediction, predicted and observed C_{max} and AUC of a formulation not used to develop the IVIVC model are compared (CR 3 in Figs. 4 and 5). If the %PE is within the FDA criteria, the

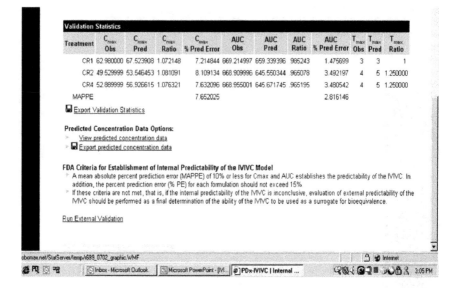

Figure 3 Output from the PDx-IVIVC software showing the internal prediction from Fig. 2. %PE and the observed and predicted C_{max} and AUC are presented.

IVIVC model can be designated a validated regulatory model with external predictability.

3. ISSUES RELATED TO DEVELOPING AN IVIVC FOR MODIFIED RELEASE PARENTAL DRUG DELIVERY SYSTEMS

Although there are many modified release parental drug delivery systems, this chapter is unable to discuss the issues associated with each delivery system. Instead, this chapter will focus on some of the general IVIVC issues associated with some of these delivery systems.

3.1. Study Design

The study design for modified release parenteral drug delivery systems should be similar to the design for oral

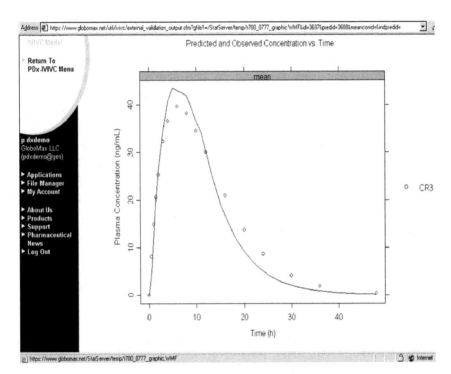

Figure 4 Output from the PDx-IVIVC software plot on plasma concentration vs. time for a formulation used for external prediction (CRB). Both observed data and the IVIVC model predicted plasma curve are presented.

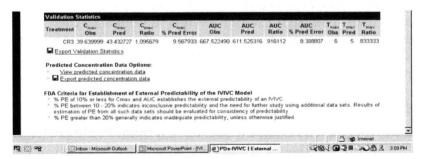

Figure 5 Output from the PDx-IVIVC software showing the external prediction from Fig. 4. Observed and predicted C_{max} and AUC are presented.

formulations, if logistically possible. Typically, two or more formulations with different release rates and formulation characteristics are administered to normal human volunteers. Patients may be used in the study if administration of the drug to normal volunteers is unsafe or the patient population significantly handles the drug and/or delivery system differently. The active drug in solution (defined here as the reference formulation) is also administered i.v. or through the same route of administration as the modified release formulation in order to perform a Level A IVIVC. Although a complete crossover study design is preferred, the logistical problems associated with running such a study may be difficult given the time-course of in vivo delivery (e.g., implant delivery over a number of months). If the complete crossover study design is not possible, incomplete block and parallel designs have also been used. Regardless of the design, every subject should receive the reference formulation as the first arm of the study in order to define the unit impulse response and to ensure that a deconvolution can be performed even if a subject drops out after receiving only one of three modified release formulations.

3.2. In Vitro Release System

Although IVR systems are well established for all types of oral formulations, standard IVR systems for modified release parenterals do not exist. The literature reports a range of systems from destructive test tube systems to the USP 4 apparatus. Although the IVR system is critical to the IVIVC modeling, this chapter will concentrate on the modeling aspects and leave any further discussion of the IVR systems to other chapters.

3.3. IVIVC Using Time Scaling and Shifting

With some of the modified release parenteral dosage forms, IVR occurs over hours or days while complete in vivo release may take days, weeks, or months. The linear IVIVC models developed in the 1970s and 1980s could not deal with this time difference between the two releases. Over the last decade, time-variant models (1,10) have been introduced and used to deal with the differences in the time course of release. A model that has provided an enormous amount of flexibility in its ability to fit time-variant and linear time-invariant IVIVC data has been the model described by Gillespie (10) and others (9,11). Both time shifting and time scaling can be described by the model, which allows the model to fit a wide variety of in vitro–in vivo profiles. The model used to describe both time shifting and scaling is presented in the following equation:

$$x_{\text{vivo}}(t) = \begin{cases} 0 & t < 0 \\ & u = t \text{ for } t \leq T \\ a_1 + a_2 x_{\text{vitro}}(-b_1 + b_2 u) & u = T \text{ for } t > T \end{cases} \quad (3)$$

where $x_{\text{vivo}}(t)$ is the cumulative amount absorbed or released in vivo, x_{vitro} the cumulative amount released in vitro, a_1 the intercept for a linear IVIVC, a_2 the slope for a linear IVIVC, b_1 the coefficient representing a time shift between in vivo and in vitro, and b_2 is the coefficient representing a time scaling between in vitro and in vivo. If $b_1 = -1$ and $b_2 = 0$, the IVIVC is the linear "point-to-point" model that has been reported in the literature over the years.

Predictable models have been developed using this approach for modified oral and parenteral drug delivery systems. An example of the impact of these type of models can be illustrated using Fig. 6. The in vivo vs. IVR of four formulations are presented in Fig. 6. Two of the formulations (K1, K2) have faster in vivo release than in vitro while two of the formulations have faster IVR (K3, K4). It would be impossible to develop one model to describe all four formulations using a conventional linear time-invariant model. However, using Eq. (2) to describe the shift and scaling, b_1 and b_2 can be estimated to obtain a single time-variant model for all four formulations. The %PE of C_{max} and AUC for each formulation was <15% and the average %PE was <10% for both C_{max} and AUC. These %PE met the FDA criteria for internal predictability.

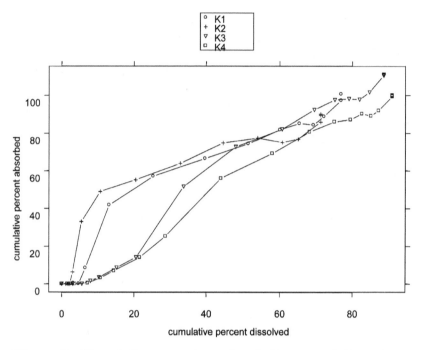

Figure 6 Cumulative percentage absorbed vs. cumulative percentage dissolved with no time shifting.

3.4. Plasma Concentration Profiles

There are two types of profiles (Fig. 7) that have been seen with modified release parenteral delivery systems: Type 1 with one peak and continuous delivery (Fig. 7, Product A), and Type 2 with an initial peak and a second peak at a later time (Fig. 7, Product B). The ideal approach to IVIVC modeling is to develop one IVIVC model for the total plasma profile. This approach has been used to develop an IVIVC for Type 1 plasma profiles but has been less successful for Type 2 profiles. Developing a single model to describe the in vitro–in vivo relationship for Type 1 formulations has been successfully accomplished for multiple formulations (i.e., representing an IVIVC) and for a single formulation. To illustrate a slightly different approach in developing a model between in vitro dissolution and the entire in vivo relationship for the

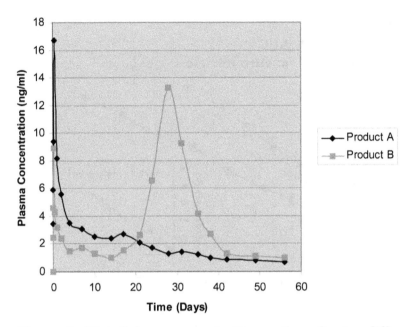

Figure 7 Plot of plasma concentration vs. time after two different microsphere products were administered, Product A with an early peak and prolonged continued release and Product B with an initial release and a second release.

Type 1 curve, Fig. 8 shows the successful correlation between the percent of total AUC at each time point and the percent released in vitro for a microsphere drug delivery system (12). Although this example does not represent a true IVIVC since only one formulation was used, it illustrates how the entire in vivo curve can be related to the in vitro curve. Other approaches (e.g., two-stage approach and compartmental modeling) have also used to develop an IVIVC model for multiple formulations but have not been reported in the literature (13).

Investigators have attempted to develop a single IVIVC model for Type 2 plasma profiles using two-stage deconvolution and compartmental modeling approaches (14). In order to develop a single model, the IVR system must be able to correlate to the very fast absorption rate of the first plasma peak

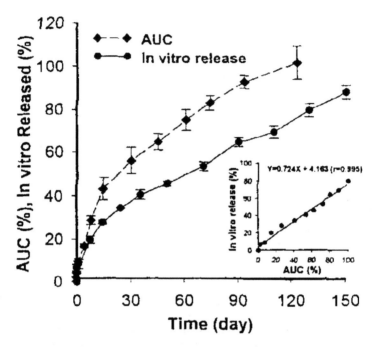

Figure 8 Comparison of in vitro leuprolide release profile with in vivo release profile, the latter plotted as cumulative area under serum peptide curve—normalized as percent of the total area (large panel) and in vitro–in vivo correlation plot (small panel).

and the slower absorption rate of the second plasma peak. Figure 9 illustrates what type of in vitro curve is required in order to develop an IVIVC for the Type 2 dual peak plasma profile (13). Although a significant amount of time has been spent in trying to develop an IVIVC and IVR system for formulations with a Type 2 profile, at the present time an example of this type of in vitro system or a validated IVIVC model describing both plasma peaks has not been reported in the literature.

For Type 2 plasma profiles, investigators have also attempted to develop an IVIVC for different parts of the plasma profile (14). This approach has also been difficult because the first peak and second peak represent different release mechanisms from the dosage form and the magnitude of the second peak appears to be related to the magnitude of

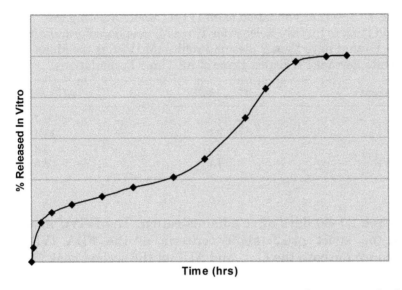

Figure 9 Plot % released in vitro vs. time that is required to develop an IVIVC with Type 2, two peak plasma concentration profiles (Fig. 7, Product B).

the first peak. Investigators have attempted to use one IVR system as well as multiple IVR systems: the accelerated release IVR systems (e.g., higher temperature) and more physiologically relevant conditions. Although acceptable IVIVC models have been developed for formulation development work, IVIVC models that meet the strict predictability criteria of the FDA IVIVC Guidance have not been reported for the complete Type 2 plasma profile.

The major problem appears to be developing the relationship between the first peak and the IVR profiles. The in vitro profile in this situation represents the complete release of drug from the microsphere over time rather than the release of drug associated with the surface of the microsphere, usually the major source of drug for the first plasma peak.

Using time-invariant models, predictable IVIVC models have been successfully developed for the second peak. Table 1 illustrates how a time-variant model successfully related the IVR over 7 days to the in vivo release of a second peak that

Table 1   Internal Validation of an IVIVC Model Relating the
Second Peak to In Vitro Release for Three Microsphere Formulations
(A, B, C) that have Type 2 plasma Profiles (MAPPE is the Mean
Absolute Percent Prediction Error of All Three Formulations)

Treatment	C_{max} $|\%PE|$	AUC $|\%PE|$
A	11.8	6.7
B	9.5	13.4
C	2.1	5.3
MAPPE	7.8	8.5

occurred 20–50 days after administration. The IVIVC model
met the strict predictability criteria of the FDA IVIVC
Guidance for both the C_{max} and AUC of the second peak (14).

4.   CONCLUSION

The FDA IVIVC Guidance has described the basic approach
that all scientists have used to develop an IVIVC for all delivery systems (1). This chapter has not presented all aspects of
IVIVC but has provided some of the specific aspects of IVIVC
modeling that are relevant to modified release parenteral
drug delivery systems. Given the number of different parenteral delivery systems, it is not possible to discuss or present
examples for each system but the basic principles presented
here apply to all parenteral delivery systems. In order to
apply IVIVC models throughout the development and regulatory cycle for all parenteral products, further IVIVC research
is still required in order to develop appropriate in vitro
systems as well as modeling the more complex relationship
between the in vitro and in vivo processes.

ACKNOWLEDGMENTS

The authors would like to thank Wendy Guy for her assistance in preparing the chapter.

REFERENCES

1. Food & Drug Administration Guidance for Industry. Extended Release Oral Dosage Forms: Development, Evaluation, and Application of In Vitro/In Vivo Correlations, CDER 09, 1997.

2. Food & Drug Administration Guidance for Industry. SUPAC-MR: Modified Release Solid Oral Dosage Forms, CDER 09, 1997.

3. Food & Drug Administration Guidance for Industry. SUPAC-IR: immediate Release Solid Oral Dosage Forms, CDER 10, 1997.

4. Food & Drug Administration Guidance for Industry. Bioavailability and Bioequivalence Studies for Orally Administered Drug Products—General Considerations, CDER 10, 2000.

5. Burgess DJ, Hussain AS, Ingallinera TS, Chen M. Assuring quality and performance of sustained and controlled release parenterals. AAPSPharmSci 2002; 4(2):7 (http://www.aaps pharmsci.org/).

6. Burgess DJ, Crommelin DJA, Hussain AS, Chen M-L. Assuring quality and performance of sustained and controlled release parenterals: EUFEPS Workshop Report. AAPSPharmSci 2004; 6(1):11. Available at http://www.aapspharmsci.org/.

7. Young D, Devane JG, Butler J, eds. Advances in Experimental Medicine and Biology—Volume 23: In Vitro–In Vivo Correlations. New York: Plenum Press, 1997.

8. PDx-IVIVC™ Tools for In Vitro–In Vivo Correlation; PDx by GloboMax.

9. Bigora S, Farrell C, Shepard T, Young D. IVIVC Applied Workshop Manual – Principles and Hands-on Applications in Pharmaceutical Development; PDx by GloboMax.

10. Gillespie WR. Advances in Experimental Medicine and Biology—Volume 23: In Vitro–In Vivo Correlations; Chapter 5, Convolution-Based Approaches for In Vivo–In Vitro Correlation Modeling. New York: Plenum Press, 1997.

11. Devane J. Advances in Experimental Medicine and Biology—Volume 23: In Vitro–In Vivo Correlations; Chapter 23, Impact

of IVIVR on Product Development. New York: Plenum Press, 1997.

12. Woo BH, Kostanski JW, Gebrekidan S, Dani B, Thanoo BC, DeLuca P. Preparation, characterization and in vivo evaluation of 120-day Ply (D,L-lactide) leuprolide microspheres. J Control Release 2001; 75:307–315.

13. Young D. Personal Communication, 2002.

14. Young D. Personal Communication, 2001.

6

Coarse Suspensions: Design and Manufacturing

STEVEN L. NAIL and MARY P. STICKELMEYER

Lilly Research Labs, Lilly Corporate Center,
Indianapolis, Indiana, U.S.A.

1. INTRODUCTION

A coarse suspension is a system in which an internal, or suspended, phase is uniformly dispersed in an external phase, which is called the vehicle. The suspended phase is solid, and the vehicle may be either aqueous or non-aqueous. Dispersions of a solid in a liquid vehicle are also categorized according to the size of the suspended particles. *Colloidal dispersions* are suspensions in which the particle size is small enough that the suspended phase does not settle under the force of gravity; that is, the particles remain suspended by Brownian motion. The particle size in a colloidal dispersion ranges from about 1 nm to an upper limit of about 1 µm.

Examples of colloidal dispersions in the pharmaceutical world include association colloids such as micellar systems and lipsomes. Macromolecules such as proteins and DNA that have at least one dimension larger than about 1 nm also exhibit properties of colloidal systems.

In contrast to colloidal dispersions, coarse suspensions typically contain dispersed solid particles in the size range of about 1 to about 50 μm. The suspending medium, or vehicle, may be either aqueous or non-aqueous. Pharmaceutical coarse suspensions fall into three categories—oral suspensions, topical suspensions, and parenteral suspensions. Ophthalmic suspensions; that is, suspensions instilled onto the eye, can be regarded as parenteral suspensions in the sense that they must be prepared as sterile dosage forms. This chapter will deal exclusively with parenteral coarse suspensions.

Parenteral suspensions are typically administered intramuscularly (into the muscle tissue), subcutaneously (into the layer of tissues between the skin and the muscle tissue), intra-articularly (into a joint), or intradermally (just beneath the outermost layer of skin). Coarse suspensions should never be administered intravenously (into a vein) or intra-arterially (into an artery), since the particles in a coarse suspension are usually larger than the diameter of capillaries. Inadvertent intravenous administration of coarse suspensions has led to fatalities, for example, when lipid emulsions were administered which contained precipitated calcium phosphate from calcium gluconate and sodium phosphate additives to the emulsion (1).

Parenteral suspensions are typically used when (i) the drug has limited aqueous solubility, and attempts to solubilize the drug would compromise safety, (ii) sustained release of the drug is needed, or (iii) when a local effect is needed. Sustained release formulations typically are either aqueous or oil-based suspensions administered intramuscularly or subcutaneously. A common example of a suspension for local effect is intra-articular (into the synovial sacs of joints) administration of steroid suspensions, which generally result in prolonged relief from the inflammatory effects of arthritis.

Most parenteral suspensions are provided as a ready-to-use product, with the solid uniformly dispersed in the vehicle. The USP nomenclature for such products is [*Drug Name*] *Injectable Suspension*. However, some parenteral suspensions are provided as sterile powders for reconstitution with water for injection. In this case, the USP nomenclature is [*Drug Name*] *for Injectable Suspension*. An example of this is Spectinomycin Hydrochloride for Injectable Suspension.

Parenteral coarse suspensions are, in general, difficult to formulate and difficult to manufacture relative to other pharmaceutical dosage forms. This arises largely because such systems tend to be physically unstable, resulting in loss of quality attributes essential to a pharmaceutically acceptable product. Particle size of the drug is critical to drug product performance; that is, the particles must be small enough to easily pass through a syringe needle, and stay that way throughout the shelf life of the product. However, because of Ostwald ripening effects, where small particles tend to become smaller and large particles tend to become larger, the particle size distribution tends to change over time. The milling process needed for particle size reduction tends to create amorphous material, and this can give rise to both physical and chemical instability. Amorphous solids have been shown to be up to an order of magnitude less chemically stable in the solid state than the same compound as a crystalline solid (2). In addition, amorphous solids tend to crystallize over time. If a substantial amount of amorphous material was present to begin with, this crystallization may not only lead to particle growth, but may also change the bioavailability of the product. Polymorphism is commonly observed in drugs, and the pharmaceutical scientist must be concerned with the potential for conversion of one crystal form to another, again changing the release characteristics of the drug from the injection site.

Manufacture of parenteral coarse suspensions is challenging because of the difficulty of carrying out the size reduction step in such a way as to provide a uniformly small particle size and narrow particle size distribution. Aseptic processing of parenteral suspensions is challenging because suspensions, in general, cannot be terminally sterilized by autoclaving. They cannot be sterile filtered once the solid phase has been dispersed into the vehicle; rather, they must be manufactured by sterile filtration of the vehicle, aseptic preparation of the solid phase, and aseptic incorporation of solid into the vehicle. Uniform dispensing of suspensions into vials can be a challenge as well, given the tendency of coarse suspensions to settle with time. Therefore, the system must be appropriately agitated, and the process carefully monitored, in order to avoid problems with vial-to-vial dose uniformity.

The purpose of this chapter is to give the reader a broad exposure to formulation and manufacturing aspects of parenteral coarse suspensions, as well as to cover some basic biopharmaceutic aspects of these dosage forms as well as basic aspects of the physical stability of coarse suspensions.

2. PREPARATION AND CHARACTERIZATION OF THE DRUG

The ease with which a drug can be formulated into a pharmaceutically acceptable coarse suspension depends largely upon the properties of the drug itself, including chemical stability in the solid and solution states, solubility, tendency to exist in the metastable amorphous state, tendency to form polymorphs, tendency to form hydrates and solvates, and wettability of the solid. Adequate characterization of the drug requires a major commitment of development resources. Two major components of this effort are: (i) crystallization studies intended to explore the number of species and crystal morphologies that can be formed under a variety of crystallization conditions, and (ii) characterization of these species, with particular emphasis on their stability, both in the solid state and as a slurry. A detailed discussion of solid-state characterization is beyond the scope of this chapter. For a broad overview, the reader is referred to a review article on general pharmaceutics by Fiese (3). For a more in-depth review, the reader is referred to Byrn et al. (4).

2.1. Crystallization Studies

Drugs for use in injectable suspensions are generally aseptically crystallized in order to avoid occlusion of microorganisms within drug crystals. A typical crystallization scheme consists of dissolving drug in a suitable solvent and sterile filtering this solution into a previously sterilized vessel that can be temperature controlled. The solution is agitated and the temperature adjusted, then a previously sterile filtered second solvent, which is a non-solvent for the drug, is added with continued agitation. The temperature may be adjusted after addition of the non-solvent. The solids are then collected, typically either by filtration or on a screen-type separator. The solid cake is typically washed with water and dried.

The purposes of laboratory scale crystallization studies are to: (i) screen for useful polymorphs, hydrates, or solvates, (ii) identify conditions under which pure phases with desirable morphology and acceptable levels of residual solvent can be prepared, (iii) isolate sufficient quantities of solid for further characterization, (iv) establish at least a preliminary assessment of robustness of the crystallization process with respect to consistently producing the same solid-state form of the drug under a given set of processing conditions. Proper laboratory scale crystallization studies generally consist of systematically varying crystallization conditions, including solute concentration, solvent and antisolvent composition, order and rates of addition, and temperatures. Holding times, particularly after addition of the non-solvent, should also be examined. The drying process should be examined in detail, since differences in drying conditions can influence the physical or chemical state of the drug, particularly if hydrates or solvates are formed.

2.2. Solid-State Characterization

A typical solid-state characterization program would consist of the following:

X-ray powder diffraction—On a fairly simple level, variability of x-ray powder diffraction patterns is a good indicator of differences in sample crystallinity. Differences in peak resolution are a reasonable indicator of differences in degree of crystallinity as long as the measurements are made under the same experimental conditions. On a more sophisticated level, x-ray powder diffraction data are used to calculate the crystal structure of the solid, including the space group and intramolecular bond lengths and angles. The ability to control the temperature of the sample as well as addition of an environmental chamber to control the relative humidity are useful enhancements for solid-state characterization.

Thermal analysis—Differential scanning calorimetry is typically combined with thermogravimetric analysis. This type of analysis yields melting temperatures as well as dehydration and desolvation temperatures. Weight loss accompanying dehydration or desolvation is used to calculate stoichiometry for hydrates and solvates. Microcalorimetry can be a useful method for quantitative estimation of the amount of amorphous material present in a solid (5–7).

Solid-state spectroscopy—Solid-state ^{13}C-NMR is a standard tool for solid characterization, and can be useful for confirming that different crystal phases exist within samples, provided that highly crystalline samples are available as reference materials. Since water is observed only indirectly by its influence on ^{13}C environments in the solid, pseudopolymorphism (hydrate formation) can be inferred by its different isotropic chemical shifts of equivalent ^{13}C nuclei relative to anhydrous forms. This can be particularly helpful for confirmation of variable hydrate formation. Appearance of solvent resonances can provide conclusive evidence that solvents are incorporated in the crystal lattice.

Optical microscopy—Optical microscopy is used for determination of the optical crystallographic properties of the solid such as the crystal system (for example, monoclinic, orthorhombic, and others) as well as to estimate particle size. The sample is usually dispersed in immersion oil. Hot-stage microscopy can be a useful additional method for assessing the physical stability of the drug, particularly at elevated temperatures. Dehydration and desolvation temperatures can be identified, particularly if the solid is immersed in oil.

Hygroscopicity measurement—A water vapor sorption microbalance is used to generate a plot of weight gain of the sample vs. relative humidity, giving a quantitative measure of the hygroscopicity of the solid. Water vapor sorption can also be a useful method for estimating the amount of amorphous material in a sample.

Solubility—Aqueous solubility as well as dissolution rate are properties that are critical to efficacy of a parenteral coarse suspension. Where multiple hydrated forms are observed, it is common that the solubility of the drug decreases as the water of hydration increases, and this can sometimes be a strategy for control of solubility. Amorphous solid is generally more soluble than the same solid in crystalline form, and this can be useful as enhanced solubility may infer the presence of amorphous material.

Physical stability—Where multiple crystal forms, hydrates, and solvates are formed under different crystallization conditions, it is important to understand the relative stability of these species and to attempt to elucidate interconversion pathways. Slurry conversion experiments are a good way to compare physical stability of different crystal forms, where both forms are slurried together in a given solvent, and the more stable crystal form will be enriched at the expense of the less stable crystal form.

Chemical stability—Chemical stability studies are generally long-term experiments where a given crystal form is stored at elevated temperature under a variety of relative humidities.

Wettability—The wettability of the solid by either water or oil (for an oil-based suspension) determines the need for a wetting agent (surfactant) in the formulation.

Particle size distribution—Particle size distribution of the bulk drug is necessary for measuring the effectiveness of subsequent particle size reduction operations.

Additional studies needed for parenteral suspension development that are not related to physical or chemical state of the solid include bioburden studies, measurement of endotoxin levels, and examination of the solid for extraneous particulate matter.

3. BIOPHARMACEUTICAL CONSIDERATIONS

When a drug is administered intravenously, the onset of action is rapid, given that the drug is injected directly into the bloodstream and there is no absorption step; rather, there is only a brief interval of 2–3 min during which the drug is "mixed" within the bloodstream. For parenteral coarse suspensions, however, the intravenous (i.v.) route is not an option. If the suspension is administered for systemic effect—generally subcutaneous or intramuscular—there is an absorption step required for the drug to reach the bloodstream.

The bioavailability of a drug from subcutaneous or intramuscular injection is dependent on both physiological factors and physical–chemical properties of the drug and the drug product. The rate of absorption of a drug from a depot site is determined by the slowest, or limiting, step in the sequence:

Solid Drug → Drug in Solution →

Drug in Systemic Circulation

Absorption of drugs from depot sites is commonly dissolution limited. The effect of the absorption process on the time coarse of the plasma drug level (C_p) is shown in Fig. 1. Note that slower rates of absorption result in lower maximum levels of drug in the blood, and the maximum drug blood level is reached after a longer period of time. It is possible for two different products to be equally *bioavailable* in the sense that the areas under the plasma drug level vs. time curve are the same, but not *therapeutically equivalent*, since delayed absorption can result in the plasma drug level never reaching the minimum effective concentration. For this reason, different dosage forms are usually compared by using $C_{p,max}$ and t_{max}, the peak plasma drug level and the time at which the peak level is reached, respectively.

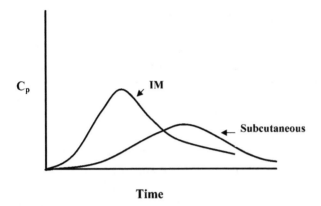

C_p

IM

Subcutaneous

Time

Figure 1 Representative plasma drug levels vs. time for intramuscular and subcutaneous injection.

3.1. Effect of Physical–Chemical Factors on Bioavailability

The rate of dissolution of drug from an injected depot of drug is influenced by the surface area of drug exposed to the interstitial fluid, which is affected by the average particle size of the drug. This relationship is given by the well-known Noyes–Whitney equation:

$$\mathrm{d}m/\mathrm{d}t - KDS(C_s - C)$$

where $\mathrm{d}m/\mathrm{d}t$ is the rate of dissolution, K is a constant, D is the diffusion coefficient of the drug in the interstitial fluid, S is the surface area of drug exposed to the medium, C_s is the equilibrium solubility of drug in the interstitial fluid, and C is the concentration of drug in interstitial fluid at a given time. Buckwalter and Dickinson (8) studied the influence of particle size of procaine penicillin G on maximum blood level following intramuscular administration. Maximum blood levels varied from 1.37 units/mL for particle sizes in the range of 150–250 μm to 2.14 units/mL for particles in the size range of 1–2 μm. Particle size of insulin zinc suspensions has been shown to be a critical factor affecting duration of response (9). A long-acting insulin (Ultralente) showed a slow decline in blood glucose levels and a depression of blood glucose level for longer than 10 hr after administration when the average particle size was 50 µm. When a suspension with an average particle size of 5 µm was administered, a much more rapid decline in blood glucose was observed over a period of 4 hr, followed by relatively rapid rise to the pre-administration glucose level.

One strategy for sustained release from depot injections of coarse suspensions is to manipulate the solubility of the drug. Preparation of a prodrug with decreased solubility should, in principle, result in longer duration of therapeutic effect. However, this approach assumes that a high enough level of drug in the serum is reached to achieve a therapeutic effect.

Solubility, and bioavailability, may also be influenced by the physical state of a drug. The amorphous form is meta-stable, and has higher solubility than crystalline forms of the same drug. Chloramphenicol and novobiocin are examples of drugs where only the amorphous form shows bioavailability. Ballard and Nelson (10) studied absorption of methylprednisolone from subcutaneous implantation of two crystal forms. The in-vitro dissolution rate was about 1.4 times faster for the less stable Form II than Form I, and the absorption rate of Form II from the implant was about 1.7 times faster than that of Form I.

When multiple polymorphs exist, the choice of a crystal form may involve a trade-off between stability and solubility. The choice of a less stable polymorph would be expected to result in higher solubility of the drug, and better bio-availability, but this approach risks conversion of the less stable form to the more stable (less soluble) form during storage.

The Noyes–Whitney equation predicts that increased viscosity of the vehicle should retard drug release because the diffusion coefficient would be expected to decrease in a more viscous medium. Thus a measure of control over the release rate may be achieved by manipulation of viscosity of the vehicle. Use of an oil, such as sesame oil, as a vehicle can result in retarded drug release from a depot injection.

3.2. Drug Absorption from Intramuscular and Subcutaneous Injections

3.2.1. Absorption Across the Capillary Wall

Once the drug is in solution in the environment of the depot, it must still cross the capillary wall in order to enter the systemic circulation. The structure of the capillary wall consists of a unicellular layer of endothelial cells on the luminal side of the capillary. The outside of the capillary is formed by a basement membrane. The total thickness of the capillary wall is about 0.5 μm. There are two small passageways connecting the interior of a capillary with the exterior. One of these passageways is the *intercellular cleft*, which is the thin gap between adjacent endothelial cells. The width of this gap varies between organs, but is normally about 6–7 nm across. In the brain, these junctions are particularly tight, and form what is commonly called the *blood–brain barrier*. The tightness of these junctions presents a major challenge in the delivery of drugs to the brain. The liver represents the opposite extreme with respect to the "tightness" of these junctions, where the clefts between endothelial cells are relatively wide open, such that almost all dissolved substances in the plasma, even plasma proteins, can pass from the blood into the liver tissue. In the kidney, small openings called fenestrae penetrate directly through the middle of the endothelial cells, so that large amounts of substances can be filtered through the glomeruli without needing to pass through the gaps between endothelial cells.

The relative surface area of the junctions between endothelial cells is small—only about 1/1000 of the total surface area of the capillary. However, water molecules, ions such as sodium and chloride, and small molecules such as glucose move freely between the interior and the exterior of the capillary. The ability of drugs to cross these junctions depends largely on molecular size. Relative permeability of capillaries in muscle tissue to different solutes is shown in Table 1 (11). It is safe to state that protein pharmaceuticals would not be absorbed significantly by passage through junctions between capillary endothelial cells.

Another passageway from the interior to the exterior of capillaries is *plasmalemmal vesicles*. These vesicles form at one surface of the cell by imbibing small volumes of plasma or extracellular fluid. These vesicles can move slowly through the cell, and it has been postulated that they can transport significant amounts of substances across the capillary wall. It has further been postulated that these vesicles can coalesce to form vesicular channels all the way through the membrane. It is doubtful, however, that transport through such channels is a significant mechanism for absorption of drugs from depot injections.

It is probably safe to state that transport of drugs through the junctions between endothelial cells is a significant route for absorption of only small molecule drugs, probably with a molecular weight of less than about 500. Regardless, a drug from a depot injection must still make its way across the basement membrane in order to even reach the tight junctions. Biological membranes are complex structures that are primarily composed of lipid and protein layers. The phospholipid bilayer generally forms the backbone of the membrane, and layers of protein give added strength to the membrane. The lipid component of such membranes is important to drug absorption in general, and the oil/water partition coefficient of a drug becomes a critical factor in drug absorption. In general, the more lipophilic a drug, the more easily it is able to diffuse across a biological membrane. Even for intravenously administered drugs, where absorption is not an issue, the distribution of drug within tissues is determined largely by the lipophilicity of the drug.

Given that many drugs are weak acids or weak bases, the lipophilicity is strongly affected by the ionization state of the drug, where the un-ionized state is more lipophilic. The degree of dissociation of the drug is determined by the ionization constant and the pH of the medium. This relationship is expressed by the familiar Henderson–Hasselbach equation which, for a weak acid, is stated as follows:

$$pH = pK_a + \log([A^-]/[HA])$$

where A^- and HA are the concentration of ionized and un-ionized states, respectively. The interrelationship between dissociation constant, pH at the absorption site, and the absorption of drugs is the basis for the *pH-partition hypothesis*.

3.2.2. The Lymphatic System

About one-sixth of the volume of the body consists of spaces between cells. These spaces collectively are called the *interstitium*, and the fluid in these spaces is the *interstitial fluid*. The solid structures in the interstitium consist mostly of collagen fiber bundles and proteoglycan filaments. The fluid in the interstitium is filtrate from capillaries. The composition of this fluid is much the same as the fluid in the capillaries except for proteins, which are too large to filter through the spaces between endothelial cells. The fluid is largely entrapped in the spaces between proteoglycan filaments. The combination of the proteoglycan filaments and the fluid has the consistency of a gel, and is often called the tissue gel. Because of the gelatinous consistency of the interstitial fluid, it flows poorly. Instead, the fluid travels largely by a process of diffusion. The diffusion of solutes through this gel occurs about 95% as rapidly as it would in a freely flowing fluid. Thus this diffusion allows relatively rapid transport of solutes, including electrolytes, nutrients, products of cellular metabolism and, in the case of a subcutaneous or intramuscular injection, drug molecules.

Almost all tissues in the body have lymphatic channels that drain excess interstitial fluid from the interstitial spaces. Most of the lymph flows through the *thoracic duct* and empties into the venous system at the junction of the subclavian veins and the jugular vein. Approximately 100 mL of lymph flows through the thoracic duct of a resting adult per hour. While the body of published research on the relevance of the lymphatic system as a route for drug absorption is very limited, it must be considered as an alternative pathway for drug absorption from intramuscular or subcutaneous injections.

3.3. Physiological Factors

In addition to physical–chemical properties of the drug and the formulation, certain physiological factors influence the absorption of drug from parenteral suspensions. As discussed above, drugs administered by the intramuscular or subcutaneous route require an absorption step in order to be bioavailable. Given that the drug is primarily absorbed by diffusion into the capillaries at the injection site, then the greater the blood flow at the site of injection, the more rapid the drug absorption. Thus, any factor that influences the flow of blood at the injection site also influences the rate of absorption. For example, epinephrine inhibits the flow of blood at the injection site when co-administered with a drug, and consequently slows absorption. Increased muscular activity causes increased blood flow, with subsequent enhanced absorption. For intramuscular injections, the site of administration, i.e., the deltoid muscle, the gluteal muscle, or the lateral thigh, can have a significant influence on bioavailability. Zener et al. (12) studied the influence of injection site on bioavailability of intramuscularly administered lidocaine solution. When patients were administered 200 mg of lidocaine into the deltoid muscle, the gluteal muscle, or the lateral thigh, the rate of absorption of drug followed the order deltoid > lateral thigh > gluteal muscle. This would be expected based on the relative rates of blood flow, particularly when considering that the patients were hospitalized, and greater muscular activity would be expected in the upper body than in the lower body. It is important to keep in mind, however, that this study was carried out using a solution formulation. When a suspension is administered, the dissolution of the drug is often the controlling resistance, in which case the effect of site of administration illustrated here would probably not be observed.

Gender has been shown to influence bioavailability of drugs from intramuscular injections. Vukovich et al. (13) studied absorption of cephradine when administered once a week for three consecutive weeks to six male and six female volunteers in either the deltoid, gluteal, or lateral thigh muscles. Serum levels were not significantly different when administered in the deltoid muscle or the lateral thigh but, when administered into the gluteal muscle, the peak cephradine concentrations were 11.1 and 4.3 µg/mL for males and females, respectively.

Significant differences in bioavailability would be expected when drugs are administered intramuscularly vs. administration into fat (intralipomatous). Cockshott et al. (14) reported results of studies showing that <5% of women and <15% of men actually received intramuscular injections when a 3.5-cm needle was used. Differences in bioavailability would be expected based on the physical–chemical nature of the injection site as well as the extent of vasculature. Hydrophobic drugs would be expected to tend to remain in the hydrophobic environment of fatty tissue. This may be part of the reason that some drugs are poorly or erratically absorbed when injected "intramuscularly." Greenblatt et al. (15) reported that oral doses of 50 mg of chlordiazepoxide are more rapidly absorbed than the same dose when administered intramuscularly. The reader is referred to the chapter by Ousseron in this book on biopharmaceutical principles.

4. PHYSICAL STABILITY OF COARSE SUSPENSIONS

Parenteral coarse suspensions are not, strictly speaking, colloidal systems, because they exhibit settling under the force of gravity. However, principles of colloidal science are useful in understanding the physical stability of these systems, particularly regarding flocculation behavior.

The interface between the suspended solid and the liquid phase plays an important role in determining the stability of suspensions. The interfacial free energy is an expression of the degree of preference of a molecule of the dispersed solid for its bulk relative to its interface. This interfacial free energy is always positive, meaning that energy must be put into the system in order to create the free energy; for example, through mechanical milling. When this energy is removed and the suspension is formulated, thermodynamics takes over and tends to drive the system toward its more stable, lower free energy state. While thermodynamics will ultimately win, appropriate manufacturing techniques and rational formulation can often result in a system that is, for practical pharmaceutical purposes, "stable."

The angle that a liquid makes with a solid surface is called the contact angle, and contact angles yield useful information about the solid surface. Generally, liquids are considered non-wetting if the contact angle is larger than 90°, and wetting if the contact angle is < 90°. Complete wetting results in a contact angle of 0°. To measure the contact angle, the solid can be pressed into a flat wafer using a Carver press, a defined volume of liquid is applied to the surface, and the contact angle is measured using a contact angle goniometer. An image of the drop is examined using a lens with adjustable cross-hairs, where one cross-hair is aligned with the surface, and the other is rotated until it forms a tangent to the drop. The contact angle is read directly. For a detailed discussion of characterization of solid surfaces by measurement of contact angle, the reader is referred to Evans and Wennerstrom (16).

The greater the percentage of molecules at the surface; that is, the smaller the particle size, the more important surface properties are in determining the stability of the system. Nature tends to reduce this free energy to zero by various means. One is reduction of the interfacial area by the growth of larger particles at the expense of smaller ones. This phenomenon is known as *Ostwald ripening*. This is expressed quantitatively by the Ostwald–Freundlich equation:

$$\ln C_1/C_2 = (2M\gamma/\rho RT)(1/R_1 - 1/R_2)$$

where C_1 and C_2 are the solubilities of particles of radius R_1 and R_2, respectively, M represents molecular weight, γ is the surface energy of the solid in contact with solution, ρ is the density of the solid, R is the gas constant, and T is the absolute temperature. Use of this equation predicts that the solubility of a 0.2-μm particle is about 13% higher than the same solid when present as a 20-μm particle.

Another way that nature tries to reduce the free energy of the system is by aggregation of dispersed particles as attractive forces overcome repulsive forces. Successful formulation of suspensions generally depends on the scientist's appreciation for the importance of surface properties of the system. For example, understanding the role of surface charge characteristics allows formulation such that a flocculated network of particles is formed that can be easily resuspended with gentle agitation. Adsorption of small hydrophilic colloids or non-ionic polymers may stabilize the system by increasing its interaction with water or sterically hindering particles from approaching closely enough that repulsive forces are replaced by attractive ones, resulting in caking of the suspended solid.

When the solid powder is added to the vehicle, it is agitated vigorously, and this agitation may be in the form of further particle size reduction by wet milling. Dispersion refers to the extent to which the solid exists as individual particles, as opposed to clumps or aggregates of particles. The extent to which a uniform distribution of particles is maintained is referred to as the dispersion stability. The particles will settle at a rate described by Stoke's law:

$$V = d^2(\rho_s - \rho_l)g/18\eta$$

where V is the settling velocity, d is the particle diameter, ρ_s and ρ_l are densities of the solid and liquid phases, respectively; g is the gravitational constant, and η is the viscosity of the liquid phase. This equation suggests several ways to reduce settling. One way is to reduce the particle diameter, another is to minimize the density difference between the liquid and solid phases by increasing the specific gravity of the vehicle, and another is to increase the viscosity of the vehicle. From the standpoint of formulating a parenteral suspension, however, increasing the density or the viscosity of a vehicle to a point where settling is prevented is not practical, since the resulting suspension would not be syringeable. Settling must be accepted, and efforts should focus on formulation conditions that result in easy resuspension of the solids with gentle shaking.

When two colloidal particles undergoing Brownian motion approach each other, they experience two types of interaction—static forces arising from attractive van der Waals forces and electrostatic interaction, and hydrodynamic forces mediated by the vehicle. The attractive static forces include dipole–dipole, dipole-induced dipole, and van der Waals forces. van der Waals forces are quantum mechanical in origin, and are always attractive, irrespective of charge effects. These are very short range forces, and vary inversely with r^6, where r is the distance between particles. Opposing these attractive forces is electrostatic repulsion arising from the fact that the surface of the suspended solid is generally charged. This charge may arise from ionization of surface ionizable groups (amines or carboxylic acid groups, most commonly), from adsorption of molecules that impart a charge, or perhaps from charge generated by particle size reduction operations. Repulsion can also be caused by adsorption of polymers that sterically inhibit close approach of particles. The layer of fixed charges at the surface, called the Stern layer, is characterized by both a charge density and a surface potential (Φ_0). Direct measurement of Φ_0 is uncertain. Instead a quantity called the *zeta potential* (ζ) is measured. The zeta potential is measured by a variety of electrokinetic methods, and represents the electrical potential at the slip plane, or the effective hydrodynamic radius of the particle.

Interaction between charges that are fixed at the surface and those that are free in solution plays an important role in the stability of colloidal systems. The electrolyte solution is characterized by the charge and concentration of electrolytes as well as the dielectric constant of the medium. The combination of the charged surface and the neutralizing layer of counter ions is said to constitute an *electrical double layer*. The thickness of the double layer is expressed by

$$1/\kappa = (\varepsilon KT/e^2 \sum_i n_i z_i^2)^{1/2}$$

where $1/\kappa$ is the Debye length, ε is the dielectric constant of the medium, K is the Boltzman constant, n_i is the number of ions of type i per unit volume near the surface, e is the charge on an electron and z_i is the valence of the electrolyte. Note the strong dependence of the Debye length on the valence of the electrolyte.

Figure 2 is a potential energy diagram representing the attractive forces, the repulsive forces, and the net interaction potential between colloidal particles. The diagram shows that

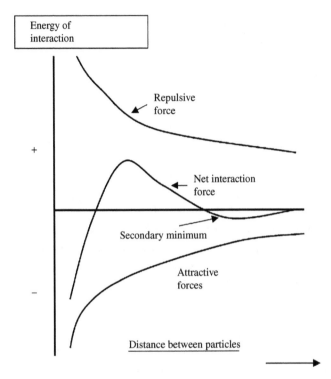

Figure 2 Forces of interaction between colloidal particles.

attractive forces predominate at very short interparticle distances. If no repulsive forces exist, the particles can come together in the primary minimum. Depending on the depth of the primary minimum, the aggregation can be either reversible or irreversible. However, for most pharmaceutical suspensions, aggregation in the primary minimum is usually irreversible; that is, the solid cannot be redispersed by simple shaking. Electrostatic repulsion creates an energy barrier opposing approach of particles closely enough to reach the primary minimum. The thickness of the electrical double layer determines the rate at which electrostatic repulsion decreases with increasing distance between particles. The net potential energy curve has a maximum. The size of this maximum, relative to the thermal energy of the system (expressed as kT, where k is the Boltzman constant), determines the ability of particles to reach the primary minimum. If the size of the potential energy barrier is very large compared with kT, then the primary minimum is inaccessible, and the system is kinetically stable. The net potential energy curve may contain a secondary minimum at a relatively large interparticulate distance. Aggregation in this secondary minimum gives rise to a loosely structured network of particles, and the aggregation is readily reversible by shaking. Such floccules are reported to display fractal properties.

Flocculation is an important property of any coarse suspension, and the pharmaceutical scientist should understand both the properties of flocculated suspensions and the forces that mediate the aggregation state of suspensions. In flocculated suspensions, particles are loosely aggregated by electrostatic forces, such that the suspension consists of a loose network of particles. This ensemble of particles, or floc, settles relatively rapidly, and forms a clear boundary between the precipitate and the supernatant. The sediment is loosely packed, and a hard, dense cake is not formed. As a result, the solid is easy to redisperse. In a deflocculated suspension, particles exist as separate entities. The rate of settling is slow, and dependent on the particle size. Since there are minimal repulsive forces between particles, eventually a hard, dense sediment is formed which is difficult, or perhaps impossible, to redisperse.

DLVO (for Derjaguin, Landau, Verwey, and Overbeek, the scientists who published the original theory of colloidal stability in the 1940s) theory states that colloidal stability is determined by a balance between electrical double layer repulsion, which increases exponentially with decreasing distance between particles, and van der Waals forces of attraction. The practical lessons to be learned from DLVO theory are primarily that: (i) ionic strength of the vehicle is a dominant factor controlling flocculation of the system, and (ii) adsorption of polymers can be used to sterically stabilize a suspension by preventing two particles from approaching closely enough to aggregate in the primary minimum. The reader is referred to the chapter by Burgess in this book on physical stability of dispersed systems.

5. FORMULATION OF PARENTERAL SUSPENSIONS

A well-formulated parenteral suspension resuspends easily upon shaking, with uniform dispersion of the drug in vehicle. The suspension is easily drawn through a needle into a syringe (syringability), and is injected without the use of excessive force (injectability). It is not irritating to the tissue into which it is injected. The suspension is both physically and chemically stable over the shelf life of the product.

Developing a sterile powder which is reconstituted with water to form a suspension at the time of use should be considered, particularly for drugs that are unstable in aqueous media. This dosage form has the advantage of avoiding physical stability issues that can be troublesome for ready-to-use suspensions, such as particle size growth and caking of the suspension, making resuspension, syringeability, or injectability troublesome. Powders for suspension also have the advantage of being more amenable to terminal sterilization than ready-to-use suspensions, either by ionizing radiation or by thermal methods. Powders for reconstitution to form a suspension are typically provided with a companion vial of vehicle containing the appropriate excipients.

5.1. Particle Size Distribution

The product development exercise should include studies to determine the role of particle size distribution on bioavailability of the drug as well as syringability and injectability. These are generally small-scale studies, where hand sieving is feasible. The experiment consists of preparing different sieve "cuts" from a single batch of powder. Prototype formulations are prepared with each of four to six particle size distributions, and syringeability and injectability are tested prior to proceeding further. Particle size distributions with a median particle size above about 50 μm are likely to produce problems with syringeability and injectability.

Bioavailability studies are generally carried out in animals, and consist of collecting plasma drug levels as a function of time after injection of the suspension. The peak

Table 1 Relative Permeability of Muscle Capillary Pores

Substance	Molecular weight	Relative permeability
Water	18	1
NaCl	58	0.96
Urea	60	0.8
Glucose	180	0.6
Sucrose	342	0.4
Insulin	5000	0.2
Albumin	69,000	0.0001

(From Ref. 20)

blood level of drug, $C_{p,max}$, and time at which the peak is reached, t_{max}, are important responses, as well as total area under the drug blood level vs. time curve.

5.2. Excipients

In addition to the drug, a typical parenteral coarse suspension contains a dispersing or suspending agent, a surfactant, a buffer, a tonicity adjusting agent, and, in the case of multiple dose containers, an antimicrobial preservative. Examples of commercial formulations of parenteral suspensions are shown in Table 2 (17). The roster of excipients is small, as it is for most parenteral products.

The DLVO theory, discussed briefly above, provides a conceptual picture of the interactions between particles that control physical properties of suspensions. The interaction curves in Fig. 2 show that close approach of particles will be opposed if the repulsive energy is high; that is, if the zeta potential is high and ionic strength is low. However, when these particles settle, this energy barrier may be overcome, and particles may interact at the primary minimum, generally resulting in caking and difficult (or impossible) redispersion.

Sedimentation volume and zeta potential measurement are useful formulation tools for helping to assure an easily redispersable suspension, where sediment volume is the height of the sediment relative to the height of the liquid. The idea is that, the higher the relative sediment volume,

Table 2 Representative Sterile Coarse Suspensions (From Ref. 17)

Drug	Excipients	Category
Aurothioglucose, 50 mg/mL	Aluminum monostearate, 20 mg/mL; propylparaben, 0.1%; sesame oil	Suspending agent; preservative; vehicle
Betamethazone sodium phosphate/betamethazone acetate (3 mg/mL each)	Sodium phosphate; EDTA sodium, 0.1 mg/mL; benzalkonium chloride, 0.2 mg/mL, pH 6.8–7.2	Buffer; chelator; preservative
Desoxycorticosterone pivalate	Methylcellulose; sodium carboxymethylcellulose; polysorbate 80; sodium chloride; thimerosal	Suspending agents; surfactant; tonicity adjustment; preservative
Dexamethazone acetate, 8 mg/mL	Sodium CMC, 5 mg/mL; polysorbate 80, 0.75 mg/mL; sodium chloride, 6.7 mg/mL; creatinine, 5 mg/mL; sodium bisulfite, 1 mg/mL; EDTA disodium, 0.5 mg/mL, pH 5.0–7.5	Suspending agent; surfactant; antioxidant; chelator
Hydrocortisone acetate, 50 mg/mL	Sodium CMC, 5 mg/mL; polysorbate 80, 4 mg/mL; sodium chloride, 9 mg/mL; benzyl alcohol, 9 mg/mL	Suspending agent; surfactant; ionic strength/tonicity; preservative
Methylprednisolone acetate, 20–80 mg/mL	PEG 3350, 30 mg/mL; polysorbate 80, 2 mg/mL; sodium chloride (isotonic); sodium phosphates, 2 mg/mL; benzyl alcohol, 9 mg/mL, pH 3.5–7.0	Suspending agent; surfactant; ionic strength/tonicity; buffer; preservative

(Continued)

Table 2 Representative Sterile Coarse Suspensions (From Ref. 17) (*Continued*)

Drug	Excipients	Category
Medroxyprogesterone acetate, 150–400 mg/mL	PEG 3350, 20–29 mg/mL; polysorbate 80, 2.4 mg/mL; sodium chloride, 8.7 mg/mL; methylparaben, 1.4 mg/mL; propylparaben, 0.15 mg/mL	Suspending agent; surfactant; ionic strength/tonicity; preservatives
Cortisone acetate, 50 mg/mL	Sodium CMC, 5 mg/mL; polysorbate 80, 4 mg/mL; sodium chloride, 9 mg/mL; benzyl alcohol, 9 mg/mL	Suspending agent; surfactant; tonicity/ionic strength; preservative
Epinephrine HCl, 5 mg/mL	Glycerin, 325 mg/mL; thioglycolic acid, 6.6 mg/mL; ascorbic acid, 10 mg/mL; phenol, 5 mg/mL	Tonicity; preservative
Triamcinolone diacetate, 20–40 mg/mL	PEG 3350, 3%; polysorbate 80, 0.2%; sodium chloride, 8.5 mg/mL; benzyl alcohol, 9 mg/mL, pH about 6	Suspending agent; surfactant; ionic strength/tonicity; preservative
Penicillin G benzathine and penicillin G procaine, 150,000–600,000 U/mL	Sodium CMC, 0.55%; lecithin; polyvinylpyrrolidone, 0.1%; methylparaben, 0.1%; propylparaben, 0.01%; sodium citrate, pH 6–8.5	Suspending agent; surfactant; suspending agent; preservatives; buffer
Triamcinolone hexacetonide, 5–20 mg/mL	Sorbitol, 50%; polysorbate 80, 0.2–0.4%; benzyl alcohol, 9 mg/mL, pH 4.5–6.5	Suspending agent; surfactant; preservative

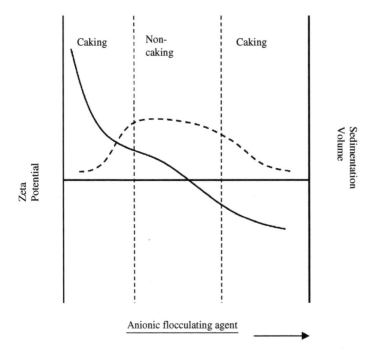

Figure 3 General relationship between sediment volume and zeta potential.

the greater the ease of redispersion. The general relationship between sediment volume and zeta potential is illustrated in Fig. 3. Addition of a flocculating agent, such as an electrolyte, causes a reduction in zeta potential, which causes changes in sediment volume. In the region where the sediment volume is maximized, there should be minimum probability of caking. Note that too much added electrolyte can result in over flocculation and subsequent caking. Measurement of sedimentation rate can also be a useful formulation tool reflecting the physical stability of the system.

5.3. Buffers

Physiological pH is always desirable for any injected product in order to minimize irritation at the injection site. However, the intramuscular and subcutaneous routes are fairly forgiving in this regard. Table 2 illustrates the wide range of formulation pH of commercially available suspension products—from as low as 3.5 to as high as 8.5. Of course, buffer capacity should be considered as well—formulations that deviate from physiological pH that have a high buffer capacity will be more irritating than those with low buffer capacity. It is good practice to keep the concentration of any buffer to a minimum.

As discussed above, the ionic strength of the vehicle is an important determinant of the effective range of the electrical double layer, which influences the physical stability of the system with respect to aggregation characteristics. Again, a flocculated system is desired, given the relative ease of resuspension of such systems. Of course, ionic strength is determined by the total number of ionic species present, including any added electrolytes.

The most commonly used buffer in parenteral suspensions is sodium phosphate. Ascorbic acid is used in a commercial formulation of epinephrine HCl, and sodium citrate is used in a commercial formulation of penicillin G. Other buffers used in commercial parenteral formulations include sodium lactate, sodium acetate, sodium succinate, histidine, and tris(hydroxymethyl) aminomethane (see Table 2).

5.4. Wetting

The tendency of a solid to be wetted by a liquid is a measure of the affinity of the substances, where hydrophilic surfaces tend to be readily wet by aqueous media. For injectable suspensions, the drug generally has limited aqueous solubility, tends to be hydrophobic, and thus tends not to be easily wet. The use of contact angle data to characterize the solid surface was mentioned above. Another method of measuring the wettability of a powder is the *wet point* method (18), which consists of determining the amount of vehicle needed to just wet all of the powder, usually by measuring the amount of liquid needed to carry a powder through a gauze. The more effective the wetting agent, the lower the "wet point" value.

Wetting agents are surfactants that lower the interfacial tension and the contact angle between the solid and the vehicle. Most parenteral suspensions use polysorbate 80 as a wetting agent. Concentrations in commercial products are given in Table 2. Other surfactants include lecithin and sorbitan trioleate.

5.5. Suspending Agents

Suspending agents usually refer to excipients used to control the viscosity of the vehicle as well as polymers that interact with the solid surface to improve physical stability. Suspending agents include sodium carboxymethylcellulose, polyvinylpyrrolidone, polyethylene glycol, and propylene glycol.

5.6. Tonicity Adjusting Agents

Tonicity adjusting agents may be either electrolytes or non-electrolytes. The type of tonicity adjusting agent used depends on the effect of ionic strength on settling properties of the suspension. Sodium chloride is commonly used for both adjustment of tonicity and adjustment of ionic strength. Sorbitol and mannitol are examples of non-electrolytes used for adjustment of tonicity.

5.7. Antimicrobial Preservative

Since many parenteral suspensions are multiple dose containers, antimicrobial preservatives are common. These include benzalkonium chloride, chlorobutanol, parabens, and benzyl alcohol.

6. MANUFACTURE OF PARENTERAL COARSE SUSPENSIONS

6.1. Ready-to-Use Suspension vs. Powder for Reconstitution

For a dry powder formulation for suspension, the manufacturing process may consist only of milling of aseptically crystallized and dried bulk drug. If terminal sterilization of the final container is not feasible, a surface sterilization of the powder blend, perhaps by ethylene oxide or irradiation, may be done to inactivate microbial contamination introduced by milling and powder handling steps. The powder is aseptically filled into vials. The suspension vehicle is prepared as a companion vial containing, for example, a surfactant, a buffer, a suspending agent, perhaps ionic strength and tonicity adjusting agents, and perhaps a preservative.

Aseptic powder filling has a number of operational challenges, including fill weight uniformity, potential concerns with cross-contamination due to airborne dust, and potential concerns with worker exposure to airborne drug if the drug is a potentially hazardous material. However, these challenges can generally be met, and may be small compared with the challenges associated with physical or chemical instability of a ready-to-use suspension.

Feasibility of terminal sterilization should be examined, both by thermal methods and by ionizing radiation (see discussion below). If feasible, this approach gives maximum sterility assurance and avoids the need for an intermediate surface sterilization during processing.

For ready-to-use suspensions, the manufacturing scheme generally consists of milling of bulk drug and surface sterilization of the powder. The vehicle is prepared and sterile filtered, and the powder is aseptically dispersed in the vehicle, followed by aseptic filling of the liquid. The process must be monitored closely, particularly during development, to assure uniformity of the bulk suspension during the filling operation.

The feasibility of terminal sterilization of a ready-to-use suspension must be determined but, generally speaking, this has low probability of success. Autoclaving usually results in changes in particle size distribution resulting from elevated temperature, where smaller particles dissolve to a greater extent than larger particles then, when the system is cooled, larger particles grow as a result of lower solubility. As for gamma or electron beam sterilization, the probability of success is low because of the generation of free radicals in solution that tend to degrade the drug.

6.2. Unit Operations in Suspension Manufacture

6.2.1. Particle Size Reduction

The most critical aspect of milling is the ability of the unit operation to produce a narrow particle size distribution of the appropriately sized drug particles. The impact of milling on the physical state of the drug can also be critical, particularly since mechanical milling can produce small but significant quantities of amorphous material (19). This can lead to subsequent problems with chemical stability, physical stability, or both.

The compatibility of the mill with aseptic operations is important. Specific design criteria include the potential for metal-to-metal contact, use of double mechanical seals, cleanability of the equipment and, for aseptic milling applications, the ability to sterilize the mill in place.

Common techniques for production of fine particles are briefly discussed below:

Air micronization is probably the most common method of particle size reduction for parenteral suspensions because of: (i) the ability to obtain very small particles with a uniform size distribution, (ii) a particle size classifier internal to the equipment that returns oversized particles to the zone where particle size reduction takes place, and (iii) the absence of any moving parts to generate extraneous particulate matter. Sterilization of air is straightforward, but dust containment can be a significant operational complication.

Spray drying can be used to produce sterile powders with uniform spherical particles by spraying a solution containing the drug into a chamber where a warm stream of air flows counter-currently to the spray. Advantages of the technique are the ability to control the droplet, therefore the particle, size by choice of the appropriate spray nozzle and feed rate, and the uniform size and particle morphology produced. The application is limited, however, since organic solvents would be needed if the intended suspension vehicle is aqueous, and problems associated with handling of large volumes of solvent vapor can be significant. Spray drying might be useful for control of particle size of drugs intended for an oil-based suspension.

Supercritical fluids, typically carbon dioxide, can be used as either a solvent or an antisolvent to achieve supersaturation and subsequent particle generation. The technique has been reviewed by Tom and Debenedetti (20). The method is generally compatible with aseptic operations, but ability to consistently produce a narrow particle size distribution is uncertain. Large particles are particularly troublesome from the standpoint of syringeabilty. While the technique holds some promise, the technology does not appear to be advanced enough to produce suitably narrow particle size distributions for injectable suspensions without an additional particle size reduction step.

6.2.2. Sterilization

This discussion is limited to terminal sterilization of product, as opposed to sterilization that is required as a part of aseptic processing. International regulatory guidelines require that the feasibility of terminal sterilization be determined; however, terminally sterilizing a ready-to-use aqueous suspension has a low probability of success. The likelihood of success for terminal sterilization of an oil-based suspension may be somewhat better.

Probability of success in terminal sterilization is best for powders for injectable suspension. This may be done by autoclave, but problems may be encountered with caking of the powder, making suspension of the solids upon reconstitution difficult. Terminal sterilization by ionizing radiation may be a better choice. Ionizing radiation for sterilization may be via electron beam, gamma irradiation, or, in a more recent development, by x-rays (21). All forms of ionizing radiation work by collision of an electron or a photon with the electrons in the outer shell of atoms to produce ions. This ionization process causes covalent bond breakage; in particular, DNA is very susceptible to depolymerization by this process. Therefore, ionizing radiation sterilizes by prevention of reproduction of microorganisms.

Electron beam—An accelerator is used to generate high-energy electrons that are focused on the product. The allowable power used in electron beam accelerators is limited because of the potential for photonuclear reactions at high power. The electron beam has limited penetration depth, but the time required for sterilization is very short—on the order of seconds to a few minutes. Scale-up of electron beam sterlization is probably move straightforward than after irradiation methods.

Gamma irradiation—^{60}Co is the most commonly used source of gamma irradiation, where the radioactive decay of ^{60}Co to ^{60}Ni produces two photons. An advantage of gamma irradiation is its inherent safety with respect to the potential to render material being sterilized radioactive, since the photons produced are not energetic enough to cause photonuclear reactions. Gamma irradiation also has a greater ability to penetrate materials than electron beam radiation. However, exposure times needed to produce an equivalent degree of sterility assurance are significantly longer than electron beam irradiation.

X-rays—If an electron beam is focused on a target material such as tungsten, x-rays are produced by the target. Sterilization by x-rays is a more recent development, and combines the penetration ability of gamma irradiation with the controllability of electron beam radiation; that is, the power of the electron beam can be controlled or turned off as needed.

One packaging-related concern pertaining to radiation sterilization is the tendency for radiation to cause glass to discolor. This, at a minimum, can cause loss of pharmaceutical elegance. Use of cerum oxide-containing glass minimizes disoderation by irradiation.

Ethylene oxide is commonly used as a surface sterilant for aseptically processed suspensions. The drug is typically crystallized aseptically, then milled under sanitary conditions. The milled drug is then sterilized by ethylene oxide prior to aseptic processing of the suspension to inactivate any microorganisms that may be present on the surface of the powder. There are many operational challenges associated with ethylene oxide sterilization. Byproducts of ethylene oxide exposure include ethylene chlorohydrin and ethylene glycol. It is important to maintain adequately low levels of these residuals. Ethylene oxide is also a potential carcinogen, and rigorous environmental monitoring is necessary to assure worker safety.

6.2.3. Filling

Filling of liquid suspensions is generally straightforward using commercially available filling equipment. The most challenging aspect of filling suspensions is maintenance of content uniformity throughout the lot. Aggressive monitoring is often required, particularly late in the filling time interval.

Filling of powders is more challenging than filling of liquids for several reasons. First, batch-to-batch and even within batch variability in mechanical properties of powders, such as compressibility, flow characteristics, and bulk density can result in variability in fill weight. Second, some dust formation is probably inevitable, and containment of dust is critical in controlling the potential for cross-contamination. Third, for potentially hazardous compounds, containment of airborne particulate matter is important from the standpoint of worker protection. An emerging process analytical technology that is pertinent to filling of dry powders is a non-contact check weighing system, based on magnetic resonance that checks the fill weight of every vial, allowing over- or under-dosed vials to be rejected.

7. EVALUATION OF PRODUCT QUALITY

Stability of the particle size distribution is a critical aspect of evaluating quality of parenteral course suspensions, since changes in the particle size distribution can affect the drug release profile from the site of injection, as well as having the potential to cause difficulty in syringeability (ability to easily withdraw the contents of a vial into a syringe) and injectability (ability to expel the contents of the syringe into the injection site using a reasonable amount of force). Clogging of the needle during administration can be traumatic to the patient. Concentrated suspensions have a greater tendency to clog the needle than more dilute suspensions. For some suspensions that display shear-induced thickening, it is appropriate to test syringeability before and after vigorous agitation.

There is no standard method for either syringeability or injectability. Use of 19–22 gauge needles is typical. A major source of uncertainty in injectability testing is the medium into which the suspension is injected. While injection into an animal is perhaps most realistic, this is often not practical. Injection into a polyurethane sponge has been reported (22). An instrument for monitoring injection force, such as an Instron device, is a significant improvement over subjective assessment of injection force.

Testing of suspensions after simulated shipping should be part of assessment of suspension quality, particularly with regard to the anticipated extremes of thermal history during shipping. Simulation of the vibration associated with shipping should also be considered, as this may affect the settling and redispersion characteristics of the suspension.

8. CONCLUSION

Formulation and manufacture of parenteral coarse suspensions, relative to other parenteral dosage forms, is not trivial. The difficulty arises primarily from the inherent thermodynamic instability of such systems, resulting in loss of such critical quality attributes as the ability to redisperse the solids sufficiently to achieve adequate uniformity of dosing and the ability to draw the suspension into a syringe and expel the product using a reasonable amount of force. However, knowledge of the principles of colloidal systems and the importance of forces at solid/liquid interfaces can be used to design formulations that, while thermodynamically unstable, can be made kinetically stable enough to retain critical quality attributes throughout the shelf life of the product.

REFERENCES

1. Shay DK, Fann LM, Jarvis WR. Respiratory distress and sudden death associated with receipt of a peripheral parenteral nutrition admixture. Infect Control Hosp Epidemiol 1997; 18:814–817.

2. Pikal MJ, Lukes AL, Lang JE, Gaines K. Quantitative crystallinity determinations for β-lactam antibiotics by solution calorimetry: correlations with stability. J Pharm Sci 1978; 67: 767–772.

3. Fiese EF. General pharmaceutics—the new physical pharmacy. J Pharm Sci 2003; 92:1331–1342.

4. Byrn SR, Pfeiffer RR, Stowell JG. Solid-State Chemistry of Drugs. 2nd ed. West Lafayette, IN: SSCI Inc., 1999.

5. Ahmed H, Buckton G, Rawlins DA. Use of isothermal microcalorimetry in the study of small degrees of amorphous content of a hydrophobic powder. Int J Pharm 1996; 130:195–201.

6. Hogan SE, Buckton G. Water sorption/desorption—near IR and calorimetry study of crystalline and amorphous raffinose. Int J Pharm 2001; 227:57–69.

7. Kawakami K, Numi K, Ida Y. Assessment of amorphous content by microcalorimetry. J Pharm Sci 2002; 91:417–423.

8. Buckwalter FH, Dickinson HL. The effect of vehicle and particle size on the absorption, by the intramuscular route, of procaine penicillin G suspension. J Am Pharm Assoc Sci Ed 1958; 47:661.

9. Feldman S. Biopharmaceutic factors influencing drug availability. In: Turco S, King R, eds. Sterile Dosage Forms: Their Preparation and Clinical Application. Philadelphia: Lea and Febiger, 1979:101–115.

10. Ballard BE, Nelson E. Physicochemical properties of drugs that control absorption rate after subcutaneous implantation. J Pharmacol Exp Ther 1972; 135:120.

11. Guyton AC. Textbook of Medical Physiology. 8th ed. New York: Harcourt Brace Jovanovich, Inc, 1976:170–182.

12. Zener JC, Kerber RE, Spivack AP, Harrison DC. Blood lidocaine levels and kinetics following high-dose intramuscular administration. Circulation 1973; 47:984.

13. Vukovich RA, Brannick LJ, Sugarman AA, Neiss ES. Sex differences in the intramuscular absorption and bioavailability of cephradine. Clin Pharmacol Ther 1975; 18:215.

14. Cockshott WP, Thompson GT, Howlett LJ, Seeley ET. Intramuscular or intralipomatous injections? N Engl J Med 1982; 307:356.

15. Greenblatt DJ, Shader RI, Koch-Weser J. Slow absorption of intramuscular chlordiazepoxide. N Engl J Med 1974; 291:1116.

16. Evans DF, Wennerstrom H. The Colloidal Domain: Where Physics, Chemistry, Biology, and Technology Meet. Chapter 2. New York, NY: VCH Publishers, Inc.

17. Strickley RG. Parenteral formulations of small molecule therapeutics marketed in the United States (1999)—Part I. PDA J Pharm Sci Technol 1999; 53:324–349.

18. Falkiewicz MJ. Theory of suspensions. In: Lieberman HA, ed. Pharmaceutical Dosage Forms: Disperse Systems. Chapter 2. New York: Marcel Dekker, 1996:1:17–52.

19. Brittain HG. Effects of mechanical processing on phase composition. J Pharm Sci 2002; 91:1573–1580.

20. Tom JW, Debenedetti PW. Particle formation with supercritical fluids—a review. J Aerosol Sci 1991; 22:555–584.

21. Fairand BP. Radiation Sterilization for Health Care Products. New York, NY: CRC Press, 2002:9–20.

22. Akers MJ, Fites AL, Robison RL. Formulation design and development of parenteral suspensions. J Parenteral Sci Technol 1977; 41:88–96.

7

Emulsions: Design and Manufacturing

N. CHIDAMBARAM

Senior Scientist,
Research & Development,
Banner Pharmacaps Inc., High Point,
North Carolina, U.S.A.

DIANE J. BURGESS

Department of Pharmaceutical
Sciences, School of Pharmacy,
University of Connecticut,
Storrs, Connecticut, U.S.A.

1. INTRODUCTION

An emulsion is a heterogeneous mixture of two or more immiscible liquids, with a third component (emulsifier) used to stabilize the dispersed phase droplets. Co-emulsifiers and other additives are often used to improve stability. The most commonly used parenteral emulsion system is for parenteral nutrition (PN). Parenteral nutrition is a means of providing intravenous nutrition to patients who are unable to absorb nutrients via the gastrointestinal tract. Infused nutrients include amino acids, dextrose, electrolytes, minerals, vitamins, fatty acids, and trace minerals (1–3). Drugs such as barbituric acid (4), nitroglycerin, and cyclandelate (5) have been incorporated into emulsions prepared for PN. In addition, emulsions have been specifically prepared as delivery vehicles for drugs with poor or no solubility in water.

1.1. Parenteral Nutrition

Parenteral nutrition is the provision of a sterile liquid containing nutrients, which is administered intravenously. This science has evolved in step with the acquisition of knowledge in chemistry, physiology, and microbiology. Provision of nutrients via parenteral administration has become an accepted method of treating nutritionally debilitated patients. Attempts to supplement nutrients intravenously were unsuccessful until the concepts of microbial infection became widely accepted. Early last century, as the principles of nutrition and metabolism became recognized, attempts were made to replenish protein deficits parenterally. Initially these efforts were hampered by allergic complications; however, in one of the first successful experiments on PN, in 1915, Henriques and Anderson (6) maintained nitrogen equilibrium in goats for over 2 weeks through the infusion of a mixture of hydrolyzed protein together with glucose and salts. Protein hydrolysates were subsequently shown a safe and effective source of nitrogen for human use.

Numerous complicating and limiting metabolic and technical issues became apparent when clinicians sought improvements in the parenteral delivery of nutrients. The provision of basal energy and nitrogen was not always sufficient. However, the infusion of excess volume resulted in pulmonary edema and high concentrations of glucose resulted in venous thrombophlebitis. In order to address this problem, the infusion of lipids was investigated as a means to effectively meet energy needs. The high caloric density and low thrombogenic potential of lipid emulsions made these substrates obvious candidates for PN. Researchers infused lipid emulsions into animals and also developed lecithin-stabilized lipid emulsion for human parenteral use. In 1945, McKibbin et al. (7) advocated the provision of calories from lipid emulsions, and shortly thereafter the first commercial preparations based on castor and cottonseed oils became available. After World War II, lipid emulsions were developed commercially to serve as an intravenous source of both calories and essential fatty acids. An i.v. cottonseed oil emulsion was marketed in 1957 (8). Since the frequency of toxic reactions with this product was high, this fat emulsion was withdrawn from the market in the 1960s. These early emulsions had toxic side effects including: fevers, back pain, and liver dysfunction, which led to their disuse and eventual withdrawal from the US market in 1964 (9). Meanwhile, in Europe, Wretlind et al. were developing an emulsion based on soybean oil stabilized with egg yolk phospholipids that were shown in clinical trials to be safe and effective as a source of calories and essential fatty acids. Supporting clinical results from the United States (10) led to FDA approval of Intralipid (Baxter & Pharmacia, Upjohn) and Liposysn (Abbott) in 1981. Shulman and Phillips (11) investigated PN in infants and children. To date in excess of 100 million units of lipid emulsion have been administered to patients as an integral part of PN. There are several commercially available fat emulsions such as Intralipid 10 and 20% (Kabi Vitrum), Lipofundin and Lipofundin S (Braun) and Liposyn (Abbott), all of which are essentially nontoxic (12). For example, Intralipid 20% has an LD_{50} of 163 mL/kg body wt. in mice (13).

1.2. Parenteral Drug Delivery

Pioneering work using lipid emulsions as drug delivery systems began in the early 1970s. Jeppsson (4) investigated the incorporation of drugs such as barbituric acid, nitroglycerin, and cyclandelate (5) into fat emulsions. Progress in the field of parenteral drug delivery using emulsions has been reported in numerous reviews and monographs (14). Singh and Ravin (15) reviewed parenteral emulsions as drug carriers. The review of Singh and Ravin (15) includes emulsion preparation, drug incorporation and specifically discusses diazepam and amphotericin B as examples of parenteral emulsions under development. In addition, water-in-oil and gelatin-in-oil emulsions have been developed with the ability to incorporate water-soluble or gelatin-soluble drugs. Although their toxicity and metabolism are less well defined compared to fat emulsions, these preparations offer untested alternative carriers for hydrophilic drugs. O'Hagan (16) summarized the application of emulsion as particulates in vaccine delivery systems. The following injectable lipid drug emulsions are commercially available:

Trade name	Manufacturer	Drug	Activity
Diazemuls®	Pharmacia	Diazepam	Sedative
Diprivam™	Pharmacia	Propofol	Anesthetic
Fluosol-DA®	Green Cross	Perfluorodecalin	Oxygen delivery
Vitalipid®	Pharmacia	Vitamins A, D, E, K_1	Parenteral nutrition
Eltanolone™	Pharmacia	Prenanolone	Anesthetic
Imagent®	Alliance Pharm.	Perflubron	Imaging agent
Intraiodol™	Pharmacia	Iodinated oil	Imaging agent
Kynacyte™	Sphinx/Eli Lilly	Dihydrosphingosine	PKC inhibitor
Oncosol™	HemaGen/PFC	Perfluorocarbon	Oxygen delivery

1.2.1. Drug Solubility

A common problem experienced in the early development of drugs intended for parenteral use, especially intravenous administration, is that many drug candidates have poor water solubility. Solubilization processes are complex and require expertise in physical chemistry to interpret and apply current theoretical models. Unfortunately, most of the literature deals with solubilization theory rather than with the practical aspects of solving solubility problems. Solubility theories deal with the conversion of a substance from one state to another, and the equilibrium phenomena that are involved. In most pharmaceutical systems, the routine application of these models to predict solubility and simplify formulation development is complex. The majority of parenterally acceptable cosolvents—such as propylene glycol, ethanol, and water—are capable of self-association through hydrogen bond formation. Such interactions may alter solvent structure and consequently, influence solubility in an unpredictable manner (17). In order for solubility models to adequately describe solubility behavior, the relative importance of competing self-associations and strong intermolecular interactions must be considered.

The first approach that is commonly used to increase the aqueous solubility of a drug is to form water-soluble salts. Berge et al. (18) wrote what is now a near classic review of salt form strategies acceptable for pharmaceuticals. If salt formation is not possible (e.g., the salt form is too unstable, or does not render the molecule sufficiently water-soluble), a series of formulation approaches may be investigated. pH adjustment may be used to increase the aqueous solubility of an ionizable drug. The next most frequently attempted approach is the use of water-miscible cosolvents. Another approach is the use of surfactants and complexing agents. The use of emulsions and other colloidal drug delivery systems for intravenous administration is becoming widely and successfully applicable (14). These delivery systems may confer to the entrapped or associated drug significantly different properties than the free drug, providing the opportunity for prolonged drug presence in the bloodstream or alteration of its disposition in the body.

Surface active agents are often incorporated into parenteral drug delivery systems to provide one of several desirable properties: (i) increase drug solubility through micellization, (ii) prevent drug precipitation upon dilution (19), (iii) improve drug stability through entrapment within a micellar structure (20), and (iv) prevent aggregation of protein formulations due to liquid/air or liquid/solid interfacial interactions.

While many different types of surfactants exist, very few have been used in parenteral products. For example, for stabilization of proteins against aggregation, polyoxyethylene sorbiton monooleate (polysorbate 80) has been used in an FDA approved product (e.g., Altepase, Genentech) (21). Other surfactants, which have been used in parenteral products, include poloxamer 188 (polyoxyethylene–polyoxypropylene copolymer), polysorbate 20 and 40 (polyoxyethylene–polyoxypropylene sorbiton monofatty acid esters), Cremophor EL®, and Emulphor EL 719® (polyethoxylated fatty acid esters and oils). Examples of products that include some of the above surfactants are multivitamins, calcitriol, teniposide, paclitaxel, and cyclosporine. The most effective as a solubilizer or stabilizer is often a matter of empirical investigation (22). Detailed reviews of micelle structure, characterization techniques, and pharmaceutical applications have been published (23,24). Attwood and Florence (24) summarized the toxicity of surfactants reported in the literature prior to 1983. Reviews on the pharmacology of polysorbate 80 and the incidence of clinical side effects of Cremophor EL have been published. Children and newborns may be particularly sensitive to these agents and administration to this population has been discussed (25).

If a drug candidate has sufficient lipid solubility, emulsions may be used as a delivery system. Typical emulsions contain triglyceride rich vegetable oils, lecithin as a surfactant and may contain nonionic surface-active agents. Insoluble drugs may be incorporated into commercial fat emulsions or an emulsion may be formed from oil-solubilized drug, surfactant and aqueous phase. The former is usually not successful since the drug can influence the stability of the commercial emulsions (26). Kreilgaard (27) summarizes the influence of microemulsions on cutaneous drug delivery.

1.3. Classification of Emulsions

Emulsion systems can be classified as either oil-in-water (O/W) or water-in-oil (W/O) depending on the nature of the dispersed phase. O/W emulsions are dispersions of oil in an aqueous continuous phase and W/O emulsions are dispersions of water in an oil continuous phase. A third type of emulsion is a multiple emulsion, which can be either oil-in-water-in-oil (O/W/O) or water-in-oil-in-water (W/O/W). Multiple emulsions are prepared by dispersion of W/O emulsions in an aqueous solution to form a W/O/W emulsion or by dispersion of an O/W emulsion in a oil to form an O/W/O emulsion.

The nature of emulsion systems (i.e., W/O or O/W) can be determined using the following methods: indicator (28), drop test (29), electrical conductance (30), and the direction of emulsion creaming (31). The indicator method utilizes a dye, which is only soluble in one phase. The stained phase can be observed visually or with the aid of a microscope (28). The drop test method involves determination of emulsion miscibility with water or oil phases (28). The emulsion will be miscible with the phase, which constitutes its continuous phase. The electrical conductance method is based on the ability of the emulsion to conduct electrical current. O/W emulsions will conduct electricity, whereas W/O emulsions will not. The direction of creaming of an emulsion depends on the density difference between the oil and aqueous phases (31). Systems with oil as the dispersed phase will cream at the top and those with water as the dispersed phase will cream at the bottom.

Emulsion systems can be further classified according to dispersed phase droplet size (macroemulsion, miniemulsion, and microemulsion) and the nature of the emulsifiers (nonionic surfactants, ionic surfactants, and nonionic polymers, polyelectrolytes and solid particles) (32). Macroemulsions are opaque with droplet sizes >400 nm. Miniemulsions are blue-white semiopaque systems with droplet sizes between 100 and 400 nm (33,34). Miniemulsion preparations often involve the addition of an ionic surfactant and a cosurfactant (a long-chain alcohol) in the form of a mixed micellar system (33). Brouwer et al. (34) reported that the driving force for O/W miniemulsion formulation is the net transfer of fatty alcohol from the aqueous to the oil phase during the mixing process. Microemulsions are transparent with droplet sizes in the order of 10–100 nm (35,36). They form spontaneously and are thermodynamically stable, since the interfacial tension of these systems approximates zero. Microemulsion systems contain large amounts of surfactant and usually have an intermediate chain length alcohol as a cosurfactant (34). Microemulsions are thermodynamically stable transparent colloidal dispersions. The advantages they have over macroemulsions are their stability and ease of manufacture. Droplet sizes are typically 10 times smaller than macroemulsions. However, the high surfactant concentration of these systems has imposed limitations on their use via the parenteral route, due to toxicity considerations.

1.4. Emulsion Destabilization

Emulsions are thermodynamically unstable systems. The phase inversion temperature is the temperature at which an emulsion inverts from either an O/W form to a W/O form or vice versa. Temperature changes such as those occurring during homogenization or sterilization procedures may cause inversion to occur. There is some evidence to suggest that relatively stable emulsion systems can be obtained when the phase-inversion temperature of O/W emulsion is approximately 20–65°C higher than the storage temperature (37).

There are several physical changes, such as creaming, flocculation, and coalescence, which may occur when a drug is added extemporaneously to a sterilized lipid emulsion (Fig. 1), some of which lead to emulsion breakage (separation of oil and water phases). While creaming and gross oil separations are visible to the trained eye, other changes such as flocculation and coalescence must be detected by light-scattering or light microscopy. An emulsion is considered stable when there is no change in certain parameters (i.e., number of particles of the disperse phase, particle size distribution, total interfacial area, mean droplet size, chemical composition of the components involved, and other related parameters) while standing undisturbed under normal conditions. Thermodynamics dictates that all emulsion systems will attempt to decrease their surface free energy, with the concomitant decrease in total interfacial area. Emulsions can exhibit both chemical and physical instability.

Chemical instability refers to the chemical changes, in liquid phases and/or the emulsifying agents that produce intrinsic emulsion instability or the tendency toward demulsification. Physical instability relates to the shelf life of the emulsion when only field forces (such as gravitational and thermal) are considered. Physical instability can be

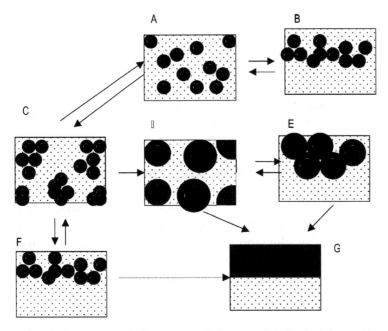

Figure 1 Physical changes possible in a lipid emulsion: (A) freshly prepared lipid emulsion; (B) creaming—readily reversible, slow flotation of lipid droplets on more dense aqueous phase; (C) flocculation—aggregated droplets are not readily redispersed by agitation; (D) coalescence—irreversible merging of smaller droplets; (E) rapid creaming of coalesced emulsion; (F) rapid creaming of flocculated emulsion; (G) broken emulsion—separation of oil and water phases.

manifested in one of two ways: reversible or irreversible instability. Creaming or sedimentation and flocculation are reversible phenomena, whereas, coalescence, demulsification, breaking, and inversion are irreversible phenomena.

Reversible instability signifies any manifestation of emulsion instability where the dispersed droplets have not lost their identity and there is a capability of restoring the emulsion to its original condition. Creaming is caused by density differences between the disperse phase and the dispersion medium. In the case of true creaming, the emulsion is separated into two emulsions: one richer in the dispersed phase and the other poorer than the original emulsion. If the dispersed phase is lighter than the dispersion medium, the phenomenon is termed "creaming"; if it is heavier, it sinks to the bottom, and this phenomenon is called "sedimentation." In creamed emulsions, the droplets do not lose their identity. Adjusting the densities of the dispersion medium and the disperse phase so that they are similar can avert creaming.

Flocculation occurs when two or more droplets approach each other and form an aggregate or "floc" (in which the droplets have not lost their identity). Flocs can be separated and the emulsion is restored to its original state (38,39). Important characteristics of emulsion flocculation are: (a) individual droplets make up a floc and they retain their identity; (b) flocs can be broken up restoring the emulsion to its original state; (c) the flocs behave as single droplets and the rate of reversible creaming is accelerated if the density of the flocs is appreciably different from that of the dispersion medium; (d) since the droplets are in close contact, flocculation may lead to coalescence and subsequent demulsification.

Irreversible instability is when the identities of the original dispersed phase droplets are changed and the emulsion cannot be restored to its original form. Once the droplets are in close contact (e.g., by flocculation), intermingling of the droplets is possible and the flocs transform into a single large droplet. The appropriate term for the process of transformation of flocs into a single large droplet is "coalescence." Coalescence brings about internal changes in the emulsion and there is no conspicuous outward manifestation. As the coalesced drops grow larger, these constitute a separate phase and the process is termed "demulsification" or "breaking."

1.5. Emulsion Characterization

Physical properties of emulsions, which can be readily quantified, include mean particle diameter, size distribution, surface (zeta) potential, interfacial tension and rheology, osmolality, and phase inversion temperature. For more details on characterization, the reader is referred to the chapter on Characterization and analysis in this book. All of these properties are important predictors of emulsion stability and biocompatibility. For intravenous use, mean droplet diameters <1 μm are required. Droplets larger than 5 μm are capable of forming emboli in small capillaries such as those that occur in the lungs. Size distribution is equally important, since a more homogenous (monodisperse) emulsion tends to exhibit less coalescence and greater resistance to phase separation. Low interfacial tension is associated with more stable emulsions, while high interfacial tension predicts short shelf life stability due to phase separation. Biocompatibility is related to the net charge on the droplet surface. In general, a more electronegative surface exhibits a reduced tendency to aggregate in the presence of blood proteins. An ideal biocompatible emulsion is also isotonic, i.e., containing 280–300 mOsm/kg. In practice, there are few physiologically acceptable tonicity agents, which may be incorporated into an emulsion without causing disruption during thermal sterilization. Isotonic saline is one such agent that does not cause emulsion disruption (40).

1.5.1. Evaluation of Emulsion Stability

According to King and Mukherjee (41), the only precise method for determining emulsion stability involves size-frequency analysis of the emulsion from time to time as the product ages. An initial frequency distribution analysis of an emulsion is not an adequate test of stability, since stability is related to the rate of change of particle size. For rapidly breaking emulsions, macroscopic observation of the separated internal phase is adequate, although the separation is difficult to read with any degree of accuracy.

Emulsion diameter can be measured microscopically. Finkle et al. (42) were the first to report the use of this method to determine emulsion stability. Brownian movement affects the smallest droplets, causing them to move in and out of focus so that they are not consistently counted. The velocity of creaming is directly proportional to the square of the droplet diameter, and therefore creaming favors the largest droplets since they move faster toward the cover glass than do the smaller droplets. Accordingly, microscopic measurement

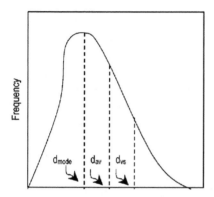

Figure 2 Droplet size distribution of an emulsion.

has a bias toward larger droplets. The particle size or diameter of the droplets in micrometers is plotted on the horizontal axis against the frequency or number of droplets in each size range on the vertical axis (Fig. 2).

Light extinction and blockage counting devices detect particles by the partial blockage of a beam of light (43). These instruments generally consist of five subsystems: an autodiluter, an optical sensor, a pulse-height analyzer, a system computer, and a software controller. Hydrodynamic chromatography is a method, which was developed for the fractionation of colloidal solutions according to their particle size. Hydrodynamic chromatography utilizes a column packed with rigid nonporous beads along the capillary tube. Light scattering, light extinction, and electrical zone sensing methods relate particle size to the equivalent spherical diameter based on surface area or volume. Jiao et al. (44) investigated the evaluation of multiple emulsion stability using a pressure balance and interfacial film strength.

Other methods used to determine emulsion stability include accelerating the separation process, which normally takes place under storage conditions. These methods employ high temperature treatment, freezing, freeze–thaw cycling, and centrifugation. Merill(45) introduced the centrifuge method to evaluate emulsion stability. Turbidimetric analysis and temperature tests have also been used in an effort to evaluate new emulsifying agents and to determine emulsion stability. Garti et al. (46) developed a method to evaluate the stability of oil–water viscous emulsions (ointments and cosmetic creams) containing nonionic surfactants. This method is based on electrical conductivity changes during nondestructive short heating–cooling–heating cycles. This method was applied in a series of emulsions with different emulsifiers, emulsifier concentrations, and oil phase concentrations.

Burgess et al. (47–56) in a series of publications have reported a method to predict emulsion stability based on evaluation of interfacial properties (elasticity, tension, and charge). These authors were able to correlate emulsion stability with the interfacial properties of the emulsifier systems used. Both small surfactants and protein emulsifiers were investigated at different concentrations and in the presence of various additives. The reader is referred to the chapter by Burgess in this book on Physical Stability of Dispersed Systems.

1.5.2. In Vitro Release

Release of drug from emulsions and subsequent bioabsorption are controlled by the physicochemical properties of the drugs and the emulsion dosage form, and the physiological properties at the site of administration. Drug concentration, aqueous solubility, molecular size, crystal form, protein binding, and pK_a are among the physicochemical factors that need to be considered. It is technically difficult to characterize in vitro drug release from emulsions due to the physical obstacles associated with separation of the dispersed and continuous phases (57). Various techniques such as sample and separate, membrane barrier, in situ and continuous flow methods have been used to characterize in vitro drug transport kinetics from submicron-sized emulsions. However, problems are associated with each of these methods.

Sample and separate techniques are not ideal since it is difficult to preserve the physical integrity of emulsion droplets during the separation process. For example, to separate the released drug present in the continuous phase from the releasing source (dispersed phase droplets), filtration and centrifugation are used (58). These techniques involve the application of external energy, which can result in emulsion destabilization and hence erroneous results. Membrane barrier techniques include cell diffusion and dialysis bag equilibrium methods (59), where the dispersed phase is separated from the receiver phase by a semiporous membrane. Submicron droplets have a large surface area compared to their volume, which can lead to rapid transport from the oil to the continuous phases and the potential for violation of sink conditions in the environment adjacent to the droplets (60,61). Barrier techniques can lead to a violation of sink conditions, due to the limited membrane surface area available for transport from the donor to the receiver chambers compared to the surface area available for transport from the dispersed phase droplets to the continuous phase. An additional constraint of the barrier method is the limited volume of continuous phase available to solubilize the released drug in the donor chamber (62–64). In situ methods involve analysis of released drug without separating from the releasing source. Another advantage of in situ methods is that the emulsions can be diluted infinitely and therefore violation of sink conditions is not a problem. However, not all compounds are suitable candidates for in situ techniques, as a method of analysis must be available which does not suffer interference from the dispersed phase droplets. Suitable drug candidates include molecules with fluorescent or phosphorescent moieties (65). Continuous flow methods involve the addition of dispersed phase material to a filtration cell, which is linked in series to an analysis cell (66,67). The sink phase is continuously circulated through the filtration and analysis cells. A limitation of this technique for emulsion systems is clogging of the filter, which causes alteration in media flow and emulsion destabilization, both of which can affect release rates.

As a consequence of the limitations of the above methods, investigators have attempted to account for the different constraints of these methods mathematically (68,69). The most commonly used technique for assessing in vitro transport from submicron emulsions is the side-by-side diffusion cell (barrier method). As stated previously, the major limitation of this method is the potential for violation of sink conditions.

To overcome the limitations of the side-by-side diffusion cell technique, a reversed dialysis bag method was developed by Chidambaram and Burgess (70). Dialysis bags containing the continuous phase (receiver phase) alone are suspended in a vessel containing the donor phase (diluted emulsion) and the system is stirred. At predetermined time intervals, each dialysis bag is removed and the contents are analyzed for released drug. The bulk equilibrium reverse dialysis bag technique overcomes the shortcomings of the side-by-side diffusion cell and bulk equilibrium dialysis bag techniques by diluting the submicron sized emulsion in the donor chamber (theoretically the emulsion can be infinitely diluted) and by increasing the surface area of the permeating membrane (dialysis bags). Consequently, violation of sink conditions can be avoided. No separation step is required since the dispersed phase does not penetrate into the dialysis bags. Thus, the possibility of emulsion destabilization during mechanical separation is avoided. Another advantage of this method is the increased efficiency in terms of manpower as a consequence of reduction in the number of steps involved. The reversed dialysis bag technique (70) mimics the in vivo situation more closely than the side-by-side diffusion method for emulsion dosage forms and other dispersed systems introduced directly into the blood stream (i.v.) or taken orally as these systems will experience infinite dilution following administration via these two routes.

1.5.3. Theoretical Models Proposed for Drug Transport in Emulsion Systems

Mass transport phenomenon is important to understand drug transport processes and has been extensively reviewed (71–73). Fick's law of diffusion may describe drug transport through membranes. Diffusion is the process of molecular movement of a permeant as a result of a concentration gradient. Fick's first law of diffusion states that in an isotropic and continuous medium, the rate of mass transport across a plane of unit area is proportional to the concentration gradient measured perpendicularly to that plane. Fick's first law of diffusion can be expressed mathematically as

$$J = D\left(\frac{dC}{dx} + \frac{dC}{dy} + \frac{dC}{dz}\right) \tag{1}$$

where J is the mass flux of permeant per unit area, C is the concentration of permeant in the diffusional medium, x, y, and z are the distances in the direction of flux in each Cartesian dimension, and D is the diffusion coefficient of the permeant in the medium. The negative sign indicates that the net mass flux is in the direction of decreasing concentration. According to the Stokes–Einstein equation (74), the diffusion coefficient of a permeant is a function of its thermal mobility in the medium, the size of the permeant and the nature of the medium. The diffusion coefficient is related to the frictional resistance and can be expressed as follows

$$D = \frac{RT}{f} \tag{2}$$

In a homogeneous liquid, the frictional resistance is dependent on the size and shape of the permeant and on the nature of the solvent. Frictional resistance (75) can be expressed by

$$f = 6\pi\eta r\left\langle\frac{2\eta + r\beta}{3\eta + r\beta}\right\rangle \tag{3}$$

where η is solvent viscosity, r is solute radius, and β is a slip factor. This equation is applicable for spherical particles in dilute solution where solute–solute interactions can be neglected. More than one barrier may be involved in the diffusion process. The rate-determining step is the slow diffusion of drug across a barrier. A membrane is defined as a layer that is distinct from the medium and restricts permeant transport. When the steady-state approximation is applied in one-dimensional diffusion, Eq. (8) can also be expressed as

$$J = \frac{KD}{h}(C_d \quad C_r) = P(C_d \quad C_r) \tag{4}$$

where K is the membrane/medium partition coefficient, h is the membrane thickness, C_d is the donor concentration of permeant, C_r is the receiver concentration of permeant, and P is the permeability coefficient. The permeability coefficient through a series of barriers is inversely proportional to the total resistance, which is the sum of the individual resistances (76). This is expressed as

$$R_T = \frac{1}{P_T} = \frac{h_1}{D_1 K_1} + \frac{h_2}{D_2 K_2} + \cdots + \frac{h_m}{D_m K_m} \tag{5}$$

where $h_1, h_2, \ldots, h_m$ are the thicknesses of the individual barriers and $D_1, D_2, \ldots, D_m$ are the respective diffusivities in each barrier. Stagnant boundary layers, poorly stirred layers, at the donor and receiver membrane surfaces, can also act as barriers to mass transport.

Madan (77) developed the simplest theoretical model for drug transport in emulsion systems. This model was based on drug partition coefficient between the two phases and utilized mass balance to determine the drug concentration in the two phases

$$Q_T = C_O V_O + C_W V_W \tag{6}$$

where Q_T is the total amount of the drug in the emulsion, subscript 'o' and 'w' are for the oil and aqueous phases, respectively, and C and V refer to concentration of the drug and volume of the phases, respectively. This model did not take into consideration the interfacial barriers as a consequence of surfactant films. Ghanem et al. (68) studied the effect of interfacial barriers on transport in emulsion systems. Interfacial transport was retarded approximately 1000 times in the presence of an adsorbed gelatin film. Ghanem developed a theoretical model for drug transport in emulsions, which included the effect of an adsorbed gelatin interfacial film. Their model for interfacial transport between an aqueous and oil environment is based on Fick's first law of diffusion as follows:

$$G = \frac{V(dC_{do})}{dt} \tag{7}$$

where G is the flux of drug, C_{do} is the concentration of the drug in the oil, and V is the volume of the oil. They did not include the interactions of the drug with the surfactant. The effect of interfacial interactions between drugs and surfactants on drug transport in emulsion systems was investigated by Lostritto et al. (78) and they developed a theoretical model assuming monolayer surfactant coverage at the interface. They assumed that the charged and neutral drug species partition differently into a hydrophobic internal phase in accordance with the pH partition hypothesis

$$Q_d = C_e V_i [F_n K_{en} + (1 - F_n) K_{ec}] + C_e V_e$$
$$+ C_e A_s [F_n (k_i + k_0 S_i) + (1 - F_n)(k_i^* + k_0^* S_i)] \tag{8}$$

where the subscripts 'd', 'e', 'i', and 's' indicate donor emulsion, external, internal, and interfacial regions, respectively. C and V are the concentration of the drug and volume of the phases, respectively. A_s is the total surface area of the O/W emulsion, k_i is the interfacial activity of the drug, k_0 is the adsorption of the drug with surfactant at the interface, S_i is the concentration of surfactant at the interface. K_{en} and K_{ec} are the partition coefficients of the neutral and charged species between the oil and water phases, respectively, and superscript '*' denotes the charged species. F_n is the neutral fraction of drug in the aqueous phase. They did not include the effect of excess surfactant.

Yoon and Burgess (69) were the first to consider the effect of the micellar phase on drug transport in emulsion systems

$$Q_d = C_e V_e [F_n K_{en} + (1 - F_n) K_{ec}] + C_W V_W + C_W A_S$$
$$+ [F_n (k_i + k_0 S_i) + (1 - F_n)(k_i^* + k_0^* S_i)]$$
$$+ C_W V_m [SAA](F_n K_{mn} + (1 - F_n) K_{mc} \tag{9}$$

where V_m is the volume of the micellar phase, K_{mn} and K_{mc} are the partition coefficients of the neutral and charged species, respectively. However, their model did not include the effect of micellar phase on surface-active model drugs where the drug may compete with the surfactant for the interface and consequently can affect emulsion stability and transport phenomenon of the model drug. Chidambaram and Burgess (79) studied this effect:

$$Q_d = C_W[K_e V_e + A_S(k_i + k_0 S_i) + V_m[SAA]K_m] \qquad (10)$$

where K_m is the micellar distribution coefficient of the drug.

1.5.4. Effect of Micelles on Drug Transport in Emulsion Systems

Emulsion systems are stabilized using surfactants, which reduce interfacial tension and/or form an interfacial film barrier. Surfactants are added at a concentration, more than is required to form a monolayer at the emulsion droplet interface in order to overcome instability caused by thinning and rupture of the interfacial film which can lead to droplet coalescence and phase separation. Excess surfactant present in the continuous phase as micelles can allow the droplets to resist coalescence and phase separation through the Marangoni and Gibbs effects. The Marangoni effect is a liquid motion caused by movement of surfactant in the interface in the direction of higher interfacial tension due to the difference between dynamic and static tensions. The Gibbs effect is a local expansion or compression of the interface due to the surfactant gradient between the interface and the bulk. Therefore, a dynamic equilibrium exists between surfactant molecules in the interfacial area and those in the bulk. Consequently, surfactant concentrations in excess of that required to form a monolayer at the emulsion droplet interface are necessary for emulsion stability. Surfactant in excess of that necessary to form a monolayer is termed "excess surfactant." Emulsions usually contain excess surfactants and therefore it is necessary to determine the effect of this excess surfactant on drug transport and emulsion stability.

Micelle formation, complex formation, and the presence of cosolvents can affect the permeability coefficient of a permeant through a membrane, since these can affect the thermodynamic activity of the permeant either in the medium or in the barrier (80). If the permeant has no affinity for the micelle, there is no significant effect of micellar phase on membrane transport. As the affinity of the permeant for the micellar phase increases, the fraction of unassociated diffusing species will be depleted and the flux will decrease proportionally. When the permeant has high micellar affinity, transport is limited by the rate of micellar diffusion and/or the driving force to transfer the diffusant from the micellar phase and into the membrane.

Complex formation, like micelle formation, affects the apparent solubility and the partition coefficient of a permeant (81). If complexation occurs in the aqueous phase, it influences transport in a manner analogous to micellization. However, if complexation is irreversible, transport of the complexed form effectively results in a parallel transport pathway. Nakano and Patel (82) studied the effect of alkylamides on the transport of *p*-nitrophenol across silicone membranes and found that dimethylacetamide did not affect the apparent permeability but that dimethylpropamide, diethylacetamide, and diethylpropamide, increased permeability by 11%, 29%, and 95%, respectively. They reported that the stability constant for the complex, the diffusivity of free and complexed drug in each phase and the partition coefficient of the drug and complex between phases must be taken into account in order to explain the transport mechanism. Bates et al. (83) reported that the apparent permeability coefficient of drug was dependent on the partition coefficient of a caffeine-drug complex between membrane and medium.

Micelles and monomer surfactants can affect membrane transport of drugs by modification of the aqueous diffusion coefficient. It is considered that the adsorption of drug, surfactant, and micelles on membranes can affect drug transport due to possible competition for adsorption sites (84).

2. MANUFACTURING AND PROCESS CONDITIONS

For all injectable lipid emulsions, oil is the internal phase, dispersed as fine droplets in the continuous phase, usually water. The emulsification process requires the addition of surfactant and mechanical energy. The two main functions of surfactants are to reduce the interfacial tension of the oil and water phases and to prevent flocculation and coalescence of the dispersed phase.

If the pharmaceutically active compound is added to a preformed, sterilized emulsion, the process is termed extemporaneous incorporation. A filtration step is commonly included to remove poorly emulsified material from the product. In pharmaceutical manufacturing, a cartridge type filter might be used, provided that it is nonpyrogenic, contains no extractables, sheds no particles and possesses a medium pore size (about 1–5 μm). The bottle-filling operation takes place with full gowning under clean room conditions or in a laminar flow hood, followed by heat sterilization. All process stages are carried out in a closed system to prevent both microbial and particulate contamination. There are a number of important variables inherent in the process. The effect of moderate increases in some of the more easily controlled parameters are summarized in Table 1, assuming that a lipid based drug emulsion is being developed. The spontaneous formation of an emulsion is a relatively rare occurrence as mentioned earlier in this chapter on microemulsions. Instead, either the condensation or the dispersion method can prepare emulsions. To obtain a metastable emulsion, with a larger number of droplets of one liquid dispersed in the other liquid, considerable ingenuity must be employed. Droplets of the required size may be obtained by two different approaches.

Either, start from very tiny nuclei, which are allowed to grow to the required size (basis for the condensation method) or break up large drops of bulk liquid into small droplets (basis for the dispersion method).

2.1. Condensation Method

In the condensation method, the vapor of one liquid is supersaturated, and, therefore, tends to deposit on the nuclei that may be present in the system. These nuclei may be natural specks of dust and smoke, or may be ions and other seeds that

Table 1 Variables and Their Effect on the Final Emulsion

Variables to be increased	Expected effect (possible reasons)
Oil concentration	Larger droplet size (without proportionate increase in surfactant concentration)
Salt concentration	Wider droplet size distribution (acts as peptizing or coagulating agent)
Drug concentration	Unpredictable due to altered surfactant solubility in the oil phase
Surfactant concentration	Smaller droplet size until optimum, then increased viscosity
Process temperature	Smaller droplet size (via reduced viscosity)
Process pressure	Smaller droplet size (increased field forces)
Homogenizer passes	Narrower droplet size distribution (increased energy)

are artificially introduced. The excess vapor is deposited on the nuclei and consequently these grow in size. This condensation procedure can be used for preparing emulsions and is best illustrated by the arrangement (Fig. 3) developed by Sumner (85).

The vapor of one liquid (the dispersed phase) is injected below the surface of the other liquid, which forms the external phase of the emulsion. In this process, the vapor becomes supersaturated and condenses as micron-sized particles. The particles are stabilized in the external liquid, which contains

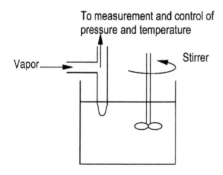

Figure 3 Principle of the condensation method.

a suitable emulsifier. The liquid to be dispersed is heated in a separate vessel and the heat input, the temperature, and the pressure of the vapor are controlled to achieve the desired emulsion characteristics. The pressure of the injected vapor, the diameter of the jet orifice, and the emulsifier added to the external phase are the principal factors affecting the size of the particles. Particle sizes of about 20 μm are easily obtained although the concentrations of the emulsions are not high.

2.2. Dispersion Method

The most common method of preparing emulsions is to apply force to break up the interface into fine droplets. When a liquid jet of one liquid is introduced under pressure into a second liquid, the initially cylindrical jet stream is broken up into droplets. The factors that enter into breakup of the liquid jet include the diameter of the nozzle, the speed with which the liquid is injected, the density and the viscosity of the injected liquid, and the interfacial tension between the two liquids. There are many variants in this method, which fall into four broad categories, viz. mixing, colloid milling, ultrasonification, and homogenization. The available commercial instruments cover a wide range of capacities, from small laboratory models to large industrial units. However, before taking this up, it is advantageous to discuss a rather naive method of emulsification, namely the intermittent shaking method.

2.3. Intermittent Shaking Method

This is a simple way of demonstrating emulsion formation and involves introducing the two liquids into a test tube, which is then shaken vigorously. It was determined by Briggs (86) that intermittent shaking, with rest periods between shakes, was vastly more effective than uninterrupted shaking. For instance, to emulsify 60% by volume of benzene in 1% aqueous sodium oleate about 3000 uninterrupted mechanical shakes (in a machine), lasting approximately 7 min were necessary. The same mixture could be completely emulsified with only five shakes by hand in 2 min if after each shake an interval of 20–30 sec was allowed. The plane interface between the two liquids becomes corrugated and deformed after each shake. The corrugations grow in size as fingers of one liquid into the other and they then disintegrate into small drops. This process takes about 5 sec for the magnitude of the parameters involved in shaking by hand (86). In hand shaking the globules are polydisperse and are in the size range 50–100 μm under favorable conditions. More vigorous shaking needs to be applied in order to get smaller droplets. To be more specific, the shaking must be on the microscale so those bigger drops, which are formed initially, are torn into smaller droplets. This will happen only if there are large velocity gradients. Mixers, colloid mills, ultrasonicators, and homogenizers are all designed to achieve large velocity gradients.

2.4. Mixers

Liquid mixing is an established operation in chemical engineering and mixing apparatues are available with capacities ranging from less than a liter to several cubic meters. The basic idea behind simple mixers is demonstrated in Fig. 4. Suppose a simple paddle is rotated in a large cylindrical vessel, then the liquid will be set in rotation and the free surface will attain a rough parabolic shape. This swirling motion often results in stratification rather than mixing, especially in large tanks. The liquid moves in large circular paths with little vertical motion.

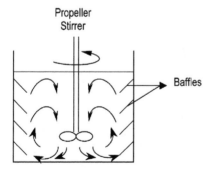

Figure 4 Flow pattern from a rotating propeller in a tank with baffles.

Mixing is best accomplished when there are lateral and vertical flows, which distribute the materials rapidly to all parts of the tank. An efficient and convenient way to achieve such mixing is to use vertical baffles near the walls, which deflect the fluid upwards. As a result of viscous and other dissipation, there is usually a small rise in temperature in most mixers. While this is tolerable in small units (whisks and churns), larger units usually have some form of cooling arrangement. Most mixers, are well suited for making low or medium viscosity emulsions, the turbine mixers tolerating somewhat higher viscosities than the propeller impellers. This effect is due to the fact that the liquid is drawn towards the center in the case of turbine mixers, whereas, in the case of propeller mixers, the liquid is forced towards the wall. The mean average particle diameters of emulsions prepared in this way are usually of the order of 5 μm.

2.5. Colloid Mills

In a colloid mill, emulsification of the liquids is carried out under strong shearing flow in a narrow gap between a high-speed rotor and a stator surface (Fig. 5).

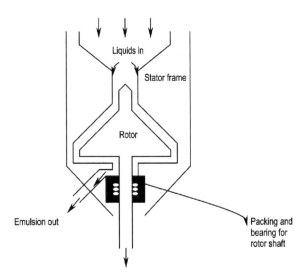

Figure 5 Section of a vertical colloidal mill.

The liquids enter at the top through suitable tubes in the stator frame, flows through the narrow clearance between the stator and the rotor and exit. The rotor is dynamically balanced and can rotate at speeds of 1000–20,000 rpm. Due to the high speed and small gap, strong shear flow is set up and the liquid interfaces are violently torn apart. Centrifugal forces also come into play. The liquids may flow either under gravity feed or under slight over-pressure. Colloid mills may be modified to suit different conditions. In colloid mills, on account of the large shear stresses and dissipation, the temperature rise is large. Consequently, some form of cooling is always used in larger units. The input materials may be liquids or pastes, with force feeding in the latter case. The throughput varies inversely with the viscosity of the emulsion. Particle diameters of the order of 2 μm are easily obtained with colloid mills.

2.6. Ultrasonifiers

The use of ultrasonic energy to produce pharmaceutical emulsions has been demonstrated, and many laboratory-size models are available. These transduced piezoelectric devices have limited output and are relatively expensive. They are useful for laboratory preparation of fluid emulsions of moderate viscosity and extremely low particle size. Commercial equipment is based on the principle of the Pohlman liquid whistle (Fig. 6).

The dispersion is forced through an orifice at modest pressures and is allowed to impinge upon a blade. The pressures required range from approximately 150 to 350 psi and cause the blade to vibrate rapidly to produce an ultrasonic note. When the system reaches steady state, a cavitational field is generated at the leading edge of the blade, and pressure fluctuations of approximately 60 tons psi can be achieved in commercial equipment.

2.7. Homogenizer

A homogenizer is a device in which dispersion of the liquids is achieved by forcing the mixture through a small orifice

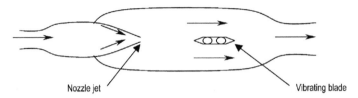

Figure 6   Principle of Pohlman whistle.

under high pressure. This results in fine particle sizes and, consequently, when highly dispersed systems with particle sizes of 1 μm or smaller are desired, a homogenizer is frequently used. Temperature rise associated with homogenizers is quite moderate and therefore cooling arrangements are usually not necessary. If the liquids are premixed, a single pass through the homogenizer will result in a fine emulsion with particles in the range of 1 μm.

A microfluidizer is an example of a homogenizer, which consists of an air driven pump and an interaction chamber connected using narrow microcolumns. High liquid pressure generated by compressed air results in dynamic interaction of the two liquids in the narrow microcolumns. Consequent laminar flow, turbulence, and cavitation generate fine emulsions with small mean diameters and narrow polydispersity values (87–91).

3. LYOPHILIZATION OF EMULSIONS

These multiphase liquid systems, being thermodynamically unstable, may require lyophilization and hence the key aspects of the lyophilization process (which maintains or regenerates the structure and size of the molecular assemblies without affecting the drug loading of the initial liquid state) will also be discussed briefly. Lyophilization of emulsions can be successfully achieved by:

1. Removal of water at such a rate that the bilayer or monolayer structure remains intact and the water interactions are replaced by (H-bonding)

interactions with cryoprotectants containing multi-OH groups.

2. On addition of water, micelles or emulsions are reformed such that:

    i. original or acceptable particle size distribution is maintained, and

    ii. drug encapsulation and location are unaffected.

Perturbations during lyophilization include:

    A. Lipid sorting, changes in the bilayer interfacial film characteristics (charge, hydration, fluidity); which include

        a. interfacial tension, and interfacial viscoelastic parameters, and

        b. zeta potential (as a result of charge reversal).

    B. Formation of multiple populations differing in charge, surface, and particle size distribution.

    C. Mechanical deformation of droplets as a result of growing crystals.

    D. Water-soluble components (protective buffers, steric stabilizing polymers, cryoprotectants) may concentrate.

4. CONCLUSIONS

The future of emulsion-based drug delivery lies in "fine tuning" the carrier to suit the requirements of both the incorporated drug and the intended therapeutic program. With a variety of nontoxic vegetable and marine oils available, this appears to be an achievable goal. For example, medium chain triglycerides from coconut oil and ethyl esters of fatty acids derived from certain vegetable oils may have potential applications in drug solubilization and emulsification. Recent reports suggest that dietary alterations in ratios of certain long chain polyunsaturated fatty acids from marine oils may suppress tumor growth in vivo and/or increase tumor sensitivity to anti-neoplastic agents (92). Future fish oil emulsions may well serve as adjuvants for cancer therapy.

An important problem with emulsions is shelf life. Although many lipid emulsions are stable for up to 2 years stored at 5°C, a more economical preparation would be stored in the dry form and reconstituted just prior to use. The early stages of research into freeze-drying and reconstitution of emulsions have begun with the investigation of various cryo-protective agents. Bachynsky et al. (93) have studied incorporation of lipophilic drugs into a liquid emulsion, which can be delivered orally. This self-emulsifying formulation was encapsulated and found to improve absorption of the lipophilic compound.

There are several problems associated with emulsion-based drug delivery systems, which must be avoided. Foremost among them is the failure to adequately characterize experimental formulations with respect to emulsion integrity including droplet size distribution, free oil, osmolality, and ζ-potential. Emulsion technology may be fine-tuned to meet the unique requirements of each drug. Indeed, this will be essential if new carrier systems are to achieve their true potential: clinically significant improvements in drug efficacy and reduction in toxicity.

REFERENCES

1. Shils MC. Guidelines for total parenteral nutrition. JAMA 1972; 220:1721.

2. Heird WC, Winters RW. Total parenteral nutrition: the state of the art. J Pediatr 1975; 86:2.

3. Van Way C, Meng H, Sandstead H. An assessment of the role of parenteral alimentation in the management of surgical patients. Ann Surg 1973; 177:103.

4. Jeppsson R. Effects of barbituric acids using and emulsion form intravenously. Acta Pharm Suec 1972; 9:81.

5. Jeppsson R, Ljungberg S. Intraaterial administration of emulsion formulations containing cyclandelate and nitroglycerin. Acta Pharm Suec 1973; 10:129.

6. Henriques V, Andersen AC. Uver parenterale Ernahrung durch intravenose Injektion, Hoppe – Seylers. Z Physiol Chem 1915; 88:357.

7. McKibbin JM, Pope A, Thayer S, Ferry RM Jr, Stare J. Parenteral nutrition: studies on fat emulsions for intravenous alimentation. J Lab Clin Med 1945; 30:488.

8. Halberg D, Holm I, Obel A. Fat emulsions for complete intravenous nutrition. Postgrad Med J 1957; 43:307.

9. Lehr HL, Rosenthal O, Rhoads JE, Sen MB. Clinical experience with intravenous fat emulsions. Metabolism 1964; 6:666.

10. Hansen LM, Hardie WR, Hidalgo J. Fat emulsion for intravenous administration: clinical experience with intralipid 10%. Ann Surg 1976; 184:80.

11. Shulman RJ, Phillips S. Parenteral nutrition in infants and children. J Pediatr Gastronenterol Nutr 2003; 36(5):587.

12. Thompson SW. The Pathology of Parenteral Nutrition with Lipids. Springfield, IL: Charles C Thomas Publishers, 1974:1–55.

13. Hogskilde S, Nielsen JW, Carl P, Sorensen MB. Pregnanolone emulsion. Anaesthesia 1987; 42:586.

14. Davis SS, Hadgraft J, Palin K. Medical and pharmaceutical applications of emulsions. In: Beecher P, ed. Encyclopedia of Emulsion Technology. Vol. 2. New York: Marcel Dekker, 1985:34.

15. Singh M, Ravin L. Parenteral emulsions as drug carrier systems. J Parenter Sci Technol 1986; 40:34.

16. O'Hagan DT. Recent developments in vaccine delivery systems. Curr Drug Targets Infect Disord 2001; 1:273.

17. Grant DJW, Higuchi T. Solubility Behavior of Organic Compounds. New York: John Wiley & Sons, 1990.

18. Berge SM, Bighley LD, Monkhouse DC. Pharmaceutical salts. J Pharm Sci 1977; 66:1.

19. Tarr BD, Yalkowsky SH. A new parenteral vehicle for the administration of some poorly water soluble anti-cancer drugs. J Parenter Sci Technol 1987; 41:31.

20. Tabibi SE, Rhodes CT. Disperse Systems, Modern Pharmaceutics. New york: Marcel Dekker, 1966:299.

21. Trissel LA. Handbook of Injectable Drugs. Washington, DC: American Society of Hospital Pharmacist, Inc., 1994.

22. Fahelelbom KMS, Timoney RF, Corrigan OI. Micellar solubilization of clofazimine analogues in aqueous solutions of ionic and nonionic surfactants. Pharm Res 1993; 10:631.

23. Florence AT. Drug Solubilization in Surfactant Systems, Techniques of Solubilization of Drugs. New york: Marcel Dekker, 1981:15.

24. Attwood D, Florence AT. Surfactant System, Their Chemistry, Pharmacy and Biology. London, UK: Chapman and Hall, 1983.

25. Danish M, Kottke MK. Pediatric and Geriatric Aspects of Pharmaceutics, Modern Pharmaceutics. New york: Marcel Dekker, 1996:804.

26. Prankerd RJ, Frank SG, Stella VJ. Preliminary development and evaluation of a parenteral emulsion formulation of penclomedine (NCS-338720; 3,5-dichloro-2,4-dimethoxy-6 -trichlormehtylpyridine): a novel, practically water insoluble cytotoxic agent. J Parenter Sci Technol 1988; 42:76.

27. Kreilgaard M. Influence of microemulsion on cutaneous drug delivery. Adv Drug Deliv Rev 2002; 54(S):S77.

28. Griffin EL, Richardson CH, Burdette RC. Relation of size of oil drops to toxicity of petroleum oil emulsions to aphids. J Agric Res 1927; 34:727.

29. Swarbrick J, Zografi G, Schott H. In: Hoover JE, ed. Coarse Dispersions, Remington's Pharmaceutical Sciences. 18th ed. Easton Pennsylavania: Mack Publishing Co., 1990:298.

30. Martin A, Swarbrick J, Cammarata A. Physical Pharmacy. 4 Philadelphia: Lea & Febiger, 1993:371, 487, 490.

31. Woodman RM. Problems connected with the preparation and application of emulsions used in agricultural spraying. Atkin WREmulsion Technology. 2 New York: Chemical Publishing Co., 1946:162.

32. Banker GS, Rhodes T. Modern Pharmaceutics. Vol. 7. New York: Marcel Dekker, Inc., 1979:329.

33. Grimm WL, Min TI, El-Asser MS, Vanderhoff JW. The role of low concentrations of ionic emulsifier-fatty alcohol mixtures in the emulsification of styrene. J Colloid Interface Sci 1983; 94:531.

34. Brouwer WM, El-Asser MS, Vanderhoff JW. Experiments on emulsions. Colloids Surf 1986; 21:69.

35. Prince LM. Microemulsions: Theory and Practice. New York: Academic Press, 1977.

36. Gillberg G. Emulsion and Emulsion Technology. Part 3. Surfactant Science Series 6. New York: Dekker, 1984:1.

37. Pandolfe W. In: Lissant KJ, ed. Homogenization and Emulsification. New York: Gaulin Corporation, Technical Bulletin No. 67, New York, NY 1982.

38. Van den Tempel M. Coagulation of oil-in-water emulsions. Proceedings of the 2nd International Congress on Surface Activation. London, UK 1957:439–446.

39. Vold RD, Groot RC. An ultracentifugal method for the quantitative determination of emulsion stability. J Phys Chem 1962; 66:1969.

40. Wani M, Taylor H, Wall M, Coggon P, McPhail A. Pant antitumor agents. VI The isolation and structure of taxol, a novel antileukemic and antitumor agent from *Taxus brevifolia*. J Am Chem Soc 1971; 93:2325.

41. King AT, Mukherjee LN. The stability of emulsions: part I. Soap-stabilized emulsions. J Soc Chem Ind 1939; 58:243.

42. Finkle P, Draper DH, Hildebrand JH. The theory of emulsification. J Am Chem Soc 1923; 45:2780.

43. Orr C. Determination of particle size. In: Becher P, ed. Encyclopedia of Emulsion Technology. Vol. 3. Basic Theory, Measurement and Applications, Marcel Dekker, NY. 1983:137.

44. Jiao J, Rhodes DG, Burgess DJ. Multiple emulsion stability: pressure balance and interfacial film strength. J Coll Interf Sci 2002; 250:444–450.

45. Merill RC Jr. Determining the mechanical stability of emulsions. Ind Eng Chem Anal Ed 1943; 15:743.

46. Garti N, Magdassi S, Rubenstein A. A novel method for rapid nondestructive determination of o/w creams stability. Drug Dev Ind Pharm 1982; 8:475.

47. Burgess DJ, Sahin ON. Influence of protein emulsifier interfacial properties on oil-in-water emulsion stability. Pharm Dev Technol 1998; 3:21.

48. Opawale FO, Burgess DJ. Influence of interfacial rheological properties of mixed emulsifier films on W/O/W emulsion stability. J Pharm Pharmacol 1998; 50:965–973.

49. Opawale FO, Burgess DJ. Influence of interfacial properties of lipophilic surfactants on water-in-oil emulsion stability. J Coll Interf Sci 1998; 197:142–150.

50. Burgess DJ, Sahin NO. Influence of protein emulsifier interfacial properties on oil-in-water emulsion stability. Pharm Develop Technol 1998; 3(1):1–9.

51. Burgess DJ, Sahin NO. Interfacial rheological and tension properties of protein films. J Coll Interf Sci 1997; 189(1):74–82.

52. Yoon JK, Burgess DJ. Interfacial properties as stability predictors of lecithin stabilized perfluorocarbon emulsions. Pharm Develop Technol 1996; 1(4):325–333.

53. Burgess DJ, Yoon JK. The influence of interfacial properties on perfluorocarbon/aqueous emulsion stability. Colloid Surf B 1995; 4(5):297–308.

54. Burgess DJ, Sahin NO. Interfacial rheology of–casein solutions. In: Herb CA, Prud'homme RK, eds. Structure and Flow in Surfactant Solutions. Chapter 27. ACS Symposium Series, 1994:380–393.

55. Burgess DJ, Yoon JK. Interfacial tension studies on surfactant systems at the perfluorocarbon/aqueous interface. Colloid Surf B 1993; 1:283–293.

56. Yoon JK, Burgess DJ. Comparison of dynamic and static interfacial tension at aqueous/perfluorocarbon interfaces. J Coll Interf Sci 1992; 151:402–409.

57. Washington C. Evaluation of non-sink dialysis methods for the measurement of drug release from colloids: effects of drug partition. Int J Pharm 1989; 56:71.

58. Armoury N, Fessi H, Devissauget JP, Puisieux F, Benita S. In vitro release kinetic pattern of indomethacin from poly (d,l-lactic) nanocapsules. J Pharm Sci 1990; 79:763.

59. Friedman D, Benita S. A mathematical model for drug release from o/w emulsions: application to controlled release morphine emulsions. Drug Dev Ind Pharm 1987; 13:2067.

60. Hashida M, Liao MH, Muranishi S, Sezaki H. Dosage form characteristics of microsphere-in-oil emulsion II: Examination of some factors affecting lymphotropicity. Chem Pharm Bull 1980; 28:1659.

61. Miyazaki S, Hashiguchi N, Hou WM, Yokouchi C, Takada M. Preparation and evaluation in vitro and in vivo fibrinogen microspheres containing adriamycin. Chem Pharm Bull 1986; 34:3384.

62. Sasaki H, Takakura Y, Hashida M, Kiamura T. Anti-tumor activity of lipophilic prodrugs of mitomycin C entrapped in liposome or o/w emulsion. J Pharm Dyn 1984; 7:120.

63. Gupta PK, Hung CT, Perrier DG. Quantitation of the release of doxorubicin from colloidal dosage forms using dynamic dialysis. J Pharm Sci 1987; 76:141.

64. Armoury N, Fessi H, Devissauget JP, Puisieux F, Benita S. Physicochemical characterization of polymeric nanocapsules and in vitro release evaluation of indomethacin as a drug model. Sci Techn Pract Pharm 1989; 5:647.

65. Desai MP, Labhasetwar V, Walter E, Levy RJ, Amidon GL. The mechanism of uptake of biodegradable microspheres in caco–2 cells is size dependent. Pharm Res 1997; 14:1568.

66. Burgess DJ, Davis SS, Tomlinson E. Potential use of albumin microspheres as a drug delivery systems. I Preparation and in vitro release of steroids. Int J Pharm 1987; 39:29.

67. Koosha F, Muller RH, Davis SS. A continuous flow system for in vitro evaluation of drug-loaded biodegradable colloidal barriers. J Pharm Pharmacol 1988; 40:131P.

68. Ghanem A, Higuchi WI, Simonelli AP. Interfacial barriers in interphase transport: retardation of the transport of diethylphthalate across the hexadecane–water interface by an adsorbed gelatin film. J Pharm Sci 1969; 58:165.

69. Yoon KA, Burgess DJ. Effect of cationic surfactant on transport of model drugs in emulsion system. J Pharm Pharmacol 1997; 49:478.

70. Chidambaram N, Burgess DJ. A novel in vitro release method for submicron-sized dispersed systems. AAPS PharmSci 1999; article II:1.

71. Flynn GL, Yalkowsky SH, Roseman TJ. Mass transport phenomena and models: theoretical concepts. J Pharm Sci 1974; 63:479.

72. Baker RW. Controlled Release of Biologically Active Agents. New York: John Wiley & Sons, 1989.

73. Kydonieus AF. Controlled Release Technologies: Methods, Theory and Application. Vol.1. Boca Raton, FL: CRC Press, 1980:1.

74. Bird RB, Stewart WE, Lightfoot EN. Transport Phenomena. New York: John Wiley & Sons, 1960:514.

75. Tanford C. Physical Chemistry of Macromolecules. New York: Wiley, 1961.

76. Flynn GL, Yalkowsky SH. Correlation and prediction of mass transport across membranes I: influence of alkyl chain length on flux-determining properties of barrier and diffusant. J Pharm Sci 1972; 61:838.

77. Madan PL. Sustained release drug delivery systems: Part V. Parenteral products. New York, NY: Pharmaceut Manufacturing, 1985:51.

78. Lostritto RT, Goei L, Silvestri SL. Theoretical considerations of drug release from submicron oil in water emulsions. J Par Sci Tech 1987; 41:214.

79. Chidambaram N, Burgess DJ. Mathematical modeling of surface active and non surface active drug transport in emulsion systems. AAPS PharmSci 2000; article 31:1.

80. Short PM, Abbs ET, Rhodes CT. Effect of nonionic surfactants on the transport of testosterone across a cellulose acetate membrane. J Pharm Sci 1970; 59:995.

81. Amidon GE, Higuchi WI, Ho H. Theoretical and experimental studies of transport of micelle-solubilized solutes. J Pharm Sci 1982; 71:77.

82. Nakano M, Patel NK. Effect of molecular interaction on permeation of organic molecules through dimethylpolysiloxane membrane. J Pharm Sci 1970; 59:77.

83. Bates TR, Galowina J, Johns WH. Effect of complexation with caffeine on the in vitro transport of drug molecules through an artificial membrane absorption model. Chem Pharm Bull 1970; 18:656.

84. Bloor JR, Morrison JC. Effects of solubilization on drug diffusion. J Pharm Pharmacol 1972; 24:927.

85. Sumner CG. On the formation, size, stability of emulsion particles. J Phys Chem 1933; 37:279.

86. Briggs TR. Experiments on emulsion shaking. J Phys Chem 1920; 24:120.

87. Gopal ESR. Rheology of emulsions. In: Sherman P, ed. Emulsion Science. Oxford: Pergamon Press, 1963:15.

88. Rushton JH. Principles of mixing in the fatty oil industries. J Am Oil Chem Soc 1956; 33:598.

89. Korstvedt H, Bates R, King J, Siciliano A. Microfluidization. Drug Cosmet Ind 1984; 135:36.

90. Silvestri SL, Lostritto RT. Theoretical evaluation of dispersed droplet radii in submicron oil-in-water emulsions. Int J Pharm 1989; 50:141.

91. Silvestri S, Ganguly N, Tabibi E. Predicting the effect of nonionic surfactants on dispersed droplet radii in submicron oil-in-water emulsions. Pharm Res 1992; 9:1347.

92. Borgeson CE, Pardini L, Pardini RS, Reitz RC. Effects of dietary fish oil on human mammary carcinoma and lipid-metabolizing enzymes. Lipids 1989; 24:290.

93. Bachynsky MO, Shah WH, Infeld MH, Margolis RJ, Malick AW, Palmer DH. Oral delivery of lipophilic drug in self-emulsifying liquid formulation. Pharm Res 1989; 6:5.

8

Liposomes: Design and Manufacturing

SIDDHESH D. PATIL

Antigenics Inc.,
Lexington, Massachusetts, U.S.A.

DIANE J. BURGESS

Department of Pharmaceutical
Sciences, School of Pharmacy,
University of Connecticut, Storrs,
Connecticut, U.S.A.

1. INTRODUCTION

Since their discovery in the mid-1960s, liposomes as well as formulations based on drug–lipid complexes have been extensively investigated for the delivery of a wide range of pharmaceutical compounds. Commercial licensing of liposomal dispersions for parenteral and topical use in humans has been fairly recent and has occurred as a result of interdisciplinary scientific advances in surface chemistry, membrane biophysics, and molecular pharmacology (1–4). These developments were supported by pioneering contributions from scientists in the fields of molecular pharmaceutics, drug delivery, and chemical engineering

(1–4). Thus, since their original simplistic use as vesicular models for studying membrane properties and transport, liposomal systems have evolved into complex yet elegant assemblies for drug delivery. Liposomes are now being developed with specific therapeutic roles for tissue and cellular targeting and have been skillfully manipulated to achieve desired clinical outcomes by altering in vivo drug biodistribution and disposition (4). There are currently more than 10 approved liposome products on the market world-wide and several more have progressed into advanced clinical trials (2).

In this chapter, some of the most important recent technological developments in liposome research are discussed. The chapter is divided into three major sections: (i) design of liposomes, their advantages and applications; (ii) methods of liposome manufacture and drug encapsulation techniques; and (iii) characterization and compendial requirements for liposomal dispersions. The first section involves an overview on liposome design, classification of vesicular systems, choice of lipids, pharmacotherapeutic applications, and marketed products. Some of the most commonly used methods of manufacturing, drug entrapment protocols, their advantages and disadvantages are evaluated in the second section. Finally, some of the routine tests for product evaluation, regulatory requirements for drug release, and sterility considerations for liposomal dispersion formulations are reviewed in the last section.

2. LIPOSOMES: DEFINITIONS AND CLASSES

Liposomes can be defined as phospholipid vesicles that are spontaneously formed by dispersion of lipid films in an aqueous environment. This phenomenon was first observed by Bangham et al. (5) during their investigations into the diffusion of ions across models of lipid bilayer membranes. Consistent with the original observations of Bangham, typically, dry lipid molecules upon hydration in an aqueous environment undergo swelling and self-assemble in a definitive and consistent orientation, known as the lipid bilayer (5–7). The dynamics and biophysical principles of bilayer formation are complex and depend on the overall lipid shape as well as the combination of lipids used (8,9). These characteristics determine the curvature and packing geometry in the resulting bilayer which in turn affects its interfacial behavior in aqueous medium (8,9).

In a conventional lipid-bilayer assembly, the hydrophobic acyl chains of a lipid molecule associate and interact with those of the neighboring molecules, and the polar head-groups orient themselves to the exterior of the assembly. This bilayer can be easily perturbed under mechanical stress to produce phospholipid vesicles which entrap an aqueous core surrounded by the lipid bilayer. If multiple lipid bilayers are arranged in a concentric (onion skin) fashion thereby generating multiple compartments, the resulting vesicles are termed multilamellar vesicles (MLV) (6). On the contrary, if a single bilayer surrounds an aqueous core space enclosing a single compartment, then the vesicles are termed unilamellar vesicles. Liposomes can be subcategorized into different types depending on their size and lamella (6,7). These different categories include: (i) small unilamellar vesicles (SUV), 25–100 nm in size that consist of a single lipid bilayer; (ii) large unilamellar vesicles (LUV), 100–400 nm in size that consist of a single lipid bilayer; and (iii) MLV, 200 nm to several microns in size, that consist of two or more concentric bilayers. Vesicles above 1 μm are also known as giant vesicles. If several vesicles are entrapped in a non-concentric fashion inside a single vesicle then the resulting assemblies are known as multivesicular vesicles (6). Depending on the number of lamellae inside them, vesicles may also be termed oligolamellar vesicles (OLV) (6).

3. VERSATILITY OF DRUGS DELIVERED USING LIPOSOMES

Since their discovery in the mid-1960s, liposomes have been extensively studied as parenteral delivery systems for a wide range of pharmaceutical therapeutic drug candidates that include traditional low molecular weight compounds, biotechnology-derived proteins and peptides as well as

DNA-based therapeutics. Liposomal delivery for low molecular weight drugs is one of the earliest and consequently one of the most clinically advanced amongst all other therapeutics.

3.1. Low Molecular Weight Drugs

Low molecular weight pharmaceutical compounds that have been investigated for parenteral delivery in liposome systems include antifungals (10,11), antibiotics (12,13), and anti cancer drugs (14). Liposomal delivery of these drugs can be used to alter drug pharmacokinetics, to assure adequate levels in tissues of interest and to reduce/avoid toxic side effects (for details, refer Sec. 4). Some of the approved low molecular weight liquid parenteral liposomal formulations include: AmBisome®, which contains amphotericin B and is indicated for systemic fungal infections (11) and visceral leishmaniasis; Amphotec® (branded as Amphocil® outside United States) and Abelcet® which also contain amphotericin B (15–17) and are indicated for systemic fungal infections (18); DaunoXome®, which contains duanorubicin and is indicated for first-line treatment of advanced Kaposi's sarcoma; DepoDur®, which contains morphine and is indicated for pain management following major surgery; DepoCyt®, which contains cytarabine and is indicated for lymphomatous meningitis; Doxil® (branded as Caelyx® outside United States), and Myocet® (formerly known as Evacet®), both of which contain doxorubicin and, whereas, Doxil is indicated for metastatic ovarian cancer and Kaposi's sarcoma, Myocet is being used for treating breast cancer; MiKasome®, which contains the aminoglycoside antibiotic amikacin and is indicated for severe infections; and Visudyne®, which contains the photosensitizer dye verteporfin and is indicated for age-related macular degeneration (19). The reader is referred to the case studies in this book on the development of Doxil by Martin and on the development of AmBisome by Adler-Moore. Currently, several liposome formulations containing low molecular weight drugs are in clinical trials for a range of pharmacologic effects (such as antibacterial, anticancer, antifungal, and anti-HIV activities) (20,21). See Table 1 for additional information on

Table 1 Marketed Liposomal Products

Marketed formulation—Company name	Pharmaceutical drug	Liposome composition	Indication
AmBisome—Gilead-Fujisawa Healthcare	Amphotericin B	HSPC, cholesterol, and DSPG	Systemic fungal infections and visceral leishmaniasis
Amphotec/Amphocil[a,b]—InterMune	Amphotericin B	Cholesteryl sulfate[a]	Aspergillosis
Abelcet—Elan	Amphotericin B	DMPC and DMPG	Fungal infections
DaunoXome—NeXstar Pharmaceuticals	Daunorubicin citrate	Cholesterol and DSPC	Kaposi's sarcoma
Doxil/Caelyx[b,c]—Ortho Biotech	Doxorubicin hydrochloride	MPEG-DSPE, HSPC, and Cholesterol	Metastatic ovarian cancer and Kaposi's sarcoma
DepoDur—SkyePharma—Endo	Morphine sulfate	Cholesterol, DOPC, and DPPG	Pain following major surgery
DepoCyt—SkyePharma—Enzon	Cytarabine	Cholesterol, DOPC, and DPPG	Lymphomatous meningitis
Myocet (formerly known and Evacet)—Elan	Doxorubicin hydrochloride	Egg PC and cholesterol	Breast cancer
MiKasome—NeXstar	Amikacin	HSPC, cholesterol, and DSPG	Bacterial infections

(Continued)

Table 1 Marketed Liposomal Products (*Continued*)

Marketed formulation—Company name	Pharmaceutical drug	Liposome composition	Indication
Visudyne—Novartis	Verteporfin	Egg PG and DMPC	Age-related macular degeneration
Epaxal[d]—Berna Biotech	Hepatitis A antigen	Hemagglutinin, neuraminidase, and lecithin	Hepatitis A
Inflexal V[d]—Berna Biotech	Hemagglutinin	Lecithin	Influenza
Pevaryl Lipogel[e]—Cilag	Econazole	Lecithin	Dermatomycosis and gynecological mycosis
L.M.X.4[e] (formerly known as ELA-Max)—Ferndale Laboratories	Lidocaine	Cholesterol and hydrogenated lecithin	Topical anesthetic

MPEG-DSPE, *N*-(carbonyl-methoxypolyethylene glycol 2000)-1,2-distearoyl-sn-glycero-3-phosphoethanolamine sodium salt; HSPC, hydrogenated soy phosphatidylcholine; DSPG, distearoyl phosphatidylglycerol; DMPC, dimyristoyl phosphatidylcholine; DMPG, dimyristoyl phosphatidylglycerol; DOPC, dioleoyl phosphatidylcholine; DPPG, dipalmitoyl phosphatidylglycerol; DSPC, distearoyl phosphatidylcholine; Egg PC, egg phosphatidylcholine; Egg PG, egg phosphatidylglycerol.
[a]Amphotec/Amphocil is a amphotericin B cholesteryl sulfate complex for injection.
[b]Brand name of the product marketed outside United States.
[c]Sterically stabilized (Stealth) liposomes.
[d]Immunopotentiating reconstituted influenza virosomes.
[e]Formulation for transdermal application.

approved liposomal products for the delivery of low molecular weight compounds.

3.2. Protein Therapeutics

Liposomes have been used as vaccine adjuvants for recombinant protein-based vaccines. Epaxal and Inflexal V are licensed liposomal protein vaccines that were globally approved in the mid-1990s for hepatitis A and influenza, respectively. Epaxal and Inflexal are immunopotentiating unilamellar vesicles known as virosomes since they contain phospholipids and proteins derived from the influenza virus (22,23). Virosome formulations are comprised of vaccine-specific antigens of interest embedded in the liposomal bilayer composed of the following two components: (i) natural and synthetic phospholipids (lecithin, cephalin, and envelope phospholipids from the influenza virus); and (ii) influenza surface glycoproteins: hemagglutinin and neuraminidase. For the Epaxal formulation (24), the antigen of interest is the hepatitis A virion, whereas, the Inflexal V (25) formulation contains a combination of surface antigens of three currently circulating strains of the influenza virus. See Table 1 for additional information on approved liposomal products for the delivery of proteins.

3.3. DNA-Based Therapeutics

In addition to the commercial licensing of a range of liposomal formulations, recent years have witnessed significant research efforts directed toward the development of liposomes as delivery vectors for DNA-based therapeutics (26–29). In vivo delivery of DNA-based therapeutics such as plasmids for gene expression, antisense oligonucleotides, siRNAs, ribozymes, etc., typically requires assistance of delivery vectors (29). Currently used gene delivery vectors are attenuated replication-defective viruses such as adenoviruses, adeno-associated viruses, polyomavirus, and retroviruses (26,30–33). Due to their natural mechanism of targeted introduction of DNA into cells, these vectors have high efficiency in DNA transfer (30). However, a number of potential problems are associated with viral vectors, including deletion of sequences during replication, recombination with endogenous sequences to produce infectious recombinant viruses, activation of cellular oncogenes, introduction of viral oncogenes, inactivation of host genes, development of toxic immune responses and inflammation that has been demonstrated to be lethal (26,34–36). Pharmaceutical formulations of viruses are expensive and difficult to manufacture, have a very low shelf life, and can lose potency upon storage (37,38). Thus, despite their superlative efficiency, many of these vectors are unsuitable for clinical use. As a consequence of the problems associated with viral gene delivery vectors, several alternative non-viral gene delivery systems are under development (29,39–41). Liposomal delivery vectors are emerging to be the most popular non-viral alternative to viral vectors for gene delivery (26,40,41). Liposomal formulations of DNA-based therapeutics protect these molecules from enzymatic inactivation in the plasma by degradative endo- and exo-nucleases, facilitate the entry of DNA into the cell cytoplasm, are relatively easy and inexpensive to manufacture, and are non-immunogenic (26). Both anionic and cationic liposomes have been utilized for gene delivery (26,42,43). DNA typically is entrapped inside anionic liposomes and is surface-complexed with cationic liposomes. These liposomal formulations have achieved tremendous success in introducing DNA-based gene therapeutics into a wide range of tissues and cells in animal as well as cell culture models. The first clinical trial for gene therapy for melanoma using cationic liposomes was conducted in 1992 by Nabel et al. (44,45). Currently, although several cationic liposomes are in advanced clinical trials for gene therapy of cystic fibrosis, and cancer, there is no commercially licensed formulation (45). In addition to their use in protein delivery, as described in detail in Sec. 3.2, virosomes have also been employed for gene transfer (46,47).

4. ADVANTAGES OF LIPOSOMAL DELIVERY SYSTEMS

Since liposomes can be constructed with a range of properties (particle size, lamellarity, drug loading, drug release characteristics, etc.) depending on the method of manufacture, the choice of lipids, and other excipients, etc., they can be easily manipulated to suit a wide range of drug delivery applications. Liposomal encapsulation of drugs offers several advantages over their conventional direct administration in vivo.

4.1. Liposomes for Solubilizing Drugs

Liposomes can serve as efficient solubilizing vehicles for drugs with poor solubility in pharmaceutically acceptable solvents. Such a strategy has been used to improve drug solubility for a wide range of compounds including alphaxalone (48), camptothecin (49), tacrolimus (50), econazole (51), and paclitaxel (52). Entrapment of drugs can also be used to reduce in vivo degradation and thus enhance the biological half-life of a drug; a feature that has been extremely useful for preserving and prolonging the pharmacological activity of nucleic acid therapeutics such as antisense oligonucleotides and plasmid DNA (26,27).

4.2. Liposomes for Tissue Targeting

Liposomes can be used for passive or active tissue targeting by virtue of their size or by the incorporation of immunorecognition motifs in the liposomal bilayer, respectively (6,7,53). Upon intravenous (i.v.) administration, liposomes larger than 200 nm are quickly cleared from the systemic circulation due to rapid uptake by phagocytic macrophages and other components of the reticuloendothelial system (53). The rate at which conventional liposomes are cleared from the circulation is dependent on their particle size, charge, and fluidity (54). Negative charge, larger size, and high fluidity all increase clearance (54,55). The adsorption of plasma proteins on the liposome surface, also known as opsonization, promotes reticuloendothelial uptake of liposomes (55,56). The negative charge of conventional liposomes facilitates opsonization (54). Due to this characteristic preferential uptake, liposomes larger than 200 nm can be used to target and deliver drugs into macrophages for diseases that involve these cells (56,57).

Liposomes smaller than 100 nm can escape phagocytic macrophage uptake and thus have higher circulation times in vivo compared to those of larger liposomes (58,59). Small liposomes can extravasate into tissues if the space in between cells surrounding the vasculature is significantly large. Under physiological conditions, extravasation is restricted to the liver and spleen tissue; however, the vasculature of tumors and tissues under inflammatory conditions is uneven and the intercellular spaces are abnormally wide (53,60,61). Very small liposomes can easily exit "leaky" vasculatures and are selectively accumulated in such tissues. This phenomenon is known as the enhanced permeability and retention (EPR) effect (53,60,61). The EPR effect of small liposomes having long circulation times in the systemic circulation has been used to achieve targeted delivery of anticancer drugs into tumors (53,62), and antifungal (63) as well as radiocontrast agents (64) into pathological sites in the liver and the spleen. Since drugs are released in desired tissues, their therapeutic potential can be significantly enhanced and side effects dramatically reduced (7). AmBisome, a liposomal antifungal product, has been shown to avoid uptake by the mononuclear phagocytic cells and thus a prolonged circulation time as a consequence of it small particle size (80 nm) (15).

Active targeting using liposomes has been accomplished by attaching target-specific moieties onto the surface of liposomes. These ligands, such as antibodies (65), immunoglobulins (66), lectins (67), transferrin (68), sterylglucoside (69), folates (70), peptides (71), and polysaccharides (mannan) (72), have specific recognition receptors in tissues that facilitate selectively internalization into target cells.

4.3. Liposomes for Immunopotentiation

Liposomes have been used as potent adjuvants to augment the immune response to recombinant protein vaccines (73,74). Immunopotentiating reconstituted influenza virosomes, used in commercially marketed vaccine formulations such as Inflexal and Epaxal, contain structural determinants that are responsible for enhanced immunogenic potential. Virosomes, due to their characteristic membrane bilayer structure, can mimic the natural mechanisms of antibody presentation and processing. Additionally, surface glycoproteins hemagglutinin and neuraminidase in virosomes promote cellular fusion with antigen-presenting cells of the immune system. Consequently, a robust T-cell and B-cell response is generated when antigens are delivered via virosomes (75,76). Liposomal adjuvants are biodegradable, have low toxicity, and do not stimulate the production of antiphospholipid antibodies (74). These adjuvants are well tolerated and safe for repeated use compared to traditionally used aluminum salt-based adjuvants in protein formulations (23,76).

4.4. Liposomes for Modified Release

Liposomal vesicles can be used to develop controlled release formulations due to delayed release of drug molecules complexed and/or entrapped within liposome compartments. Controlled release liposomal formulations have been developed for many drugs such as progesterone (77) and cisplatin (78) among others. The commercially licensed formulations DepoDur and DepoCyt are based on the DepoFoam technology (79) that involves encapsulating drugs into multivesicular liposomes composed of mixtures of cholesterol, triolein, 1,2-dipalmitoyl-*sn*-glycero-3-phosphocholine (DOPC), and 1,2-dipalmitoyl-*sn*-glycero-3-[phospho-*rac*-1-glycerol] (DPPG). These lipid bilayer membranes can serve as an efficient barrier to the permeation of entrapped drugs and can be programmed to release the drug over extended periods of time.

4.5. Transdermal Drug Delivery Using Liposomes

Although the focus of this book is on i.v. delivery, transdermal applications of liposomal formulations merit discussion due to the significant advances that have occurred recently in this rapidly developing field (80–82). Liposomal encapsulation for transdermal use is intended for localized delivery and has the following two major advantages: (i) ability to circumvent systemic administration and thus increase local activity and prevent toxic side effects, a feature that is particularly important for potent glucocorticosteroids and retinoids (81), and (ii) targeted site-specific delivery that is especially useful for selective treatment of tissues, a feature practical in pharmacotherapy of psoriasis, acne, and genital warts, as well as stimulation of hair growth. Liposomes can be a non-toxic substitute for dermal penetration enhancers such as dimethylsulfoxide to improve localized drug transport into skin (83). They are believed to release entrapped drugs upon interaction with cells, by fusion and/or endocytosis (80,83). Some of the liposomal formulations commercially available for transdermal applications include Pevaryl Lipogel® (51), which contains econazole and is indicated for dermatomycosis and gynecological fungal infections; and L.M.X.4® (formerly known as ELA-Max®) (84), which contains lidocaine and is indicated for local anesthesia and itch relief. See Table 1 for additional information on approved dermatologic liposomal products.

5. LIPOSOME COMPOSITION: CHOICE OF LIPIDS

The choice of lipids for drug encapsulation into liposomes is dependent on the drug characteristics and intended applications. Liposomal composition determines the properties (including surface charge, rigidity, and steric interactions) and the in vitro and in vivo performance of liposomes. The specific properties of the liposomes are determined by the chemistry of the head and tail groups of the constituent lipids. Selection is often decided on a case-by-case basis, since product performance can be drastically affected by slight changes in the liposomal vesicle composition.

5.1. Conventional Liposomes

Liposomes were traditionally prepared from a variety of neutral and anionic lipids. Examples of some of these lipids include lecithins (85), sphingomyelins (86), phosphatidylcholines (87) and phosphatidylethanolamines (88) (neutral) and phosphatidylserines (89), phosphatidylglycerols (90) and phosphatidylinositols (91) (anionic). These liposomes have non-specific interactions with their environment and are often referred to as conventional or unmodified liposomes. These conventional liposomes are recognized by the mononuclear phagocytic cells and are removed from the circulation within a few minutes to several hours (59). They are subsequently taken up by the liver and the spleen and therefore are very effective in targeting therapeutic agents to treat diseases of these organs (59). The addition of acidic phospholipids to mixtures of zwitterionic phospholipids imparts a negative charge that helps to reduce liposomal aggregation and improve their stability (7).

Specific liposome characteristics can be altered by incorporation of various liposomal components that have different properties (7). An example of an important liposomal property that is affected by lipid composition is the phase transition temperature (T_m) (7,92,93). This is the temperature of a liposome at which the membrane changes from ordered solid to disordered fluid and is dependent on the length and degree of saturation of the hydrocarbon chains (93,94). T_m may be dependent on the acyl chain length (94), composition of the lipid bilayer (95), and the entrapped drug (96). For example, for lipids composed of phosphatidylcholine polar head group, the T_m can vary from $-15°C$ for dioleoyl chains to $55°C$ for distearoyl chains (94). The fluid state of the membrane is relatively unstable, more elastic, can form transient hydrophilic channels and is permeable to the transport of materials. Consequently membranes above their T_m tend to be "leaky" to entrapped drug substances (97). Conventional liposomes particularly those of high fluidity may disrupt on contact with the plasma, as a result of interactions with plasma components (7,97). The addition of cholesterol causes an ordering of the disordered fluid phase and therefore increasing amounts of cholesterol eventually lead to an elimination of the phase transition (95). Consequently, liposomes containing cholesterol are more cohesive and have high stability against proteins in vivo and against leakage of encapsulated materials (98). Addition of cholesterol to conventional liposome formulations increases their stability in the plasma (7,98).

5.2. Sterically Stabilized Liposomes

Sterically stabilized liposomes constitute an important class of second generation liposomal vesicles that are engineered to have extended circulation times in vivo compared to conventional liposomes (99–101). Due to their ability to circumvent immune surveillance and recognition by the body as foreign and hence avoid opsonization and phagocytic uptake, sterically stabilized liposomes are also popularly known as stealth liposomes (99,101,102). Stealth liposomes are composed of lipids that have covalently linked polymers with hydrophilic head groups such as poly(ethylene glycol) (PEG) on their surface (100). The process of PEG conjugation to conventional lipids is known as pegylation and the lipids with covalently attached PEG can be included in the formulation at a desired ratio. Pegylation prevents the opsonization of proteins on the surface of stealth liposomal vesicles and prevents phagocytic uptake by the reticuloendothelial system, thus leading to their long circulating times in the systemic circulation (101,103). Commercially approved Doxil is a pegylated liposomal product, with surface grafted segments of the hydrophilic polymer methoxypolyethylene glycol (MPEG) (104). The MPEG segments extend from the surface of the liposomes, reducing interactions between the lipid bilayer membrane and plasma components (103). The reader is referred to the case study on Doxil by Martin et al. in this book for more information. In addition to surface grafting with hydrophilic polymers, such other molecules as ganglioside and phosphatidylinositol have also been used to have a stealth effect (7,100). Addition of specific immunorecognition motifs such as integrin antibody segments on stealth liposomes can couple the advantages of cell targeting and improved circulation times, respectively (105).

5.3. Liposomes for Gene Delivery

Liposomes for gene delivery are typically composed of combinations of cationic and zwitterionic lipids (42,106–108). Cationic lipids commonly used are 1,2-dioleoyl-3-trimethylammonium propane (DOTAP), 2,3-dioleoyloxy-*N*-[2-(sperminecarboxamido) ethyl]-*NN*-dimethyl-1-propanaminium (DOSPA), 3,[*N*-(N^1N-dimethylethylenediamine)- carbamoyl]cholesterol (DC-chol), *N*-[1-[2,3-dioleyloxy]propyl]-*NNN*-trimethylammonium chloride (DOTMA), and dioctadecyl amido glycil spermine (DOGS) (109). Commonly used zwitterionic lipids, also known as helper lipids, are 1,2-dioleoyl-*sn*-glycero-3-phosphoethanolamine (DOPE) and cholesterol (109).

Cationic liposomes upon electrostatic attraction with the anionic DNA backbone form a cationic complex also known as a lipoplex that is capable of transferring DNA molecules into cells by a process known as transfection (110). The cationic lipids in the formulation facilitate DNA complexation and condensation in the lipoplex (111,112). The zwitterionic lipids help in membrane perturbation and fusion. The overall positive charge of the lipoplex facilitates cellular association and transfection (113). Excess cationic lipids also help to stabilize the liposomes in vivo and prevent release of DNA by anionic molecules in the serum.

Liposomal gene delivery vectors are believed to achieve transfection through the following sequence of events (114–116): (i) interaction with the cell membrane; (ii) receptor mediated endocytosis; (iii) release from the endosome into the cytoplasm, usually through destabilization and disruption of the endosome membrane; and (iv) uptake from the cytoplasm into the cell nucleus. X-ray diffraction studies have also indicated that cationic lipoplexes are successful in transfection because of the formation of the H_{II}^c (hexagonal) phase instead of the Lα (lamellar) phase that is typically observed in liposomal bilayers (114). Formation of the hexagonal phase is attributed to due to the small less hydrated inverted-cone shape of the DOPE molecule (114,117).

The first use of cationic liposomes for gene delivery was demonstrated by Felgner et al. (110) when they successfully introduced a plasmid DNA encoding the chloramphenicol acetyltransferase enzyme into mammalian cells using cationic liposomes composed of DOTMA. Since then numerous synthetic cationic lipids and their formulations have been successfully used for gene delivery in a wide range of cell culture and animal models (118–120). Some of the commercially available cationic liposomal formulations used for in vitro gene delivery include: LipofectAmine® (Invitrogen, Carlsbad, CA, USA); Effectene® (Qiagen, Valencia, CA, USA); and Tranfectam® (Promega, Madison, WI, USA) (119,121).

The human in vivo gene therapy trial for melanoma conducted by Nabel et al. (44,122). used cationic liposomes composed of cationic DC-chol and zwitterionic DOPE to transfer a gene encoding the foreign major histocompatibility complex protein, HLA-B7 into cancer nodules. The clinical trial demonstrated the feasibility of cationic liposomes for gene delivery in humans. Currently, cationic liposomes have progressed into clinical trials for several indications that include cystic fibrosis (123), metastatic head and neck carcinoma (124), breast cancer (124), and ovarian cancer (125). There is, however, no commercially approved cationic liposomal product on the market.

Despite this progress, cationic liposomes suffer from several undesirable issues that reduce their overall potential of DNA delivery. These include inactivation in the presence of serum, instability upon storage (109), and cytotoxic effects on cells, both in vitro (126,127) and in vivo (128–132). Cytotoxicity of cationic lipids has been demonstrated in a variety of cell types including phagocytic macrophages (127), pulmonary intratracheal tissue (128,129,132), and arterial cell walls (130). Toxicity is attributed to the production of reactive oxygen intermediates (128), induction of apoptosis (133), or stimulation of proinflammatory cytokines (129) in response to the administration of cationic lipids. It is evident that there is a need for efficient and well-tolerated delivery systems to exploit the benefits of gene medicine. As a non-toxic alternative to cationic lipids, anionic liposomal formulations for the delivery of DNA-based therapeutics have been recently developed (27,28,134,135). The endogenous negative charge of these naturally occurring lipids is thought to be responsible for their low toxicity (27,28).

5.4. Lipid Specifications

Lipids in liposomal formulations can be synthetic, semisynthetic, or derived from natural sources such as egg yolk or soybeans (136). For FDA approval of liposomal formulations, strict control of lipid excipients is mandated and specific information is required to be submitted prior to product approval (7,136). Natural lipids contain a mixture of lipid chains with different head groups, whereas synthetic lipids can be pure. For completely synthetic products, the source and process specifications should be supplied (7). Formulations comprised of mixtures of natural lipids or natural starting products for semi-synthetic lipids are required to specify individual lipid composition, degree of saturation, and relative percentages of fatty acids. Lipids if obtained from genetically modified plant and animal sources have also to be indicated. In addition, lipids in human formulations are also mandated to be free of contamination from animal proteins and viruses. Typically lipids and their impurities (synthetic by-products if applicable and/or degradants) can be identified using spectroscopic techniques, which can be used to determine the acceptance criteria for starting materials (2,136). Some of the quality-determining specifications are adapted from the egg yolk phospholipid monograph (7).

6. MANUFACTURE OF LIPOSOMES

As their clinical potential for diverse drug delivery applications has begun to be realized, the last few decades have witnessed the development of a large number of techniques for the manufacture of liposomal formulations (3,137–139). Early protocols were suitable for small laboratory scale liposome production; however, newer protocols are more sophisticated and amenable to expedited large-scale industrial manufacture and processing under cGMP conditions (137). The selection of a particular protocol is primarily dictated by the nature of the therapeutic in the liposomal formulation and should ensure preservation of its stability and biological activity during processing. Protocols that necessitate prolonged exposure to organic solvents or high temperature are unsuitable for protein therapeutics. In addition, the production method should maximize drug entrapment in liposomal vesicles. The following are some of the commonly used methods for the preparation of liposomes:

6.1. Liposomes Preparation from Lipid Films

Preparation of liposomes by hydration and agitation of lipid films is one of the oldest and most widely used laboratory scale methods. This method exploits the natural self-assembly process of bilayer membranes and involves the formation of MLV from dried lipid films upon their exposure to an aqueous medium (6,92). This is usually achieved by the dissolution of lipids (in the desired ratio) in an organic solvent such as chloroform followed by its complete evaporation which leads to the deposition of a thin lipid film (7,140). Evaporation of the organic solvent can be assisted using a steady stream of nitrogen gas over the lipid surface (7). Use of inert nitrogen prevents oxidation of lipids and prevents chemical instability (for details refer Sec. 8.3). The resulting dried film of lipids is then dispersed in a solution of the material to be encapsulated. As the lipids hydrate, they assemble and form a suspension of MLV. Mechanical agitation and sonication during hydration can assist the formation of MLV from lipid films (7). However, sonication produces unstable SUV that are susceptible to physical degradation-related fusion (3). Drug entrapment in liposomal MLV is dependent on the volume of their enclosed aqueous compartments. The trapped aqueous volume of MLV is very small ($<1\ \mu L/\mu mol$ lipid), thereby reducing the entrapping efficiency (140). Liposome preparation from lipid films yields large polydisperse vesicles and typically MLV are further treated to achieve vesicles with desired and consistent properties (6,92).

6.2. Liposome Preparation by Freeze–Thaw Cycling of MLV

To improve the drug entrapment efficiencies, frozen and thawed multilamellar vesicles (FATMLV) were developed by Mayer et al. (141,142). FATMLV are generated from MLV by repeated freeze–thaw cycling of MLV. This procedure involves rapid freezing of MLV suspensions using liquid nitrogen followed by thawing at 40°C. Microscopic investigations have revealed that subjecting MLV to freeze–thaw cycling leads to breakdown of the characteristic concentric lamellae of MLV. Although the exact mechanism still remains unknown, formation of ice crystals is speculated to be a contributing factor to MLV disruption. It is interesting to note that the average size (>1 μm) and over all size distribution (high polydispersity index) of FATMLV remain similar to those of MLV (140). The choice of lipids and their concentration along with the nature of the drug to be encapsulated in the formulation determine the entrapment efficiencies of FATMLV (141,142). Typical drug entrapment efficiencies of FATMLV are higher than those of MLV and can range from 2 to 17 μL/μmol lipid (140).

6.3. Liposome Preparation by Extrusion Techniques

Though FATMLV improve the entrapment efficiency of drugs in liposomes, such liposomes are large in size and generate non-homogenous suspensions (7,143). Extrusion techniques have been used to produce SUV from MLV (143). These techniques typically involve passage of an MLV suspension through polycarbonate membranes or filters of definite size. Typically, smaller size vesicles are obtained by the sequential passage of the MLV suspension through a series of progressively smaller pore size filters. In addition to yielding liposomes with homogenous populations, extrusion techniques can also handle higher lipid concentrations, as high as 400 mg/mL lipids. The entrapped volumes of liposomes generated by the extrusion technique range from 1 to 3/μmol lipid and are higher than those of conventional MLV (7,138,140). Extrusion techniques can easily be adapted to industrial production and can be compliant to cGMPs and other regulatory requirements (7).

6.4. Liposome Preparation by Dehydration/ Rehydration

In this technique, SUVs and the solute to be entrapped are dispersed in buffer and the solution is frozen and dehydrated by the passage of liquid nitrogen till complete evaporation of the aqueous medium takes place (74). The dried film of lipids and solutes is then reconstituted by rehydration with the necessary buffer. Upon rehydration of this solid mixture MLV are produced. These MLV can be further subjected to size reduction using microfluidization or sequential passage through membranes. Liposomes prepared by the dehydration/rehydration technique have been used in the production of recombinant protein vaccines due to high entrapment efficiencies (74). Antigens have also been entrapped using such techniques; however, instead of SUVs, giant vesicles that can incorporate larger particulates have been used in the original freeze-drying step (74).

6.5. Liposome Preparation by Reverse Phase Evaporation

Preparation of liposomes using reverse phase evaporation technique involves the introduction of an aqueous medium (buffer) containing the solute to be entrapped into a solution of lipids in an organic solvent (7,138,144). The two-phase system is sonicated to form a temporary unstable emulsion. The organic solvent is eventually removed by evaporation under reduced pressure. The resultant suspension of lipids can also be processed further as discussed in Sec. 6.3 till the vesicles are in the desired size range and with similar lamellar characteristics. The volume of the aqueous component of the initial emulsion and the concentration of the lipids are some of the factors that affect the characteristics of vesicles prepared by reverse phase evaporation (145). In general, since these vesicles have substantially larger entrapped internal aqueous volumes they have significantly higher entrapment efficiencies compared to MLV (7,138,144).

Liposomes generated by reverse phase evaporation can be subclassified into two types based on modifications to the evaporation process made in the general method: stable plurilamellar vesicles (SPLV) (146) and multilayered vesicles prepared by the reverse-phase evaporation method (MLV-REV) (145,147). SPLV are prepared by simultaneous sonication and concurrent evaporation of the initial emulsion, whereas MLV-REV does not involve the sonication process. MLV-REV have uniformly dispersed homogenous lamellae and higher entrapment efficiencies compared to SPLV (145,147).

6.6. Preparation of Liposomes Based on Lipid–Alcohol–Water Injection Technology

This technique of liposome preparation involves the injection of alcoholic solutions of lipids into aqueous media (6,92,148). The lipids undergo precipitation in the form of polydisperse unilamellar vesicles. The size and polydispersity is affected by lipid concentration, relative percentage of alcohol to the aqueous phase as well as the dilution effect (148). Newer methods utilizing the same general principles of alcohol injection technology with minor modifications have been recently developed to yield SUVs with greater homogeneity (6,92,149). One such method involves introduction of the aqueous phase into the lipid ethanol solution followed by complete evaporation of ethanol using evaporation (149). First described by Batzri and Korn (150), the ethanol injection technique has been developed for entrapment of pharmaceutical proteins (151) and used for the commercial manufacture of Pevaryl Lipogel ®, the first approved dermatological liposomal formulation (51,152).

6.7. Liposome Preparation Using Detergent Dialysis

Developed by Weder et al. (153) for topical liposomal formulations, detergent dialysis technique is similar to the alcohol injection technology in terms of general principles of facilitated lipid solubilization. Instead of using alcohol, this technique accomplishes lipid solubilization in the form of mixed micelles in aqueous media using detergents (80,153). The detergents are then removed by dialysis, which leads to the disruption of the mixed micelles and the solubility of the phospholipids is lowered in the aqueous medium. Consequently, the mixed micelles are converted into liposomes. Commonly used detergents in this method of liposome manufacture include bile salts such as sodium cholate, sodium taurocholate, and sodium deoxycholate, and other ionic and non-ionic tensides such as sodium dodecyl sulfate and dodecyl maltoside (154). This technique has been used in the preparation of protein liposomes (155) and stabilized plasmid-lipid particles (156). However, detergent dialysis techniques are protracted and incomplete removal of the residual detergent may compromise liposome stability (3).

6.8. Freeze-Drying of Liposomes

Though lyophilization or freeze-drying itself cannot generate liposomes, it is included as a manufacturing process since liposomes prepared by any method described above can be converted into dry solid formulations using this technique. It can be appended as a continuation to any manufacturing process after liposomes with desired characteristics have been produced. Lyophilization of liposomes is one of the best ways to circumvent many of the stability problems associated with liquid liposome suspensions (for details refer Sec. 8.3) (157,158). Lyophilization involves three major processes: (i) freezing; (ii) primary drying; and (iii) secondary drying (157,158). The freezing process involves cooling of the liposome suspension at very low temperatures such that the water component of the liposomal suspension is frozen into solid ice and the viscosity of the suspension is significantly reduced by the formation of an amorphous glass. The frozen matrix is then subjected to the second phase of primary drying. Primary drying involves removal of ice by sublimation under high vacuum and low temperature. At the end of the primary drying process, the frozen matrix is converted into a freeze-dried porous cake. This resulting porous cake is then brought to shelf temperature (usually 25°C) and subjected to secondary drying to facilitate complete removal of water in the formulation. The headspace in the vials is replaced by nitrogen to minimize phospholipid oxidation. The process thus yields an elegant dry formulation that is reconstituted with a recommended buffer prior to administration (157,158).

It should be noted that during the lyophilization process, the liposomal bilayer structure may be disrupted or punctured due to the temperature stresses generated or due to the ice crystals formed during the initial freezing phase. This damage may lead to leakage of entrapped components, liposome fusion, and aggregation. Freeze-drying may also affect the stability of some of the sensitive entrapped molecules such as proteins. To minimize this structural damage, lyophilization is conducted in the presence of cryoprotectants such as sorbitol, mannitol, trehalose, lactose, and sucrose (158–160). Cryoprotectants decrease vesicle fusion and aggregation and improve liposome stability by forming a low mobility amorphous glass surrounding the vesicles during the freezing phase as well as due to interactions between them and the phospholipid head groups during the freezing cycle (161,162). The temperature of formation of this amorphous glass is characteristic of each cryoprotectant and is known as the glass transition temperature (T_g). From a regulatory perspective, commonly used saccharide-based cryoprotectants currently qualify as generally recognized as safe (GRAS) food ingredients and thus can easily be incorporated into parenteral formulations.

7. LIPOSOME DRUG ENCAPSULATION TECHNIQUES

The physicochemical characteristics of the drug as well as those of the lipids used determine drug loading into liposomal vesicles (6,7,92,163). Some of the commonly used techniques for loading of liposomal vesicles are passive and active encapsulation and complexation. The choice of the entrapment process is determined by the nature of the drug as well as that of the lipids (6,92,164). Water soluble drugs can be easily entrapped within the aqueous compartment of SUV or within the interlamellar spaces of MLV (6,7,92). Hydrophobic drugs, primarily as a result of their affinity towards lipids, associate with the hydrocarbon chains. DNA-based therapeutics such as oligonucleotides and plasmids, due to their anionic charge, complex with cationic liposomes predominantly due to electrostatic interactions (111,112). Entrapment of DNA-based therapeutics into anionic liposomes is enhanced using divalent cations (e.g., Ca^{2+}) or polycations (27,28). Such cations can facilitate electrostatic interactions between negatively charged liposomes and DNA-based therapeutics (27,28,165).

On the basis of their interaction with the liposome bilayer, entrapment drug candidates can be classified into three major types (92,163):

1. Drugs with low oil/water and low octanol/water partition coefficients;
2. Drugs with low oil/water partition coefficients but high or variable octanol/water coefficients;
3. Drugs with high oil/water and high octanol/water partition coefficients.

Drugs in the first class are typically hydrophilic and freely water soluble, due to which they can be encapsulated into liposomes using passive encapsulation strategies (7,92). Class 2 drugs are usually amphiphilic whose membrane permeability is dependent on the pH in the aqueous medium; thus they can easily be encapsulated into liposomes using active encapsulation techniques (7,92). Class 3 drugs are hydrophobic in nature and tend to strongly associate with lipid bilayers. Class 3 drugs are unsuitable for encapsulation into liposomes since they can phase separate easily, due to which such drug compounds are delivered using oil-in-water emulsions (7,92).

7.1. Passive Encapsulation

Passive entrapment of drugs in liposomes involves preferential partitioning of the drug either in the aqueous compartment or by association with the lipids (7). Passive entrapment of drug molecules in MLV typically takes place when these vesicles are formed in aqueous solutions of drugs (6,7,92). Passive entrapment in SUV is also facilitated during their extrusion and sequential passage through filters. Retention of drug inside liposomes is typically low and is determined by membrane permeability, the stronger the membrane association the better the drug retention in the vesicles. For example, drugs such as methotrexate and hydroxyzine tend to remain in the liposomes for a long time due to their strong association with zwitterionic members, whereas, charged drugs such as adriamycin typically do not interact with the lipids and therefore are released rapidly (7). Membrane association and thus passive entrapment of drugs can be improved by synthesis of their lipophilic derivatives that have a higher oil/water partition coefficient. This derivatization approach has been used to improve the passive entrapment efficiency of 6-mercaptopurine from 1.92% for the parent drug to 91.8% for the lipophilic derivative (glyceryl monostearate drug-conjugate) (86); and that of triamcinolone from 5% for the parent drug to 85% for triamcinolone acetonide 21-palmitate (166). The commercially available product AmBisome has been developed using passive loading technologies (163).

7.2. Active Encapsulation

Active loading of liposomal vesicles was pioneered by Cullis et al. (3,6,167). This method is based on the pH-dependent differential membrane permeability of ionization states of drugs (3). Active loading consists of initial suspension of empty liposomes with a pH gradient with respect to the external aqueous medium containing the drug of interest (168,169) and the entrapped aqueous core. Depending on the pK_a of the drug, the external pH is manipulated so the drug exists in a predominantly non-ionized state. In response to the concentration gradient of the drug, which is developed across the bilayer membrane of the liposome, the non-ionized species is transported to the internal aqueous space of the liposome. However, the internal pH of the liposomes is maintained at a value so that the drug is reverted to its ionized state and reverse flux of the drug into the medium is prevented. pH gradients across membranes can also be generated using ammonium sulfate to facilitate active loading (6,170). Active loading is suitable for amphiphilic weak acids or weak bases and can be used to have very high-loading efficiencies compared to passive encapsulation techniques. Active loading can also be used for liposomal entrapment of metal ions (6). This variation of the active loading strategy involves lipophilic carrier-mediated transmembrane transport of metal ions into aqueous cores of vesicles that contain metal ion chelators. After being transported into the aqueous core, chelation prevents the reverse flux of the metal ions (6). The commercially available products Doxil, Myocet, and DaunoXome have been developed using active loading technologies (171). The reader is referred to the case study in this book on the development of AmBisome by Adler-Moore.

7.3. Drug Complexation

Drug entrapment in liposomes by complexation with their surface is based on the electrostatic interaction of the drug and the lipid component in the formulation. Due to the characteristic nature of this interaction and association of the drug with the lipids compared to classical drug entrapment in the aqueous core of liposomes; these products are termed drug-lipid complexes (26–28). Although many of the currently developed formulations for small molecular weight drug candidates are based on drug loading into the liposomes, one of the original commercially licensed liposomal drug product Abelcet is a small molecular weight drug (amphotericin B)–lipid complex (15). The Abelcet formulation is amphotericin B interdigitated and complexed with lipids in a 1:1 drug to lipid ratio. The lipid component in the formulation is composed of dimyristoyl phosphatidylglycerol and dimyristoyl phosphatidylcholine in a 7:3 molar ratio. The formulation is based on the strong binding of this complex until drug release in the fungal cytoplasm.

Currently, drug complexation for loading drug into liposomes is most commonly used for cationic liposomes intended to deliver DNA-based therapeutics (111,112,165). Cationic liposomes interact with the anionic backbone of DNA-based therapeutics to form complexes as a result of electrostatic attraction (for details, refer Sec. 5.3). Complexation of DNA-based therapeutics on liposome surfaces protects them from nuclease degradation and can also facilitate their entry into cells (172). Cationic lipids have been used to deliver a wide range of DNA-based therapeutics such as oligonucleotides and gene therapy plasmids in cell culture and animal models (118). Several cationic liposomal formulations of DNA-based therapeutics are currently in human clinical trials (120,173).

8. LIPOSOME CHARACTERIZATION AND COMPENDIAL REQUIREMENTS

Although potential applications of liposomes as drug carriers have been described in the literature since the early 1970s, it has only been a few years since such products have been introduced for human therapy. Most of the marketed products are manufactured to meet internal specifications set by individual manufacturers and thus far there have been no comprehensive compendial guidelines exclusively specified for liposomal products. Furthermore due to the potential diversity of liposomal preparations in terms of their physico-chemical properties (such as vesicle charge, size, lamellarity, and composition), intended applications, and variations thereof; it has been extremely difficult to establish generalized guidelines that can be collectively applied for all liposomal products. This conundrum is further complicated when liposomes are dispersed in non-aqueous media such as creams and lotions for topical transdermal applications. Since liposomes are typically administered parenterally, some of the standards that universally apply to parenteral products are extended to liposomal preparations (174–176). For a more detailed discussion on the biopharmaceutical aspects and regulatory guidance for liposomal dispersions, the reader is referred to the chapter by Chen in this book.

Recently, there have been some initiatives by the US Department of Health and Services, Food and Drug Administration (US-FDA) to develop specifications for liposomal products (136). The Center for Drug Evaluation and Research (CDER) of the US-FDA developed a preliminary draft for a guidance document on the industrial manufacture of liposome products in August 2002 (136). The draft recommends extensive characterization of liposomal drug products to ensure product quality and prevent batch-to-batch variations. These include definition and analysis of the following formulation-related issues: (i) morphological and biophysical characteristics of liposomes; (ii) drug loading and release characteristics; (iii) liposome stability; and (iv) liposome sterilization.

8.1. Morphological and Biophysical Characterization

Morphological and biophysical characterization of liposomal vesicles has been a subject of exhaustive research ever since these drug carrier systems were discovered (2,6,92). To ensure dependable performance in vivo, it is desired to produce liposomes with consistent electrochemical and biophysical properties. According to the US-FDA guidance document, for commercial manufacturing of liposomes, the following biophysical properties of the final drug product are recommended to be analyzed: gravimetric analysis of lipids in the formulation; lamellarity; particle size and size distribution; phase transition temperature; vesicle charge; osmotic and pH properties; and light scattering index (136). Exact specifications of these properties can be used to stipulate quality control end-points to ensure minimal batch-to-batch variation during production (6,92). In addition, these tests can also serve as process control indicators for changes in manufacturing protocols or sites of manufacture of the drug product (136).

Lamellarity of liposomes can be detected using nuclear magnetic resonance (NMR) spectroscopy, small angle x-ray scattering, and cryo-electron microscopy. Particle sizing, size distribution, and light scattering indices of liposome formulations can be determined using an array of techniques depending on the expected size range of the vesicles (2,6). For submicron size particles, the following techniques can be used: dynamic light scattering (DLS); electron microscopy using either cryo-fixation techniques or negative staining; atomic force microscopy (AFM) (177); and ultracentrifugation (6). Coulter counter, gel exclusion chromatography, laser diffraction, and light microscopy are some of the techniques that could be used for particle size analysis of particles in the micron range (6). Phase transition temperature of liposomes and overall thermotropic phase behavior can be identified using differential light scattering (DSC), fluorescence methods, and NMR (6,178). Zeta potential measurements can be used to determine the electrophoretic mobility (microelectrophoresis) of liposomal vesicles and thus identify their surface charge density (6,179).

In addition to these properties, other product-specific tests may be needed depending on the chosen liposome manufacturing or drug loading process. Examples of such tests include: determination of residual alcohol or detergent in the final formulation if alcohol injection technology or detergent dialysis, respectively; estimation of stability of cryoprotectants in lyophilized liposomes; and assessment of compounds used for generating pH gradients when active loading strategies are used for drug encapsulation into liposomes.

8.2. Drug Loading and Release Characterization

Drug loading and in vitro release from liposomal vesicles are some of the most critical parameters for estimating the in vivo performance and therapeutic efficacy of these drug delivery systems (175,176). Although, release testing methodologies for solid parenteral dosage forms have been very well characterized and acceptance criteria and regulatory guidance have been established, not much has been accomplished in these areas for liposomal and other novel drug carriers (175). According to the preliminary draft of the guidance document developed by the US-FDA, manufacturers seeking regulatory approval for liposomal products are recommended to submit the following information with regard to drug loading and release testing: (i) quantification of the entrapped and unentrapped drug in the liposomal formulation; (ii) determination of volume of entrapment in liposomes; (iii) in vitro release testing of the drug candidate from liposomal vesicles; and (iv) in vivo stability assessment and release testing (136). Though the actual techniques used to determine these properties are not specified in the guidance draft (136), these issues have been discussed in literature (174–176).

Drug loading is dependent on the method of liposome manufacture as well as the technique used for drug entrapment. Since most liposomal products are administered via the i.v. route, it is necessary to have high drug loading capacities for the carriers, so that maximum drug can be delivered with minimum lipidic excipients. It is critical to remove as much unentrapped drug from the formulation as possible since high amounts of free drug will affect the overall pharmacokinetics of the formulation (136). The free drug fraction may also be responsible for toxicity (e.g., doxorubicin and amikacin) or compromise stability of the formulation (92).

Drug loading in the liposome carrier and the ratio of encapsulated to unentrapped drug can also be used to evaluate efficacy of the manufacturing process used in the preparation of the liposomes. Active loading procedures such as use of pH gradients can generate high entrapment efficiencies compared to the passive loading strategies (140,142). The complexation process for cationic liposomes with DNA-based therapeutics also has high drug loading capacities. Techniques that have low encapsulation efficiencies may be cost-prohibitive since they may necessitate additional steps for drug recovery from the liposomal dispersion.

The amount of unentrapped drug in the final formulation can be estimated by separation of liposomal vesicles from the dispersion. Commonly used separation methods include: gel permeation chromatography; ultracentrifugation; and dialysis (92). Free drug can also be separated from the liposome dispersion using ion exchange chromatography. This technique is used for drugs that are of opposite charge compared to their vesicles. Following separation of the liposomal vesicles, spectroscopic techniques can be used to estimate the free drug fraction.

Entrapment volume of liposomes is defined as the intra-liposomal volume or the "milieu interne" of the aqueous compartment of the liposome vesicle (139,180,181). The entrapment volume is dependent on the manufacturing process used for liposomes (for details, refer Sec. 6). Entrapment volume is estimated by liposomal encapsulation of water-soluble marker ions that have minimal interaction with lipid bilayers such as radioactive solutes (^{22}Na or ^{14}C/^{3}H-inulin) (139), fluorescent molecules (5,6-carboxy-fluorescein), or salt ions (Cl$^-$) (181). The vesicles can be separated using centrifugation, dialysis, or gel chromatography (180) and the internal volume can be assayed by solute exclusion, solute entrapment, or by solvent distribution combined with solute exclusion techniques (181).

In vitro release testing of liposomes can be used to estimate the in vivo performance, to assess the quality, and for process control of liposome drug products (174–176). Typical in vitro release testing USP apparatuses have been designed for oral and transdermal products and are unsuitable for parenteral liposomal dispersions. Although some modifications have been attempted on these systems to facilitate their use for liposomes and other controlled release dosage forms, they still possess several deficiencies that include: (i) improper sample containment; (ii) requirements of large volumes; (iii) violation of sink conditions; and (iv) liposome aggregation. USP apparatus 4 (flow-through cell) and small sample vials and chambers, with or without agitation, are some of the alternative release testing equipments used for liposomes. The USP apparatus 4 can handle samples without aggregation, is amenable for use of small volumes, and can maintain sink conditions throughout the testing process (175). Release testing is typically performed in simulated physiological media with or without plasma (136). To characterize the release profile adequately, release of approximately 80% of the loaded drug for liposomal carriers is sought (175). Since some of the release testing can be conducted over days or weeks, adequate provisions should be made to ensure minimal loss of water due to evaporation and prevent microbial contamination of the samples (174). Addition of commonly used preservatives such as methyl paraben, propyl paraben, cetrimonium bromide, and benzalkonium chloride to the release medium is recommended. However, it must be guaranteed that these preservatives are compatible with the drug product and the lipid excipients and do not hamper or interfere in their analytical assays (174). Furthermore, if real-time in vitro release testing necessitates long durations, liposomes are also recommended to be evaluated under accelerated stress tests (175).

In vivo drug release characteristics are critical determinants of the overall feasibility of using liposomes for durg delivery applications. These characteristics are dependent on the biological stability of these drug delivery vehicles which can be affected by plasma proteins and the dilution effect upon administration (54,55). The ratio of the unencapsulated and encapsulated drug can be used as an indicator of the in vivo stability. Constant values of this ratio prior to and after administration of a single dose of the liposome drug product in an animal model would indicate stability of the liposomal formulation.

8.3. Liposome Stability

Determination of stability of liposomal formulations is critical to identify their storage conditions and shelf life (182). Liposomal drug products need to be evaluated not only for the degradation of the pharmacotherapeutic entity but also for the liposomal excipients. Some of the product specifications related to stability issues according to the US-FDA-developed guidance document for liposome manufacture include: (i) assay for encapsulated and free drug substance; (ii) report on chemical degradation products related to lipids and the encapsulated drug; (iii) assay of lipid components; and (iv) characterization for physical instabilities (136). The draft document also recommends conducting accelerated stress testing of liposomes and stability testing of unloaded vesicles (136). Accelerated stability testing is recommended to estimate the effect of high stress conditions such as pH and temperature fluctuations on the formulation (7). Overall stability of liposomal formulations can be assessed on two levels: chemical and physical (182,183). Although they are apparently discrete, both chemical and physical instabilities can influence one another (184). Chemically degraded lipids can lose their ability to form bilayers and thus affect the physical stability and eventually the performance of the formulation (184).

8.3.1. Chemical Stability

Chemical aqueous stability of lipids is affected by two major degradation pathways: hydrolysis and oxidation (7,183). Both saturated and unsaturated lipids can undergo hydrolysis to produce fatty acids and lysolipids, which may have completely different physicochemical properties from their parent compounds. Lysolipids increase membrane permeability and lead to destabilization of liposomal bilayers (183). The formation of lysolipids is of particular concern since they have been implicated in cardiological toxic reactions in vivo in animal models (185,186). Lipid oxidation predominantly affects lipid molecules with polyunsaturated acyl chains. Free radical mediated-peroxidation of acyl chains is a complex process that produces a diverse group of degradative by-products which may include hydroperoxides, alkanes, diene conjugates (187), and 4-hydroxy-2(E)-nonenal (6,188). Oxidation of lipids can be accelerated by radiation (189) and production of reactive oxygen species during apoptosis (190). Oxidation can be minimized by the incorporation of other compounds in the formulation such as carotenoids (191), caroverine (192), and catechols (183,193). However, high concentrations of antioxidant excipients in liposomal formulations may not be permitted due to potential complications caused by their pharmacological properties. Chemical degradation of lipids can also be stimulated by γ-irradiation (194). Presence of lysolipids, fatty acids, and other chemical degradant products in liposomal formulations can be identified using high-performance liquid chromatography (HPLC) in tandem with mass spectrometry (189), thin layer chromatography (TLC), and gas liquid chromatography (GLC) (6).

8.3.2. Physical Stability

Physical instability of liposomes can be manifested in the form of aggregation and fusion that may cause leakage of the entrapped components (7,195). Aggregation or flocculation is characterized by a thermodynamically reversible assembly of discrete units of liposomes into larger colloidal entities (7,195). Aggregation is typically observed in neutral liposomes of large size due to their small curvature (relative flatness) and high contact area. Aggregation is promoted by trace elements and sedimentation, both of which promote temporary electrostatic binding of the vesicles. Due to its transient nature, aggregated vesicles can be disengaged by shaking. Addition of charged lipid molecules such as phosphatidylglycerol and cholesterol hemisuccinate in the formulation may induce a negative charge on the liposomes which may

cause electrostatic repulsion and prevent aggregation (196).
However, as discussed in Sec. 8.2, negative charge on the vesicles may compromise their biological stability (54,55).

In contrast to aggregation or flocculation, fusion of liposomes is an irreversible process that leads to the formation
of larger vesicles that cannot revert back to their original
form. Fusion of vesicles is commonly observed with very small
vesicles (<20 nm diameter) that have excessive high stress
curvature which promotes this phenomenon (195). Fusion
can be prevented choice of liposome manufacturing method
that has good control on liposome size.

Although, aggregation and fusion may result in release of
entrapped materials from liposomes, un-intended escape of
drug products from vesicles also known as leakage, by itself,
is a form of physical instability. Leakage is commonly observed
in passively loaded liposomal vesicles. In addition to its occurrence in vitro, leakage may also be caused in vivo by liposomal
interactions with serum proteins (197). Leakage of components is inversely proportional to the acyl chain length of the
lipids incorporated in the formulation. Increased stability of
liposomes with long acyl chains is attributed to their high
transition temperature and decreased membrane fluidity
(198). Leakage may also be prevented by incorporation of cholesterol into membranes which may result in a tighter packing
arrangement of the bilayers (7). Pegylation of liposomes also
minimizes leakage (199).

Aggregation and fusion of SUVs can be determined by
particle size characterization using photon correlation spectroscopy (200). Analytical techniques such as fluorescence
polarization measurement (199) and permeability measurement using carboxyfluorescein (199) have been used to
characterize liposomal bilayers and their biophysical properties and determine the effects of changes in the membrane
composition on them.

8.4. Sterilization of Liposomal Products

Since many liposomal products are developed to be administered via the i.v. route, sterilization of these products is

mandatory. Terminal sterilization using steam, routinely employed for several pharmaceutical products, may not be suitable for liposomal formulations, since high temperature may disrupt the liposome architecture, lead to physical destabilization, and may be completely prohibitive to liposomal formulations of proteins, peptides, and antibodies due to their thermolabile properties (201). Another commonly used pharmacopeial sterilization technique, γ-irradiation, may also be unsuitable for liposomal dispersions, since radiation compromises their chemical stability (194,201). Although aseptic manufacturing can be an alternative, it is not commonly used due to the expense and difficulty in validation. Liposomal formulations are typically manufactured with raw materials with low microbial burden and terminally sterilized using microbial retentive filtration. Filtration sterilization of the final product can be challenging due to the structural complexity of these vesicles. Since liposomal components can affect microbial interaction with filters used in the sterilization process and thus permit their passage into the filtered product, microbial retentivity tests for specific products are recommended. Furthermore, lipids may be lost by non-specific adsorption of lipids to filters. It is recommended to develop strict validated quality control assays for assessing sterility of commercial liposomal formulations. Sterility of the final liposomal product can be confirmed using recommended pharmacopoeial protocols such as aerobic and anaerobic bottle cultures (6). In addition to sterility testing, liposomal dispersions can also be assessed for pyrogenicity using the limulus amebocyte lysate (LAL) test (2,6,92).

9. CONCLUSIONS

Liposomal drug products have progressed beyond the experimental stage and appear to be a reliable and clinically viable strategy for the delivery of a wide range of pharmacotherapeutics. They also constitute some of the most promising non-viral gene delivery vectors developed in the last few decades. Liposomal delivery systems offer both exceptional formulator control on their design and an excellent option for safer disease management due to consistent clinical performance. As with any novel drug delivery system, global regulatory approval of several products has separated hype from realistic expectations and reinforced confidence in the scientific basis and principles of liposomes. With the anticipated entry of more biotechnology-based drugs and the observed trend of small molecular weight drug candidates with poor water solubility in pharmaceutical pipelines in the near future, the utility and efficacy of liposome-derived formulations will become even more evident and much appreciated. As clinical use of these systems becomes prevalent, such issues that cannot be completely investigated initially as chronic human exposure, potential drug interactions, and effect on endogenous lipid and cholesterol levels will be brought to light. These factors will eventually play a crucial role in evaluating the bene-fit/risk ratio of liposomal formulations and determine the fea-sibility of long-term acceptance of these drug carriers into mainstream pharmaceutical dosage forms.

REFERENCES

1. Bangham AD, ed. Liposome Letters. New York: Academic Press, 1983.

2. Crommelin DJA, Storm G. Liposomes: from the bench to the bed. J Liposome Res 2003; 13(1):33–36.

3. Cullis PR. Commentary: liposomes by accident. J Liposome Res 2000; 10(2/3):IX–XXIV.

4. Lasic DD, Barenholz Y. Liposomes: past, present, and future. In: Lasic DD, Barenholz Y, eds. Handbook of Nonmedical Applications of Liposomes. Vol. 4. Boca Raton: CRC Press, 1996:299–315.

5. Bangham AD, Standish MM, Watkins JC. Diffusion of univalent ions across the lamellae of swollen phospholipids. J Mol Biol 1965; 13(1):238–252.

6. Crommelin DJA, Schreier H. Liposomes. In: Kreuter J, ed. Colloidal Drug Delivery Systems. New York: Marcel Dekker Inc., 1994:73–190.

7. Brandl M. Liposomes as drug carriers: a technological approach. Biotechnol Annu Rev 2001; 7:59–85.

8. Israelachvili JN. Thermodynamic and geometric aspects of amphiphile aggregation into micelles, vesicles and bilayers, and the interactions between them. In: Degiorgio V, Corti M, eds. Physics Amphiphiles: Micelles, Vesicles and Micro-emulsions. Amsterdam: North-Holland, 1985:24–58.

9. Israelachvili JN, Marcelja S, Horn RG. Physical principles of membrane organization. Q Rev Biophys 1980; 13(2):121–200.

10. Arikan S. Lipid-based antifungal agents: a concise overview. Cell Mol Biol Lett 2002; 7(3):919–922.

11. Rapp RP. Changing strategies for the management of invasive fungal infections. Pharmacotherapy 2004; 24(2):4S–28S.

12. Ng AWK, Wasan KM, Lopez-Berestein G. Development of liposomal polyene antibiotics: an historical perspective. J Pharm Sci 2003; 6(1):67–83.

13. Robinson RF, Nahata MC. A comparative review of conventional and lipid formulations of amphotericin B. J Clin Pharmacy Ther 1999; 24(4):249–257.

14. Drummond DC, Kirpotin D, Benz CC, Park JW, Hong K. Liposomal drug delivery systems for cancer therapy. In: Brown DM ed. Drug Delivery Systems in Cancer Therapy. Totowa, NJ: Humana Press, 2004; 191–213.

15. Martino R. Efficacy, safety and cost-effectiveness of amphotericin B lipid complex (ABLC): a review of the literature. Curr Med Res Opin 2004; 20(4):485–504.

16. Boswell GW, Buell D, Bekersky I. AmBisome (liposomal amphotericin B): a comparative review. J Clin Pharmacol 1998; 38(7): 583–592.

17. Kayser O, Olbrich C, Croft SL, Kiderlen AF. Formulation and biopharmaceutical issues in the development of drug delivery systems for antiparasitic drugs. Parasitol Res 2003; 90(suppl 2): S63–S70.

18. Linden PK. Amphotericin B lipid complex for the treatment of invasive fungal infections. Expert Opin Pharmacother 2003; 4(11):2099–2110.

19. Ansel AC, Allen LV, Popovich NG. Novel dosage forms and drug delivery technologies. In: Ansel AC, Allen LV, Popovich NG, eds. Pharmaceutical Dosage Forms and Drug Delivery Systems. Philadelphia, PA:, MD: Lippincott Williams & Wilkins, 1999:535–551.

20. Presant CA, Crossley R, Ksionski G, Proffitt R. Design of liposomes clinical trials. In: Gregoriadis G, ed. Liposome Technology. Vol. II. 2nd ed. Boca Raton: CRC press, 1993:307–317.

21. Cattel L, Ceruti M, Dosio F. From conventional to stealth liposomes: a new frontier in cancer chemotherapy. Tumori 2003; 89(3):237–249.

22. Glueck R, Moser C, Metcalfe IC. Influenza virosomes as an efficient system for adjuvanted vaccine delivery. Expert Opin Biol Ther 2004; 4(7):1139–1145.

23. Gluck R, Metcalfe IC. New technology platforms in the development of vaccines for the future. Vaccine 2002; 20 (suppl 5): B10–B16.

24. Gluck R, Walti E. Biophysical validation of Epaxal Berna, a hepatitis A vaccine adjuvanted with immunopotentiating reconstituted influenza virosomes (IRIV). Dev Biol (Basel) 2000; 103:189–197.

25. Mischler R, Metcalfe IC. Inflexal V a trivalent virosome subunit influenza vaccine: production. Vaccine 2002; 20(suppl 5): B17–B23.

26. Patil SD, Rhodes DG, Burgess DJ. DNA-based therapeutics and DNA delivery systems: a comprehensive review. AAPS Journal 2005; In press.

27. Patil SD, Rhodes DG, Burgess DJ. Anionic liposomal delivery system for DNA transfection. The AAPS Journal 2005; 6(4):E29.

28. Patil SD, Rhodes DG, Burgess DJ. Biophysical characterization of anionic lipoplexes. Biochimica et Biophysical Acta–Biomembranes 2005; In press.

29. Luo D, Saltzman WM. Synthetic DNA delivery systems. Nat Biotechnol 2000; 18(1):33–37.

30. Kay MA, Glorioso JC, Naldini L. Viral vectors for gene therapy: the art of turning infectious agents into vehicles of therapeutics. Nat Med (NY) 2001; 7(1):33–40.

31. Kamiya H, Tsuchiya H, Yamazaki J, Harashima H. Intracellular trafficking and transgene expression of viral and non-viral gene vectors. Adv Drug Deliv Rev 2001; 52(3):153–164.

32. Mah C, Byrne BJ, Flotte TR. Virus-based gene delivery systems. Clin Pharmacokinet 2002; 41(12):901–911.

33. Lotze MT, Kost TA. Viruses as gene delivery vectors: application to gene function, target validation, and assay development. Cancer Gene Ther 2002; 9(8):692–699.

34. Flotte TR, Laube BL. Gene therapy in cystic fibrosis. Chest 2001; 120(suppl 3):124S–131S.

35. Tenenbaum L, Lehtonen E, Monahan PE. Evaluation of risks related to the use of adeno-associated virus-based vectors. Curr Gene Ther 2003; 3(6):545–565.

36. Raper SE, Chirmule N, Lee FS, Wivel NA, Bagg A, Gao G-p, Wilson JM, Batshaw ML. Fatal systemic inflammatory response syndrome in a ornithine transcarbamylase deficient patient following adenoviral gene transfer. Mol Genet Metab 2003; 80(1/2):148–158.

37. Nyberg-Hoffman C, Aguilar-Cordova E. Instability of adenoviral vectors during transport and its implication for clinical studies. Nat Med (N Y) 1999; 5(8):955–957.

38. McTaggart S, Al-Rubeai M. Retroviral vectors for human gene delivery. Biotechnol Adv 2002; 20(1):1–31.

39. Davis Mark E. Non-viral gene delivery systems. Curr Opin Biotechnol 2002; 13(2):128–131.

40. Lollo CP, Banaszczyk MG, Chiou HC. Obstacles and advances in non-viral gene delivery. Curr Opin Mol Ther 2000; 2(2): 136–142.

41. Liu F, Huang L. Development of non-viral vectors for systemic gene delivery. J Control Release 2002; 78(1–3):259–266.

42. Godbey WT, Mikos AG. Recent progress in gene delivery using non-viral transfer complexes. J Control Release 2001; 72(1–3):115–125.

43. Brown MD, Schatzlein AG, Uchegbu IF. Gene delivery with synthetic (non viral) carriers. Int J Pharm 2001; 229(1–2): 1–21.

44. Nabel GJ, Nabel EG, Yang ZY, Fox BA, Plautz GE, Gao X, Huang L, Shu S, Gordon D, Chang AE. Direct gene transfer with DNA-liposome complexes in melanoma: expression, biologic activity, and lack of toxicity in humans. Proc Natl Acad Sci USA 1993; 90(23):11307–11311.

45. Huang L, Viroonchatapan E. Introduction. In: Huang LH, Hung LM, Wagner E, eds. Nonviral Vectors for Gene Therapy. San Diego, CA: Academic Press, 1999:3–22.

46. Ponimaskin EG, Schmidt MFG. Fusogenic viral envelopes as potent vehicles for gene transfer. Curr Genomics 2001; 2(3): 261–267.

47. Hodgson CP, Solaiman F. Virosomes: cationic liposomes enhance retroviral transduction. Nat Biotechnol 1996; 14(3): 339–342.

48. Dean TP, Hider RC. Incorporation of alphaxalone into different types of liposomes. J Pharm Pharmacol 1993; 45(11): 990–992.

49. Cortesi R, Esposito E, Maietti A, Menegatti E, Nastruzzi C. Formulation study for the antitumor drug camptothecin: liposomes, micellar solutions and a microemulsion. Int J Pharm 1997; 159(1):95–103.

50. Lee M-J, Straubinger RM, Jusko WJ. Physicochemical, pharmacokinetic and pharmacodynamic evaluation of liposomal tacrolimus (FK 506) in rats. Pharm Res 1995; 12(7): 1055–1059.

51. Naeff R. Feasibility of topical liposome drugs produced on an industrial scale. Adv Drug Deliv Rev 1996; 18(3):343–347.

52. Singla AK, Garg A, Aggarwal D. Paclitaxel and its formulations. Int J Pharm 2002; 235(1–2):179–192.

53. Maruyama K. Passive targeting with liposomal drug carriers. Drug Deliv Syst 1999; 14(6):433–447.

54. Hsu MJ, Juliano RL. Interactions of liposomes with the reticuloendothelial system. II. Nonspecific and receptor-mediated

uptake of liposomes by mouse peritoneal macrophages. Biochim Biophys Acta 1982; 720(4):411–419.

55. Juliano RL, Hsu MJ, Regen SL. Interactions of polymerized phospholipid vesicles with cells. Uptake, processing and toxicity in macrophages. Biochim Biophys Acta 1985; 812(1):42–48.

56. Patel HM. Serum opsonins and liposomes: their interaction and opsonophagocytosis. Crit Rev Ther Drug Carrier Syst 1992; 9(1):39–90.

57. Tempone AG, Perez D, Rath S, Vilarinho AL, Mortara RA, Franco de Andrade H Jr. Targeting *Leishmania* (L.) chagasi amastigotes through macrophage scavenger receptors: the use of drugs entrapped in liposomes containing phosphatidyl serine. J Antimicrob Chemother 2004; 54(1):60–68.

58. Yu HY, Liu RF. Hepatic uptake and tissue distribution of liposomes: influence of vesicle size. Drug Dev Ind Pharm 1994; 20(4):557–574.

59. Litzinger DC, Huang L. Amphipathic poly(ethylene glycol) 5000-stabilized dioleoylphosphatidylethanolamine liposomes accumulate in spleen. Biochim Biophys Acta 1992; 1127(3):249–254.

60. Kasaoka S, Maruyama K. Stealth liposome for delivery system. Maku 2003; 28(3):135–144.

61. Unezaki S, Hosoda J-i, Maruyama K, Iwatsuru M. Passive targeting to solid tumor by long-circulating liposomes. Organ Biol 1996; 3(3):61–68.

62. Tsukioka Y, Matsumura Y, Hamaguchi T, Koike H, Moriyasu F, Kakizoe T. Pharmaceutical and biomedical differences between micellar doxorubicin (NK911) and liposomal doxorubicin (Doxil). Jpn J Cancer Res 2002; 93(10):1145–1153.

63. Sivak O, Bartlett K, Risovic V, Choo E, Marra F, Batty DS Jr, Wasan KM. Assessing the antifungal activity and toxicity profile of amphotericin B lipid complex (ABLC; Abelcet) in combination with caspofungin in experimental systemic aspergillosis. J Pharm Sci 2004; 93(6):1382–1389.

64. Jendrasiak GL, Frey GD, Heim RC Jr. Liposomes as carriers of iodolipid radiocontrast agents for CT scanning of the liver. Invest Radiol 1985; 20(9):995–1002.

65. Leserman LD. Immunologic targeting of liposomes. In: Nicolau C, Paraf A, eds. Liposomes, Drugs and Immunocompetent Cell Functions 1981; 109–122.

66. Toonen PA, Crommelin DJ. Immunoglobulins as targeting agents for liposome encapsulated drugs. Pharmaceutisch Weekblad 1983; 5(6):269–280.

67. Abu-Dahab R, Schafer UF, Lehr CM. Lectin-functionalized liposomes for pulmonary drug delivery: effect of nebulization on stability and bioadhesion. Eur J Pharm Sci 2001; 14(1): 37–46.

68. de Ilarduya CT, Arangoa MA, Duzgunes N. Transferrin-lipoplexes with protamine-condensed DNA for serum-resistant gene delivery. Methods Enzymol 2003; 373:342–356.

69. Hwang SH, Hayashi K, Takayama K, Maitani Y. Liver-targeted gene transfer into a human hepatoblastoma cell line and in vivo by sterylglucoside-containing cationic liposomes. Gene Ther 2001; 8(16):1276–1280.

70. Leamon CP, Cooper SR, Hardee GE. Folate-liposome-mediated antisense oligodeoxynucleotide targeting to cancer cells: evaluation in vitro and in vivo. Bioconjug Chem 2003; 14(4): 738–747.

71. Yu W, Pirollo KF, Yu B, Rait A, Xiang L, Huang W, Zhou Q, Ertem G, Chang EH. Enhanced transfection efficiency of a systemically delivered tumor-targeting immunolipoplex by inclusion of a pH-sensitive histidylated oligolysine peptide. Nucleic Acids Res 2004; 32(5):e48/1–e48/10.

72. Cui Z, Han S-J, Huang L. Coating of mannan on LPD particles containing HPV E7 peptide significantly enhances immunity against HPV-positive tumor. Pharm Res 2004; 21(6):1018–1025.

73. Moser C, Metcalfe IC, Viret J-F. Virosomal adjuvanted antigen delivery systems. Expert Rev Vaccines 2003; 2(2): 189–196.

74. Gregoriadis G, McCormack B, Obrenovic M, Saffie R, Zadi B, Perrie Y. Vaccine entrapment in liposomes. Methods (San Diego, California) 1999; 19(1):156–162.

75. Bungener L, Huckriede A, Wilschut J, Daemen T. Delivery of protein antigens to the immune system by fusion-active viro-

somes: a comparison with liposomes and ISCOMs. Biosci Rep 2002; 22(2):323–338.

76. Zurbriggen R. Immunostimulating reconstituted influenza virosomes. Vaccine 2003; 21(9–10):921–924.

77. Knepp VM, Hinz RS, Szoka FC Jr, Guy RH. New liposomal delivery system for controlled drug release. In: Lee PI, Good WR, eds. Controlled Release Technology: Pharmaceutical Applications. Washington, DC: American Chemical Society, 1987:267–272.

78. Xiao C, Qi X, Maitani Y, Nagai T. Sustained release of cisplatin from multivesicular liposomes: potentiation of antitumor efficacy against S180 murine carcinoma. J Pharm Sci 2004; 93(7):1718–1724.

79. Ye Q, Asherman J, Stevenson M, Brownson E, Katre NV. DepoFoam technology: a vechile for controlled delivery of protein and peptide drugs. J Control Release. 2000; 64(1–3): 155–166.

80. Korting HC, Schafer-Korting M. Topical liposome drugs. In: Korting HC, Schafer-Korting M, eds. The Benefit/Risk Ratio: A Handbook for the Rational use of Potentially Hazardous Drugs. Boca Raton: CRC Press, 1999:333–357.

81. Schmid MH, Korting HC. Therapeutic progress with topical liposome drugs for skin disease. Adv Drug Deliv Rev 1996; 18(3):335–342.

82. Korting HC, Blecher P, Schafer-Korting M, Wendel A. Topical liposome drugs to come: what the patent literature tells us. A review. J Am Acad Dermatol 1991; 25(6):1068–1071.

83. Redziniak G. Liposomes and skin: past, present, future. Pathol Biol 2003; 51(5):279–281.

84. Goldman RD. ELA-max: a new topical lidocaine formulation. Ann Pharmacother 2004; 38(5):892–894.

85. Fiume MZ. Final report on the safety assessment of lecithin and hydrogenated lecithin. Int J Toxicol 2001; 20(suppl 1): 21–45.

86. Taneja D, Namdeo A, Mishra PR, Khopade AJ, Jain NK. High-entrapment liposomes for 6-mercaptopurine—a prodrug approach. Drug Dev Ind Pharm 2000; 26(12):1315–1319.

87. Matos C, De Castro B, Gameiro P, Lima JLFC, Reis S. Zeta-potential measurements as a tool to quantify the effect of charged drugs on the surface potential of egg phosphatidylcholine liposomes. Langmuir 2004; 20(2):369–377.

88. Litzinger DC, Huang L. Phosphatidylethanolamine liposomes: drug delivery, gene transfer and immunodiagnostic applications. Biochim Biophys Acta 1992; 1113(2):201–227.

89. Huong TM, Ishida T, Harashima H, Kiwada H. The complement system enhances the clearance of phosphatidylserine (PS)-liposomes in rat and guinea pig. Int J Pharm 2001; 215(1–2):197–205.

90. Wassef NM, Swartz GM, Berman JD, Wilhelmsen CL, Alving CR. Toxic effects of antileishmanial reverse-phase evaporation liposomes containing diacetyl phosphate in monkeys. Drug Deliv 1995; 2(3/4):181–189.

91. Ramsammy LS, Kaloyanides GJ. Effect of gentamicin on the transition temperature and permeability to glycerol of phosphatidylinositol-containing liposomes. Biochem Pharmacol 1987; 36(7):1179–1181.

92. Barenholz Y, Crommelin DJA. Liposomes as pharmaceutical dosage forms. In: Swarbrick J, Boylan JC, eds. Encyclopedia of Pharmaceutical Technology Vol. 9. New York: Marcel Dekker, Inc., 1994:1–39.

93. Blok MC, van Deenen LL, De Gier J. Effect of the gel to liquid crystalline phase transition on the osmotic behaviour of phosphatidylcholine liposomes. Biochim Biophys Acta 1976; 433(1):1–12.

94. Blok MC, van der Neut-Kok EC, van Deenen LL, de Gier J. The effect of chain length and lipid phase transitions on the selective permeability properties of liposomes. Biochim Biophys Acta 1975; 406(2):187–196.

95. Blok MC, Van Deenen LL, De Gier J. The effect of cholesterol incorporation on the temperature dependence of water permeation through liposomal membranes prepared from

phosphatidylcholines. Biochim Biophys Acta 1977; 464(3): 509–518.

96. Ueda I, Tashiro C, Arakawa K. Depression of phase-transition temperature in a model cell membrane by local anesthetics. Anesthesiology 1977; 46(5):327–332.

97. Yatvin MB, Weinstein JN, Dennis WH, Blumenthal R. Design of liposomes for enhanced local release of drugs by hyperthermia. Science 1978; 202(4374):1290–1293.

98. Gotfredsen CF, Frokjaer S, Hjorth EL, Jorgensen KD, Debroux-Guisset MC. Disposition of intact liposomes of different compositions and of liposomal degradation products. Biochem Pharmacol 1983; 32(22):3381–3387.

99. Lasic DD. Stealth liposomes. Drugs and the pharmaceutical sciences. In: Benita S, ed. Microencapsulation: Methods and Industrial Applications. Vol. 73. New York: Marcel Dekker, 1996:297–328.

100. Ceh B, Winterhalter M, Frederik PM, Vallner JJ, Lasic DD. Stealth liposomes: from theory to product. Adv Drug Deliv Rev 1997; 24(2/3):165–177.

101. Lasic DD, Martin FJ, Gabizon A, Huang SK, Papahadjopoulos D. Sterically stabilized liposomes: a hypothesis on the molecular origin of the extended circulation times. Biochim Biophys Acta 1991; 1070(1):187–192.

102. Moghimi SM, Szebeni J. Stealth liposomes and long circulating nanoparticles: critical issues in pharmacokinetics, opsonization and protein-binding properties. Progr Lipid Res 2003; 42(6):463–478.

103. Moribe K, Maruyama K. Reviews on PEG coated liposomal drug carriers. Drug Deliv Syst 2001; 16(3):165–171.

104. Lasic DD. Doxorubicin in sterically stabilized liposomes. Nature (London) 1996; 380(6574):561–562.

105. Abra RM, Bankert RB, Chen F, Egilmez NK, Huang K, Saville R, Slater JL, Sugano M, Yokota SJ. The next generation of liposome delivery systems: recent experience with tumor-targeted, sterically-stabilized immunoliposomes and active-loading gradients. J Liposome Res 2002; 12(1/2):1–3.

106. Felgner JH, Kumar R, Sridhar CN, Wheeler CJ, Tsai YJ, Border R, Ramsey P, Martin M, Felgner PL. Enhanced gene delivery and mechanism studies with a novel series of cationic lipid formulations. J Biol Chem 1994; 269(4):2550–2561.

107. Hofland HEJ, Shephard L, Sullivan SM. Formation of stable cationic lipid/DNA complexes for gene transfer. Proc Natl Acad Sci USA 1996; 93(14):7305–7309.

108. Pedroso de Lima MC, Neves S, Filipe A, Duzgunes N, Simoes S. Cationic liposomes for gene delivery: from biophysics to biological applications. Curr Med Chem 2003; 10(14):1221–1231.

109. Marshall J, Yew NS, Eastman SJ, Jiang C, Scheule RK, Cheng SH. Cationic lipid-mediated gene delivery to the airways. In: Wagner E, ed. Nonviral Vectors for Gene Therapy. San Diego, CA: Academic Press, 1999:39–68.

110. Felgner PL, Gadek TR, Holm M, Roman R, Chan HW, Wenz M, Northrop JP, Ringold GM, Danielsen M. Lipofection: a highly efficient, lipid-mediated DNA-transfection procedure. Proc Natl Acad Sci USA 1987; 84(21):7413–7417.

111. Wiethoff CM, Gill ML, Koe GS, Koe JG, Middaugh CR. The structural organization of cationic lipid-DNA complexes. J Biol Chem 2002; 277(47):44980–44987.

112. Raeder JO, Koltover I, Salditt T, Safinya CR. Structure of DNA-cationic liposome complexes: DNA intercalation in multilamellar membranes in distinct interhelical packing regimes. Science (Washington, DC) 1997; 275(5301):810–814.

113. Kawakami S, Harada A, Sakanaka K, Nishida K, Nakamura J, Sakaeda T, Ichikawa N, Nakashima M, Sasaki H. In vivo gene transfection via intravitreal injection of cationic liposome/plasmid DNA complexes in rabbits. Int J Pharm 2004; 278(2):255–262.

114. Hope MJ, Mui B, Ansell S, Ahkong QF. Cationic lipids, phosphatidylethanolamine and the intracellular delivery of polymeric, nucleic acid-based drugs. Mol Membr Biol 1998; 15(1):1–14.

115. Xu Y, Szoka FC Jr. Mechanism of DNA release from cationic liposome/DNA complexes used in cell transfection. Biochemistry 1996; 35(18):5616–5623.

116. Szoka FC Jr, Xu Y, Zelphati O. How are nucleic acids released in cells from cationic lipid-nucleic acid complexes? J Liposome Res 1996; 6(3):567–587.

117. Safinya CR. Structures of lipid-DNA complexes: supramolecular assembly and gene delivery. Curr Opin Struct Biol 2001; 11(4):440–448.

118. Gao X, Huang L. Cationic liposome-mediated gene transfer. Gene Ther 1995; 2(10):710–722.

119. Simberg D, Hirsch-Lerner D, Nissim R, Barenholz Y. Comparison of different commercially available cationic lipid-based transfection kits. J Liposome Res 2000; 10(1):1–13.

120. Audouy SAL, de Leij LFMH, Hoekstra D, Molema G. In vivo characteristics of cationic liposomes as delivery vectors for gene therapy. Pharm Res 2002; 19(11):1599–1605.

121. Kang S-H, Zirbes EL, Kole R. Delivery of antisense oligonucleotides and plasmid DNA with various carrier agents. Antisense Nucleic Acid Drug Dev 1999; 9(6):497–505.

122. Gao X, Huang L. A novel cationic liposome reagent for efficient transfection of mammalian cells. Biochem Biophys Res Commun 1991; 179(1):280–285.

123. Stern M, Alton EWFW. Use of liposomes in the treatment of cystic fibrosis. Albelda SM, ed. Gene therapy in lung disease. New York: Marcel Dekker, 2002:383–396.

124. Yoo GH, Hung M-C, Lopez-Berestein G, LaFollette S, Ensley JF, Carey M, Batson E, Reynolds TC, Murray JL. Phase I trial of intratumoral liposome E1A gene therapy in patients with recurrent breast and head and neck cancer. Clin Cancer Res 2001; 7(5):1237–1245.

125. Hortobagyi GN, Ueno NT, Xia W, Zhang S, Wolf JK, Putnam JB, Weiden PL, Willey JS, Carey M, Branham DL, Payne JY, Tucker SD, Bartholomeusz C, Kilbourn RG, De Jager RL, Sneige N, Katz RL, Anklesaria P, Ibrahim NK, Murray JL, Theriault RL, Valero V, Gershenson DM, Bevers MW, Huang L, Lopez-Berestein G, Hung M-C. Cationic liposome-mediated E1A gene transfer to human breast and ovarian cancer cells and its biologic effects: a phase I clinical trial. J Clin Oncol 2001; 19(14):3422–3433.

126. Lappalainen K, Jaaskelainen I, Syrjanen K, Urtti A, Syrjanen S. Comparison of cell proliferation and toxicity assays using two cationic liposomes. Pharm Res 1994; 11(8):1127–1131.

127. Filion MC, Phillips NC. Toxicity and immunomodulatory activity of liposomal vectors formulated with cationic lipids toward immune effector cells. Biochim Biophys Acta 1997; 1329(2):345–356.

128. Dokka S, Toledo D, Shi X, Castranova V, Rojanasakul Y. Oxygen radical-mediated pulmonary toxicity induced by some cationic liposomes. Pharm Res 2000; 17(5):521–525.

129. Scheule RK, St George JA, Bagley RG, Marshall J, Kaplan JM, Akita GY, Wang KX, Lee ER, Harris DJ, Jiang C, Yew NS, Smith AE, Cheng SH. Basis of pulmonary toxicity associated with cationic lipid-mediated gene transfer to the mammalian lung. Hum Gene Ther 1997; 8(6):689–707.

130. Armeanu S, Pelisek J, Krausz E, Fuchs A, Groth D, Curth R, Keil O, Quilici J, Rolland PH, Reszka R, Nikol S. Optimization of nonviral gene transfer of vascular smooth muscle cells in vitro and in vivo. Mol Ther 2000; 1(4):366–375.

131. Freimark BD, Blezinger HP, Florack VJ, Nordstrom JL, Long SD, Deshpande DS, Nochumson S, Petrak KL. Cationic lipids enhance cytokine and cell influx levels in the lung following administration of plasmid:cationic lipid complexes. J Immunol 1998; 160(9):4580–4586.

132. Dokka S, Malanga CJ, Shi X, Chen F, Castranova V, Rojanasakul Y. Inhibition of endotoxin-induced lung inflammation by interleukin-10 gene transfer in mice. Am J Physiol 2000; 279(5):L872–L877.

133. Takano S, Aramaki Y, Tsuchiya S. Physicochemical properties of liposomes affecting apoptosis induced by cationic liposomes in macrophages. Pharm Res 2003; 20(7):962–968.

134. Fillion P, Desjardins A, Sayasith K, Lagace J. Encapsulation of DNA in negatively charged liposomes and inhibition of bacterial gene expression with fluid liposome-encapsulated antisense oligonucleotides. Biochim Biophys Acta 2001; 1515(1):44–54.

135. Lakkaraju A, Dubinsky JM, Low WC, Rahman Y-E. Neurons are protected from excitotoxic death by p53 antisense oligonucleotides delivered in anionic liposomes. J Biol Chem 2001; 276(34):32000–32007.

136. Guidance for Industry, Liposome Drug Products, Chemistry, Manufacturing, and Controls; Human Pharmacokinetics and Bioavailability; and Labeling Documentation. Rockville, MD: US Department of Health and Services, Food and Drug Administration, Center for Drug Evaluation and Research (CDER), 2002.

137. Walde P. Preparation of vesicles (liposomes). In: Nalwa HS, ed. Encyclopedia Nanoscience of Nanotechnology. Vol. 9. Stevenson Ranch, CA: American Scientific Publishers, 2000: 43–79.

138. Betageri GV, Kulkarni SB. Preparation of liposomes. In: Microspheres Microcapsules & Liposomes-Preparation and Chemical Applications. Vol. 1. London: Citrus Books, 1999:489–521.

139. Hope MJ, Bally MB, Mayer LD, Janoff AS, Cullis PR. Generation of multilamellar and unilamellar phospholipid vescicles. Chem Phys Lipids 1986; 40:89–107.

140. Bally MB, Hope MJ, Mayer LD, Madden TD, Cullis PR. Novel procedures for generating and loading liposomal systems. In: Gregoriadis G, ed. Liposomes as Drug Carriers, Recent Trends and Progress. New York: John Wiley & Sons Ltd, 1988:841–853.

141. Mayer LD, Hope MJ, Cullis PR, Janoff AS. Solute distributions and trapping efficiencies observed in freeze-thawed multilamellar vesicles. Biochim Biophys Acta 1985; 817(1): 193–196.

142. Cullis PR, Mayer LD, Bally MB, Madden TD, Hope MJ. Generating and loading of liposomal systems for drug-delivery applications. Adv Drug Deliv Rev 1989; 3(3):267–282.

143. Hope MJ, Nayar R, Mayer LD, Cullis PR. Reduction of liposome size and preparation of unilamellar vesicles by extrusion techniques. In: Gregoriadis G, ed. Liposome Technology. Vol. I. 2nd ed. Boca Raton: CRC Press, 1993: 123–139.

144. Szoka F Jr, Papahadjopoulos D. Procedure for preparation of liposomes with large internal aqueous space and high capture by reverse-phase evaporation. Proc Natl Acad Sci USA 1978; 75(9):4194–4198.

145. Pidgeon C, McNeely S, Schmidt T, Johnson JE. Multilayered vesicles prepared by reverse-phase evaporation: liposome structure and optimum solute entrapment. Biochemistry 1987; 26(1):17–29.

146. Gruner SM, Lenk RP, Janoff AS, Ostro MJ. Novel multilayered lipid vesicles: comparison of physical characteristics of multilamellar liposomes and stable plurilamellar vesicles. Biochemistry 1985; 24(12):2833–2842.

147. Pidgeon C. Preparation of MLV by the REV method: vesicle structure and optimum solute entrapment. In: Gregoriadis G, ed. Liposome Technology. Vol. I. 2nd ed. Boca Raton: CRC Press, 1993:99–110.

148. Domazou AS, Luisi PL. Size distribution of spontaneously formed liposomes by the alcohol injection method. J Liposome Res 2002; 12(3):205–220.

149. Maitani Y, Soeda H, Junping W, Takayama K. Modified ethanol injection method for liposomes containing b-sitosterol b-D-glucoside. J Liposome Res 2001; 11(1):115–125.

150. Batzri S, Korn ED. Single bilayer liposomes prepared without sonication. Biochim Biophys Acta 1973; 298(4):1015–1019.

151. Wagner A, Vorauer-Uhl K, Katinger H. Liposomes produced in a pilot scale: production, purification and efficiency aspects. Eur J Pharm Biopharm 2002; 54(2):213–219.

152. Kriftner RW. Liposome production: the ethanol injection technique and the development of the first approved liposome dermatic. In: Braun-Falco O, Korting HC, Maibach HI, eds. Liposome Dermatics: Griesbach conference. New York: Springer Verlag, 1992:91–100.

153. Weder HG. Liposome production: the sizing-up technology starting from mixed micelles and the scaling-up procedure for the topical glucocorticoid betamethasone dipropionate and betamethasone. In: Braun-Falco O, Korting HC, Maibach

HI, eds. Liposome Dermatics: Griesbach conference. New York: Springer Verlag; 1992:101–109.

154. Schubert R. Liposome preparation by detergent removal. Methods Enzymol 2003; 367:46–70.

155. Pirkl V, Jaroni HW, Schubert R, Schmidt KH. Liposome-encapsulated hemoglobin as oxygen-carrying blood substitute. Life Support Syst 1986; 4(suppl 2):408–410.

156. Saravolac EG, Ludkovski O, Skirrow R, Ossanlou M, Zhang YP, Giesbrecht C, Thompson J, Thomas S, Stark H, Cullis PR, Scherrer P. Encapsulation of plasmid DNA in stabilized plasmid-lipid particles composed of different cationic lipid concentration for optimal transfection activity. J Drug Targeting 2000; 7(6):423–437.

157. Van Winden ECA. Freeze-drying of liposomes: theory and practice. Methods Enzymol (Liposomes, Part A) 2003; 367:99–110.

158. Crommelin DJA, Van Bommel EMG. Stability of liposomes on storage: freeze dried, frozen or as an aqueous dispersion. Pharm Res 1984; 4:159–163.

159. Cortesi R, Esposito E, Nastruzzi C. Effect of DNA complexation and freeze-drying on the physicochemical characteristics of cationic liposomes. Antisense Nucleic Acid Drug Dev 2000; 10(3):205–215.

160. Toliat T, Arab N, Rafiee-Tehrani M. Effect of various cryoprotectants on freeze-drying of liposomes containing Na methotrexate. Proc Int Symp Control Release Bioactive Mater 1998; 25:926–927.

161. Wolkers WF, Oldenhof H, Tablin F, Crowe JH. Preservation of dried liposomes in the presence of sugar and phosphate. Biochim Biophys Acta 2004; 1661(2):125–134.

162. Crowe JH, Leslie SB, Crowe LM. Is vitrification sufficient to preserve liposomes during freeze-drying? Cryobiology 1994; 31(4):355–366.

163. Barenholz Y. Relevancy of drug loading to liposomal formulation therapeutic efficacy. J Liposome Res 2003; 13(1):1–8.

164. Mayer LD, Bally MB, Hope MJ, Cullis PR. Techniques for encapsulating bioactive agents into liposomes. Chem Phys Lipids 1986; 40(2–4):333–345.

165. Patil SD, Rhodes DG. Conformation of oligodeoxynucleotides associated with anionic liposomes. Nucleic Acids Res 2000; 28(21):4125–4129.

166. Goundalkar A, Mezei M. Chemical modification of triamcinolone acetonide to improve liposomal encapsulation. J Pharm Sci 1984; 73(6):834–835.

167. Maurer N, Fenske DB, Cullis PR. Developments in liposomal drug delivery systems. Expert Opin Biol Ther 2001; 1(6): 923–947.

168. Mayer LD, Bally MB, Cullis PR. Uptake of adriamycin into large unilamellar vesicles in response to a pH gradient. Biochim Biophys Acta 1986; 857(1):123–126.

169. Madden TD, Harrigan PR, Tai LCL, Bally MB, Mayer LD, Redelmeier TE, Loughrey HC, Tilcock CPS, Reinish LW, Cullis PR. The accumulation of drugs within large unilamellar vesicles exhibiting a proton gradient: a survey. Chem Phys Lipids 1990; 53(1):37–46.

170. Haran G, Cohen R, Bar LK, Barenholz Y. Transmembrane ammonium sulfate gradients in liposomes produce efficient and stable entrapment of amphipathic weak bases. Biochim Biophys Acta 1993; 1151(2):201–215.

171. Li C, Deng Y. A novel method for the preparation of liposomes: freeze drying of monophase solutions. J Pharm Sci 2004; 93(6):1403–1414.

172. Ruozi B, Forni F, Battini R, Vandelli MA. Cationic liposomes for gene transfection. J Drug Targeting 2003; 11(7):407–414.

173. Eastman SJ, Scheule RK. Cationic lipid:pDNA complexes for the treatment of cystic fibrosis. Curr Opin Mol Ther 1999; 1(2):186–196.

174. Siewert M, Dressman J, Brown CK, Shah VP, Aiache J-M, Aoyagi N, Bashaw D, Brown C, Brown W, Burgess D, Crison J, DeLuca P, Djerki R, Foster T, Gjellan K, Gray V, Hussain A, Ingallinera T, Klancke J, Kraemer J, Kristensen H, Kumi K, Leuner C, Limberg J, Loos P, Marguilis L, Marroum P,

Moeller H, Mueller B, Mueller-Zsigmondy M, Okafo N, Ouderkirk L, Parsi S, Qureshi S, Robinson J, Shah V, Uppoor R, Williams R. FIP/AAPS guidelines to dissolution/in vitro release testing of novel/special dosage forms. AAPS Pharm Sci Tech 2003; 4(1):43–52.

175. Burgess DJ, Crommelin DJA, Hussain AS, Chen M-L. Assuring quality and performance of sustained and controlled release parenterals. Eur J Pharm Sci 2004; 21(5):679–690.

176. Burgess DJ, Hussain AS, Ingallinera TS, Chen M. Assuring quality and performance of sustained and controlled release parenterals: workshop report. AAPS PharmSci 2002; 4(2):E7.

177. Kanno T, Yamada T, Iwabuki H, Tanaka H, Kuroda SI, Tanizawa K, Kawai T. Size distribution measurement of vesicles by atomic force microscopy. Anal Biochem 2002; 309(2): 196–199.

178. Kinnunen P, Alakoskela J-M, Laggner P. Phase behavior of liposomes. Methods Enzymol (Liposomes, Part A) 2003; 367: 129–147.

179. Cohen JA. Electrophoretic characterization of liposomes. Methods Enzymol (Liposomes, Part A) 2003; 367:148–176.

180. Perkins WR, Minchey SR, Ahl PL, Janoff AS. The determination of liposome captured volume. Chem Phys Lipids 1993; 64(1–3):197–217.

181. Gruber HJ, Wilmsen HU, Schurga A, Pilger A, Schindler H. Measurement of intravesicular volumes by salt entrapment. Biochim Biophys Acta 1995; 1240(2):266–276.

182. Grit M, Crommelin DJA. Chemical stability of liposomes: implications for their physical stability. Chem Phys Lipids 1993; 64(1–3):3–18.

183. Heurtault B, Saulnier P, Pech B, Proust J-E, Benoit J-P. Physico-chemical stability of colloidal lipid particles. Biomaterials 2003; 24(23):4283–4300.

184. Zuidam NJ, Gouw HK, Barenholz Y, Crommelin DJ. Physical (in) stability of liposomes upon chemical hydrolysis: the role of lysophospholipids and fatty acids. Biochim Biophys Acta 1995; 1240(1):101–110.

185. Caldwell RA, Baumgarten CM. Plasmalogen-derived lysolipid induces a depolarizing cation current in rabbit ventricular myocytes. Circ Res 1998; 83(5):533–540.

186. Sobel BE, Corr PB. Biochemical mechanisms potentially responsible for lethal arrhythmias induced by ischemia: the lysolipid hypothesis. Adv Cardiol (New Approaches Diagn Manage Cardiovasc Dis) 1979; 26:76–85.

187. Yurkova IL, Shadyro I, Davydov VY, Kisel MA. The effect of phosphatidic acid on the radiation-initiated peroxidation of phosphatidylcholine in liposomes. Radiatsionnaya Biologiya Radioekologiya 2004; 44(2):142–145.

188. Siu FKW, Lo SCL, Leung MCP. Effectiveness of multiple pre-ischemia electro-acupuncture on attenuating lipid peroxidation induced by cerebral ischemia in adult rats. Life Sci 2004; 75(11):1323–1332.

189. Vitrac H, Courregelongue M, Couturier M, Collin F, Therond P, Remita S, Peretti P, Jore D, Gardes-Albert M. Radiation-induced peroxidation of small unilamellar vesicles of phosphatidylcholine generated by sonication. Can J Physiol Pharmacol 2004; 82(2):153–160.

190. Jiang J, Serinkan BF, Tyurina YY, Borisenko GG, Mi Z, Robbins PD, Schroit AJ, Kagan VE. Peroxidation and externalization of phosphatidylserine associated with release of cytochrome c from mitochondria. Free Radic Biol Med 2003; 35(7):814–825.

191. Woodall AA, Britton G, Jackson MJ. Carotenoids and protection of phospholipids in solution or in liposomes against oxidation by peroxyl radicals: relationship between carotenoid structure and protective ability. Biochim Biophys Acta 1997; 1336(3):575–586.

192. Nohl H, Bieberschulte W, Dietrich B, Udilova N, Kozlov AV. Caroverine, a multifunctional drug with antioxidant functions. BioFactors 2003; 19(1/2):79–85.

193. Borisova NV, Zhigal'tsev IV, Bogomolov OV, Kaplun AP, Iurasov VV, Kucherianu VG, Nikushkin EV, Kryzhanovskii GN, Shvets VI. Oxidation in liposomes from egg phosphatidylcholine loaded with L-3,4-dihydroxyphenylalanine (DOPA)

and dopamine: mutual effect of components. Bioorganiches-
kaia khimiia 1997; 23(4):284–289.

194. Samuni AM, Barenholz Y, Crommelin DJA, Zuidam NJ. g-
Irradiation damage to liposomes differing in composition
their protection by nitroxides. Free Radic Biol Med 1977;
23(7):972–979.

195. Casals E, Galan AM, Escolar G, Gallardo M, Estelrich J.
Physical stability of liposomes bearing hemostatic activity.
Chem Phys Lipids 2003; 125(2):139–146.

196. Senior J, Gregoriadis G. Methodology in assessing liposomal
stability in the presence of blood, clearance from the circulation of injected animals, and uptake by tissues. In: Gregoriadis G, ed. Liposome Technology. Vol. III. Boca Raton, FL:
CRC Press, 1984:263–282.

197. Comiskey SJ, Heath TD. Serum-induced leakage of nega-
tively-charged liposomes at nanomolar lipid concentrations.
Biochemistry 1990; 29(15):3626–3631.

198. Anderson M, Omri A. The effect of different lipid components
on the in vitro stability and release kinetics of liposome
formulations. Drug Deliv 2004; 11(1):33–39.

199. Hashizaki K, Taguchi H, Itoh C, Sakai H, Abe M, Saito Y,
Ogawa N. Effects of poly(ethylene glycol) (PEG) chain length
of PEG-lipid on the permeability of liposomal bilayer
membranes. Chem Pharm Bull 2003; 51(7):815–820.

200. Armengol X, Estelrich J. Physical stability of different liposome compositions obtained by extrusion method. J Microencapsul 1995; 12(5):525–535.

201. Zuidam NJ, Talsma H, Crommelin DJA. Sterilization of
liposomes. In: Lasic DD, Barenholz Y, eds. Handbook nonmedical applications of liposomes. Vol. 3. Boca Raton: CRC
Press, 1996:71–80.

9

Microspheres: Design and Manufacturing

DIANE J. BURGESS

Department of Pharmaceutical Sciences, School of Pharmacy, University of Connecticut, Storrs, Connecticut, U.S.A.

ANTHONY J. HICKEY

School of Pharmacy, University of North Carolina, Chapel Hill, North Carolina, U.S.A.

Many new candidate therapeutic agents are extremely potent and must be dispensed in a controlled and accurate fashion, preferably at the site of action. These compounds may be delivered by a number of routes of administration each offering advantages and limitations based on the application.

Microspheres can be utilized to achieve drug targeting and/or controlled release. The manufacture and production of suitable microsphere systems requires consideration of issues including the physicochemical properties of the drug and other components of the system, conditions of manufacture, unit operations, scale-up conditions as well as production scale manufacturing resources and equipment. The following is an overview of the physical, chemical, biological, and engineering principles underlying development of microsphere systems.

1. INTRODUCTION

Microspheres are defined as solid, approximately spherical particles ranging in size from 1 to 1000 μm (1). Sometimes, the definition of microspheres is extended into the nano-size range. Microspheres are usually made of polymeric or other protective materials. Drugs and other substances may be incorporated within microspheres either as an encapsulated core (microcapsules) or homogeneously dispersed throughout the microspheres (micromatrices) (Fig. 1) (1). Refer to Burgess and Hickey (1) for a historical and contextual perspective on the application of microsphere technology to the pharmaceutical and other industries.

Microspheres have been proposed as a parenteral delivery system for drugs and other tissue response modifiers (TRMs). There are four microsphere drug products on the market (2,3) and several in clinical trials (134). TRMs include: traditional small molecule drugs, enzymes, proteins, DNA, vaccines, and cells. Biopharmaceutical and physicochemical advantages of parenteral microsphere delivery are numerous. Microsphere preparation may enhance the chemical stability of the TRM, extend its residence time at the site of administration, result in physical targeting, protect the TRM from

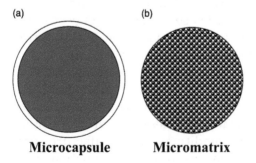

(a)   (b)

Microcapsule   **Micromatrix**

Figure 1 Schematic diagram of (a) microcapsule and (b) micromatrix.

biological degradation, result in controlled delivery, protect the in vivo environment from the TRM (e.g., immune response to protein therapeutics), and increase safety (subdivision of the dose can avoid dose dumping problems that may occur on failure of a single unit implant). In addition, microspheres used for vaccine delivery may act as an immune adjuvant. Microspheres are also useful as the dosing frequency can be reduced when a controlled release product is administered. This method of delivery is important for drugs for which there is no other satisfactory dosing technique, e.g., proteins that are rapidly degraded and cleared when administered by standard injection. However, microspheres are also under consideration for classes of small molecular weight drugs that require localized therapy and controlled release.

1.1. Small Molecular Weight Drugs

Examples of small molecular weight drugs that have been investigated for parenteral delivery in microsphere systems include: narcotic antagonists, antibiotics, local anesthetics, anti-malarials, anti-cancer drugs, narcotic antagonists, and steroids (1,4–12). Local delivery of antibiotics can assure adequate tissue levels at the local site of infection. For example, localized delivery of gentamicin in microsphere systems has been utilized successfully in animal models for acute and chronic bone infections (13). This approach is also used to prevent bone infections that could arise following surgery. Some local sites are particularly problematic in terms of achieving adequate tissue levels. Examples include isolated tissues, such as the inner ear and cancerous tissue. Controlled release microspheres have been used for localized delivery of steroids to joints for the treatment of arthritis (14,15). These delivery systems allow therapeutic concentrations to be maintained at the local site for an extended period of time compared to drug solutions or suspensions. Hence dose frequency can be reduced which is important for patient compliance as intra-articular injections are particularly painful. Localized delivery in microsphere formulations can also be used to avoid side effects that are often associated with systemic delivery. For example, systemic administration of high concentrations of antibiotics can result in organ toxicity, such as kidney damage (16). Corticosteroids and anti-cancer drugs are two other examples of drug classes with high systemic toxicity. The systemic delivery of anti-cancer drugs may cause toxicity to all rapidly dividing cells. Consequently, targeted microsphere delivery systems have been applied to anti-cancer drug delivery to reduce systemic side effects and ensure adequate drug levels in the tumor (17).

1.2. Protein Therapeutics

Over the past two decades, the therapeutic products of biotechnology (proteins, peptides, and DNA) have been investigated as candidate compounds for delivery in microsphere systems for parenteral delivery (18,19). For example, leutenizing hormone releasing hormone (LHRH) analogs have been investigated and there is currently one microsphere product containing leuprolide acetate (Leupron Depot). There are currently four microsphere products on the US market (Lupron Depot, Sandostatin LAR, Nutropin Depot, and Trelstar Depot) (134).

Biotechnology therapeutics are usually administered parenterally since these molecules are susceptible to degradation in the gastrointestinal tract (GIT) as a consequence of low pH and the high concentrations of peptidases. Localized delivery of these agents is desirable as they are expensive and may give rise to side effects at other sites. Parenteral delivery is also problematic since these pharmacological agents need to be protected from the environment (e.g., peptidases present at the local delivery site). Consequently the in vivo half-lives of these therapeutics are usually very short. An additional problem is that immune responses may occur. Delivery systems, such as microspheres, are currently under investigation to overcome such problems. Microspheres offer an advantage over other dispersed systems (e.g., liposomes and emulsions) in that microspheres are more stable, can carry higher drug loadings and can result in longer release profiles. Extended release is advantageous in reducing dosing frequency, as otherwise daily injections are often required for chronic therapy. For example, release rates in the order of months can be achieved when relatively hydrophobic polymers such as poly-lactic-co-glycolic acid (PLGA) are used.

Vascular endothelial growth factor (VEGF) delivered in PLGA microspheres has resulted in new blood vessel growth at the subcutaneous (s.c.) injection site in rats over a 1 month study period (20,135,136). This can be compared to no growth of new blood vessels on injection of unencapsulated VEGF. Like most proteins, VEGF has a short half-life in the body and is rapidly cleared from the injection site. These problems are overcome using the slow release microencapsulated form of the drug.

1.2.1. Live Cells

Microencapsulation of cells as a means of achieving artificial organs and for tissue transplant purposes has been investigated extensively over the past four decades. This area was extensively reviewed by Burgess and Hickey (1). Other publications on this subject have appeared more recently (21–23). Key issues are the use of a biocompatible polymer systems, which provides a semipermeable membrane for nutrient exchange, and a method of manufacture which is non-destructive to the cells. The alginate-polylysine microencapsulation process involving divalent calcium ion gelling of the alginate has been a popular method for live cell encapsulation (24). The processing involved is not destructive to the cells as a calcium alginate gel is formed initially which protects the cells. A "permanent" alginate-polylysine membrane is then formed which is semipermeable, allowing nutrients into the microcapsules but excluding substances that might be harmful (such as high molecular weight antibodies which would be destructive to implanted cells). For example, insulin-secreting cells have been encapsulated in alginate-polylysine membranes to prevent rejection. Purified alginate with high mannuronic acid content has been shown to have good biocompatibility and is able to form stable microcapsules.

Recent advances in the area of cell encapsulation include the use of polymer scaffolds for tissue engineering purposes. Cell growth rates should be considered in cell encapsulation, as rapidly growing cells will "grow through" the microcapsule membrane (25). This may be acceptable in some cases of tissue engineering once cell growth is established, where encapsulation is used for initial protection of the cells.

Disadvantages of microspheres for controlled release parenterals include: difficulty of removal from the site, low drug loading, possible drug degradation within the microspheres, and changes in drug crystallinity or polymorphic form during microsphere processing. Generally, the amount of drug that can be incorporated into microspheres is limited to approximately 50% at the high end (1). However, 20% or less is more common, therefore microsphere delivery is particularly relevant for highly potent drugs such as peptides and steroids otherwise the total amount that would need to be injected could be prohibitive.

Parenteral administration involves the injection of solutions, or dispersed systems through the skin usually to a tissue or body cavity from which distribution to other locations in the body occurs. The most frequently used approach is to deliver drugs directly to the venous blood supply for systemic distribution. For dispersed systems, this is usually followed by organ disposition. The reticulo-endothelial system (RES) of the liver is particularly suited to the removal of small circulating dispersed droplets or particulates (26). The capillary system of the lung, by virtue of the narrow passage presented to circulating blood, is also capable of filtering particulates. This can be a limiting factor for any intravenously (i.v.) injected dispersed system as pulmonary embolism may result from lung deposition in this manner. Subcutaneous (below the dermis) and intradermal (between the dermis and the epidermis) injection results in local depots of material, which have greater or lesser, respectively, access to local blood supply. Consequently, systemic bioavailability of materials administered by these routes is dependent upon the physicochemical properties of the drug including solubility, partition coefficient, and diffusion coefficient (27). Intramuscular (i.m.) and intraperitoneal injections deliver drug to heavily vascularized areas. Intramuscular injections form a well-defined depot in a mass of tissue to which a good blood supply is required since this is the major location for energy consumption involved in locomotion. Intraperitoneal injections place materials in the peritoneum, which represents a large surface area served by maintaining blood supply in the form of the peritoneal wall, and the mesenteric blood supply to the gastrointestinal tract. For more information on biopharmaceutical aspects of microsphere delivery, refer to Chapter 2.

Biopharmaceutical and physicochemical advantages of microspheres are numerous. Microsphere preparations may enhance the chemical stability of the drug, extend its residence time at the site of administration, result in physical targeting, protect the drug from biological degradation, and result in rate controlled delivery of the drug. The multi-unit nature of dispersed systems offers an advantage in terms of safety, since failure of individual microspheres would not cause a serious concern compared to the failure of a single unit implant which would result in total dose dumping. In addition, different populations of microspheres may be administered to achieve a desired effect. For example, a cocktail of different drugs may be given or microspheres containing the same drug but with varying release rates may be blended to achieve a desired release profile. Some drugs, when injected in solution or suspension form, may result in tissue irritation at the injection site. In the case of solution formulations, this may be a consequence of the solvent system used and/or precipitation at the injection site, refer to the case study in this book by Brazeau. The use of microspheres or similar controlled release systems, such as liposomes, can overcome this problem by slow controlled release at the site. These controlled release systems can also improve drug bioavailability by releasing the drug at a rate that is comparable to or less than the rate of absorption at the site. Absorption rates depend on blood flow at the site as well as drug physicochemical factors such as hydrophobicity/hydrophilicity and the potential for partitioning into fatty tissue that may be present at the site. Refer to Chapter 2 for more information on bioavailability issues for dispersed systems. Microspheres may assist in targeting therapeutics to specific disease states and, as a consequence, toxic side effects can be avoided as well as reducing the drug concentration required, which is important for expensive therapeutic agents such as proteins.

Microencapsulation can protect drugs from physical and chemical degradation. For example, a microencapsulated drug may be protected from degradation by a variety of pathways such as light, hydrolysis, and oxidation. For protein drugs, microencapsulation may also protect against physical degradation such as aggregation. However, the reader should be aware that the various methods of microsphere manufacture can be harmful to protein drugs in particular. The most common methods of microsphere manufacture involve processing conditions such as high heat, use of organic solvents and agitation, all of which can result in physical instability of protein therapeutics. Refer below for microsphere manufacture.

2. MICROSPHERE DISPERSIONS: METHODS AND CHARACTERIZATION

2.1. Formulation/Composition

2.1.1. Typical Polymers

Table 1 shows a list of polymers employed in microsphere manufacture and their intended purpose. These consist of natural and synthetic polymers with a variety of properties. Among the properties of interest are the rate of degradation and nature of the erosion process. In this regard, PLGA microparticles may be contrasted with polyanhydride. PLGA erodes from every available wettable surface and after an initial surface erosion develops a porosity which results in the final structural collapse of the particle. Polyanhydride in contrast erodes from the surface in a manner dictated by the particle geometry and as such offers predictable degradation throughout its lifespan.

Particle Engineering

Each method of manufacture involves a number of processing variables that can be adjusted to achieve objectives of drug load, particle size, and distribution, porosity, tortuosity, and surface area. Since each method involves different variables, it is sufficient to note that key physicochemical

Table 1   Selected Papers Reflecting Use of Polymers for Parenteral Administration (1995–2003)

Component (matrix/ encapsulator)	Application	Literature source
PLA	Chemoembolization	28, 29
	Testosterone delivery system	30
	Adjuvants for adsorbed influenza virus	31
	Encapsulation of 5-fluorouracil for treatment of liver cancer	32
	Encapsulation of cisplatin for direct intratumoral injection with reduction of acute renal toxicity	33
PLA/PLGA	Antigen entrapment for diphtheria and tetanus vaccines	34, 35
PLA/PGA	Long-lasting ivermectin delivery system for control of livestock pests (larval horn flies, hematobia irritants)	36
PLGA	Intraocular delivery of guanosine	37
	DNA vaccine and encapsulation	38, 39
	Intracerebral treatment of malignant glioma	40
	Encapsulation of ultrasonographic contrast agents for differentiation of coagulation necrosis in adenocarcinoma tumors	41
	Microencapsulation of an influenza antigen for a single dose vaccine delivery system	42
	Pulsatile single immunization for HIV	43
Poly-(caprolactone)	Potential drug delivery system	44, 45
Polyanhydride	Encapsulation of rhodium (II) citrate, an anti-tumor agents	46
Poly(anhydride-co-imides)	Controlled delivery of vaccine antigens	47
Chitosan	Device for gadolinium neutron-capture therapy by intratumoral injection	48
	Delivery system for steroids (progesterone)	49

(Continued)

Table 1 Selected Papers Reflecting Use of Polymers for
Parenteral Administration (1995–2003) (*Continued*)

Component (matrix/ encapsulator)	Application	Literature source
	Delivery system for antibiotic agents	50
	Encapulation of bisphosphonates delivered by local implantation or injection for site-specific therapy in pathological conditions associated with bone destruction	51
	Controlled and localized delivery system of endothelial cell growth factor for stimulation of vascularization	52
Gelatin	Carrier matrix of basic fibroblast growth factor (bFGF) to enhance the vascularization	53
Gelatin/chondroitin 6-sulfate	Microspheres for implantation as an embolization material	54
	Microspheres as immunological adjuvant	55
	Intraarticular delivery system of therapeutic proteins	56
Gelatin/alginate	Sustained-release microparticles for delivery of interferon-α	57
Alginate	Delivery of TGF-β to inhibit fibrosis of the corpus cavernosum	58
Alginate/polylysine	Intrahepatic implantations	59
Magnetic dextran	Targeted drug delivery system for brain tumors	60
Hydroxyethyl-starch	Potential drug delivery system	61, 62
	Imaging	63

factors include starting concentrations of drug and polymer, partitioning, and solubility; and that processing variables include the presence of surfactant (for emulsion formulations), the presence of polymerizing agent (initiator or ionic species for synthetic or natural polymer cross-linking) stirring conditions, vessel geometry, nozzle geometry, and feed conditions (spray drying, spray freezing, and supercritical fluid manufacture) and in general, fluid flow, heat and mass transfer.

2.2. Desired Performance Characteristics and Relative Performance Measurements

Desirable performance characteristics must be considered in developing a formulation for a therapeutic agent. The first consideration is the target disease state or desired therapeutic intervention. This will dictate relevant routes of administration, dosage, and period and/or frequency of delivery. Thus, therapeutic, pharmacokinetic, and pharmacodynamic considerations may be used as criteria to judge the merits of dosage forms following characterization of relevant physicochemical properties.

The goal in preparing any drug delivery system is to maximize the therapeutic effect while minimizing toxicity. Contemporary concepts in drug delivery and disposition focus on specific delivery to the site of action in the absence of effects at any other site. This may be achieved by direct or indirect local delivery. Indirect targeted delivery can be brought about by using the capacity of the dosage form to preferentially localize in certain organs or tissues or by molecular modifications to the drug to achieve high receptor/enzyme/protein affinities.

Matrix or encapsulating components and processing parameters should be selected to produce particles with the desired drug load, particle size, surface area, surface and dissolution properties to achieve acceptable in vitro and ultimately in vivo dose delivery.

2.2.1. Release

Factors governing release from microsphere systems have been discussed by Burgess and Hickey (1). Mechanisms of release are from the microsphere surface, through pores in the microspheres, diffusion from swollen microspheres, and following erosion and/or bulk degradation. Polymer type is of importance, hydrophilic polymers tend to release drug much more rapidly than hydrophobic polymers as a result of penetration of aqueous media into the microspheres. In order to achieve extended release profiles of weeks or months, it is usually necessary to use hydrophobic polymers, such as polylactic and polyglycolic acid. Amorphous polymer structures facilitate absorption of water compared to crystalline polymers and therefore amorphous polymers and polymers with a high proportion of amorphous regions tend to have faster release rates compared to crystalline polymers or polymers with a high proportion of crystalline regions. Crystalline regions decrease the diffusion rate of drug molecules by increasing the diffusional path length. Crystalline regions of polymers have a higher density and a lower specific volume than amorphous regions (64). Polylactic acid is an example of a polymer with a high proportion of crystalline regions, whereas polyglycolic acid has a high proportion of amorphous regions. These two polymers are usually used in combination to achieve desired release rates.

Other factors affecting release rates include the drug solubility, diffusivity, molecular weight, and particle size; the microsphere particle size, the percentage loading of the drug, the dispersion of the drug in the microspheres, any drug/polymer interactions, the rate of biodegradation of the polymer and the stabilities of the polymer matrix and the drug in the polymer matrix before and after injection. The larger the microsphere particle size, the longer the diffusional path length and the smaller the surface to volume ratio; therefore, release rates of larger microspheres are usually slower. For large drug molecules, the release rate is usually dominated by the polymer properties. If the microspheres have low porosity, then the polymer must degrade in order to create channels for drug release.

PLGA microspheres have been successfully used to achieve extended release profiles. Controlled release from PLGA microspheres will be discussed since this system has been widely investigated; there are five products on the market and several are in clinical trials. Although PLGA might not be the most appropriate polymer for all applications, this system is very attractive since it has been used in sutures for many years. Investigators have used a number of methods to increase and decrease release rates from this polymer system (65). Release can be diffusion and/or degradation controlled. The more crystalline the PLGA co-polymer, the higher the molecular weights of both the polymer and the drug and the less porous the microspheres, the slower the release rate. The slower the diffusion rate of the drug, the greater the contribution that polymer degradation will make to release (64). Porosity can be controlled by adding water soluble excipients such as NaCl which rapidly dissolves and diffuses away from the microspheres, leaving pores. High concentrations of albumin have been added to increase release rates of high molecular weight drugs such as protein therapeutics. The albumin releases and leaves behind large pores and the protein therapeutic agent can then diffuse out along these pores (64). Release is also ionic strength dependent. High ionic strength inside the microspheres results in the influx of water. Similarly, decreased release rates elevate the ionic strength of the medium (66). The addition of plasticizers should give enhanced release due to less resistance to diffusion. However, the use of plasticizers can result in a less porous microsphere matrix and therefore drug release can be reduced (64). Since PLGA degradation is acid catalyzed, the addition of salts such as calcium carbonate and magnesium hydroxide confers an alkaline pH and reduces PLGA degradation rates (67).

All of the above factors can be manipulated to achieve the desired release rates. Mixed populations of microspheres may also be used to this end (137,138). As a consequence of the manufacturing process, microspheres often have surface-associated drug which can contribute to a burst release effect. Microspheres can be washed with water before release testing to determine surface associated drug. It is also possible to use predegraded microspheres to avoid the initial burst release effect (137,138).

2.3. In Vitro Release Testing

The reader is referred to the chapter in this book by Clark et al., for additional information on in vitro release testing.

In vitro testing of microsphere release rates is often conducted by suspending the microspheres in the release media under conditions of mild stirring. The samples must be processed to remove any suspended microspheres prior to analysis to avoid interference with the analytical method and so that the microspheres are returned to the release media to continue the release process. This is usually achieved by filtration or centrifugation. This in vitro release methodology is referred to as sample and separate. The continuous flow method (USP apparatus 4) has also been used for microspheres. Here the microspheres are maintained in a compartment and the release media is continuously circulated through this compartment (closed system) or fresh media is continuously pumped through (open system). Filters are used to isolate the microspheres in the compartment. Glass beads can be added to avoid microsphere aggregation and to alter the flow pattern so as to avoid the production of flow channels within the microsphere bed that would lead to inaccurate release profiles due to uneven contact between individual microspheres and the release media. The USP 4 flow-through method has been recommended for microspheres as it can avoid problems of microsphere aggregation and sink conditions can be easily maintained (134). Problems may arise due to filter blockage and this should be monitored periodically.

USP apparatus 1 has been used to test in vitro release rates from microspheres. Typically microspheres are suspended in 900 mL of release medium placed in dissolution apparatus and stirred using an overhead stirrer at 100 rpm (68). The amount of microspheres used is dependent on microsphere drug loading and sink conditions. The relatively slow stirring rates used can be problematic as the microspheres may settle in the large USP vessels under slow stirring. The volumes used are another problem particularly for expensive biotech drugs. In addition, large volumes are not representative of the parenteral situation with the exception of the i.v. route and as previously discussed this route is generally not applicable for microspheres.

Release media can be standard, pH 7.4, phosphate buffer or other suitable media depending on the drug. Since microspheres usually have hydrophobic surfaces, dispersing agents (such as Tween surfactants) are often added to assure dispersion of the microspheres in the media. Ethanolic phosphate media has been used for PLGA microspheres as this causes plastization of the polymer, simulating the plastization effect of lipids in vivo. Non-ethanolic degradation of PLGA microspheres in vitro has been reported, in some cases, to be two to six times slower than occurs in plasma.

Several variations on miniaturized release methods have been reported for systems intended for parenteral use. These include: a miniaturized version of the standard USP apparatus 1 method in a scaled down beaker using 50–100 mL of media; a dialysis sac method using volumes in the order of 50–100 mL; magnetic stirring in place of overhead stirring that is used in the USP method; small sample vial method, using 1–10 mL volumes and shaking or rotating rather than stirring. If a sufficiently miniaturized method is used then each vial can be used for a single sample. The vial is centrifuged and the whole supernatant is taken for the sample. The pellet may then be discarded or analyzed for polymer degradation or percent drug remaining.

A problem that can arise during in vitro release testing of microspheres is floating of microspheres composed of hydrophobic materials (such as, PLGA) due to difficulty in wetting. To overcome this problem, Tween 80 or similar surfactants can be added to the dissolution media. Care should be taken as surfactants will solubilize hydrophobic drugs at concentrations above the surfactant CMC. This can be used as a method of obtaining sink conditions without having to dilute the sample too much and therefore can help in analysis to obtain concentrations within the detectable range. Microsphere hydrophobicity can also result in aggregation of the particles at the bottom of the dissolution vessel. This can result in irreproducible data. The surface area exposed and hence the release rates may be reduced as a consequence of aggregation, again resulting in irreproducible data.

Another problem is drug degradation during release. For microspheres designed to release drug over periods of 1 month or more, this can be a significant problem for drug released early and then left in the dissolution media during the study. Degradation can be accounted for mathematically, if the degradation rate in the dissolution media is first calculated (69).

It is also important to determine if any drug remains at the end of the release study. Samples can be filtered, the polymer dissolved and the drug extracted as appropriate (refer Sec. 2.6). In vitro release studies reported in the literature usually are incomplete and this may be due to drug degradation or to incomplete release form the microsphere system.

In vitro–in vivo correlation (IVIVC) is the ultimate goal and therefore in vitro dissolution methods should reflect the in vivo situation as much as possible. It should be possible to establish guidelines for IVIVC for controlled release parenterals through a systematic evaluation of in vivo and delivery system factors. The exact in vivo situation need not be reproduced in vitro, but a situation that results in the same outcome. For example, in vivo release from microspheres may be enhanced compared to in vitro release as a result of enzymatic degradation of the polymer. Polymer degradation could be enhanced to the same degree in vitro by alteration of pH or some other variable that affects polymer degradation (e.g., PLGA degradation is enhanced by reduction in pH). Until recently, there have only been rank order correlations between in vitro and in vivo drug release from microsphere systems (14,68). However, in the last 2 years, there have been IVIVC reports for controlled release parenteral microsphere systems and the reader is referred to the chapters in this book by Clark et al., by Young, and by Chen for more information.

2.4. In Vivo Release Testing

In vivo release rates are usually determined indirectly from drug plasma levels. Animal studies have been conducted where release has been measured directly through serial sacrifice experiments where the tissue is excised, homogenized and the amount remaining at the site is determined (13,139). These animal studies are useful in determining the release mechanism and in vivo factors that affect release rates.

2.4.1. Polymer Degradation

Gel permeation chromatography (GPC) has been utilized to determine the molecular weight distribution of PLA and PLGA before and after different incubation times (70). Scanning electron microscopy (SEM) has been used to assess the extent of microsphere degradation.

2.5. Particle Size

Particle size is dependent on the method of microsphere manufacture. For the commonly used emulsion methods, microsphere particle size can be altered by decreasing the concentration of polymer, decreasing the volume fraction of the dispersed phase, increasing the rate of agitation, increasing the surfactant concentration, and changing the type of surfactant. For methods that involve atomization, such as spray drying, the particle size is dependent on the atomization pressure, the orifice size, as well as the viscosity of the polymer solution or suspension and the flow rate.

Microspheres are in the micron size range and, therefore, both resistance and light blockage methods are appropriate particle sizing methods (e.g., HIAC, accusizer particle sizing systems, and coulter counter), refer to Chapter 3 for dispersed system characterization. Problems in particle size analysis can arise due to the tendency of microspheres to aggregate in that two or more particles could be counted as one. Size can be determined directly by microscopy, although this method is more time consuming and is subjective.

2.6. Drug Loading

Drug loading in hydrophobic polymers such as PLGA can be up to approximately 50% for insoluble materials such as steroids, but typically is closer to 20% to obtain satisfactory spheres with the desired release characteristics. For hydrophobic drugs, loading is dependent on the relative solubility of the drug in the organic solvent. Drugs with lower solubilities compared to the polymer may precipitate out of the polymer solvent system during solvent evaporation, resulting in relatively low loadings (71). For water-soluble materials, loadings are usually not more than 10%, since these drugs are rapidly lost to the external aqueous phase during manufacture by the O/W emulsification technique through partitioning into the external aqueous phase. Loading of water soluble drugs into PLGA microspheres can be enhanced using W/O/W, W/O/O, and solid S/O/W techniques (72–74). In addition, drug loading of microspheres prepared by the W/O/W method can be increased by improving the stabilization of the primary emulsion (75). This helps to prevent loss of drug to the external phase during solvent evaporation. Higher internal aqueous phase, cool temperatures, and short processing times have also been used to increase drug loading using the W/O/W method of manufacture (76). Other methods to increase loading of water-soluble drugs are to complex the drug with a more hydrophobic macromolecule or with the polymer itself (77). PLGA COO^- binds with positively charged drugs such as peptides. Chemical modification of hydrophilic materials may also be employed as a means to enhance loading as well as to modify release rates (70). It has been reported that the higher the PLGA concentration the greater the entrapment efficiency (78). Increase in drug concentration can decrease entrapment efficiency (79). In addition, high drug concentrations can result in fragmented microspheres (79).

Drug loading can be detected directly by disruption/dissolution of the polymer and subsequent release of the drug. A solvent for the polymer (e.g., methylene chloride for PLGA) is added and the mixture is ultracentrifuged to separate any precipitated polymer. Drug levels are determined in the supernatant. In some cases, direct determination is difficult. Then, loading can be calculated based on the percentage of drug in the supernatant fluid following collection of the microspheres at the end of the manufacturing process.

It is important to ensure that all of the loaded drug is determined. Several extraction methods have been reported for PLGA microspheres. These include shaking the microspheres overnight in 0.1 m NaOH with 5% SDS and measuring the released drug (80). A combination of polymer solubilization and drug extraction has also been employed for PLGA microspheres loaded with protein (66). The polymer was dissolved in methylene chloride and extracted into pH 4 acetate buffer to remove the protein. Exhaustive extraction in distilled water has also been utilized (81). DMSO has been used as an alternative to methylene chloride to dissolve PLGA (78). Drug loading of albumin microspheres has been determined by digestion with 0.5% acetic acid, followed by centrifugation and extraction of hydrophobic drugs using appropriate organic solvents (68).

2.7. Porosity

Large pores, or megaporosity (10–75 µm), can be measured by air permeability applying the Kozeny–Carmen equation (82). Macroporosity (smaller than 10 µm) can be measured by mercury porosimetry to obtain total porous volume, specific surface area, average pore radius, and pore size distribution. The conventional method of surface area determination is based on the Brunauer–Emmett–Teller equation (83) that takes into account lateral interaction energies, multilayer formation and condensation in pores and conforms to one of the five isotherms described by Brunauer (83). The porosity of particles plays a role in the dissolution and release of drug and in the erosion of the polymer matrix (84).

2.8. Sterility Testing

Sterilization of microspheres is usually achieved by aseptic processing since the final product may not be able to undergo terminal sterilization. Manufacturing methods that are single step and can be performed in an enclosed chamber, such as spray drying, are ideally suited to aseptic processing. Terminal sterilization at high temperature is likely to melt the polymer, cause alteration of drug release rates and may destroy any targeting moiety that may be attached. The glass transition temperature of PLGA is around 44°C (79) and therefore PLGA microspheres would melt and agglomerate during autoclaving for terminal sterilization. Entrapped drug may also be degraded at sterilization temperatures. Sterility assurance is another problem, as, although it is relatively easy to determine whether the exterior of the microspheres is sterile using conventional plating methodology, it is difficult to determine whether the interior of the microspheres are sterile. Methods that can be used to break or dissolve the microspheres such as crushing, grinding, or the use of organic solvents may introduce false-positive or false-negative results. A method of determining the presence of viable organisms in the interior of microspheres without breaking or dissolving microspheres was developed (85). This method involves detecting organism metabolism.

2.8.1. Chemical Stability and Protection from Degradation

Protein drugs are susceptible to breakdown by peptides at the local site and hence their in vivo half-lives are generally very short. Encapsulation in a microsphere system can protect these molecules from degradation until release at the site. This effectively extends the in vivo half-life for elimination and thus the dosing frequency can be reduced. The microenvironment within the polymer may also cause degradation of the drug. Microspheres composed of polyesters of glycolic and lactic acid present a problem for drugs that are not stable under acidic conditions as these polymers degrade to glycolic and lactic acid and present an acidic microenvironment during dissolution. The degradation of proteins and peptides can be increased in the presence of polyesters. This problem may not be serious if the release rates are much faster than the degradation rates.

Low pH environments may be deleterious to many proteins, nucleic acids, and cells. Consequently, the selection of matrix or coating materials that are chemically compatible with these molecules is important.

3. TISSUE TARGETING

Targeting of the RES of the liver has been achieved by selecting dispersed systems of an appropriate size ($<5\,\mu m$) for uptake by the Kuppfer cells. This has been achieved utilizing particulates (86), liposomes (87), and emulsion systems. Particles larger than approximately $7\,\mu m$ will occlude the capillaries of the lungs (86). Surface modifications have been adopted to target dispersed systems to bone and kidney.

Particle engineering has been used for targeted delivery: size, surface charge, surface hydrophobicity, and steric stabilization can all be manipulated to this end. Size and surface characteristics can be used to target the lung, liver, spleen, general circulation, and bone marrow following i.v. administration. Mechanical filtration by size is used to target the lung. Recognition of the particle surface by the RES is used to target the liver. Subcutaneous and peritoneal administration are used to target the regional lymph nodes. Particles below 100 nm delivered by the s.c. route are taken up by the regional lymph nodes. This is useful for anti-cancer and immunomodulating agents. Hydrophilic coatings (e.g., poloxomer) are used to keep microspheres within the systemic circulation. The presence of the hydrophilic layer minimizes uptake of opsonic factors, reduces particle–cell interaction and therefore prevents uptake by phagocytic cells. Specific polymers can target specific sites, e.g., poloxamer 407 can target the bone marrow (88). This is speculated to occur via a receptor-mediated pathway. Both particle size and surface characteristics are critical for bone marrow uptake (size range 60–150 nm). Magnetic targeting is used to overcome clearance by RES (89) and achieve target site specificity. Coated particles avoid uptake by the liver Kuppfer cells and are captured by the spleen via filtration.

Different methods can be used to achieve targeting of microspheres to specific sites in the body. The theory behind this has been discussed by (1). Some parenteral targeting methods that have been proposed are not practical, e.g., blockage of the RES to avoid RES uptake and distribution according to particle size following i.v. injection since this may lead to undesirable blockage of vessels and as discussed above microspheres are not recommended for i.v. administration. The use of antibodies to direct microspheres has met with limited success. Success has been achieved in targeting the lymph system according to size and surface characteristics. Direct injection at the site, e.g., intra-articular injection, to treat arthritic joints and i.m. or s.c. injections at the local site to treat localized infection and inflammation, have been successful. Targeting using an external magnetic field to localize microspheres has also been successful but this approach does not appear to be practical.

4. TARGETING DISEASES

Many diseases are characterized by anatomically localized abnormalities. In general, these are most suited to targeted controlled drug delivery. Diseases of a more diffuse and systemic nature are more readily treated with conventional dosage forms. Some cancers, infectious and hereditary diseases, may be considered suitable for treatment with dispersed systems.

Pharmaceutical journals are replete with examples of the "next wave" of targeted drugs. As biological molecules such as peptides (90), proteins (56), and nucleic acids (38), with highly specific mechanisms of action, are discovered and their site of action and role in disease identified, they may act at the molecular level as immunological, endocrine or neurological mediators or influence replication, translation, transcription or expression of genes. Development of new therapies depends upon delivery systems compatible with the drug and the purpose for which it is intended. Hence there are new delivery systems for antagonists for inflammatory mediators, cytokines, oligonucleotides, and vaccines (31,42).

5. COMMERCIAL PROSPECTS

- Overview of approaches to scale-up, manufacturing and anticipated.
- New products.

5.1. Unit Operations

The unit operations involved in processing microsphere or microcapsule products are based on each of the three fundamental phenomena in chemical engineering (91). That is, heat and mass transfer in the formation of the stable emulsion and fluid flow from the mixing process. The practical unit operations include solution preparation, mixing under various conditions of flow and temperature, drying, filtration or centrifugation depending upon the product (92). In addition, an in situ polymerization step may also be adopted.

The key element in the production of dispersed systems is the stability of the final product. In most cases, these products are in kinetic and not thermodynamic equilibrium. This requires that all of the potential forces of interaction are considered. These forces include, polar forces, i.e., dipole moment and polarizability, hydrogen bonding, steric hindrance, ionic interactions, van der Waals, and gravitational forces. It is the balance between these forces that allows a product to exhibit stability suitable for long-term storage. Some of these forces can be evaluated by considering bulk properties such as dielectric constant, interfacial tension, and sedimentation rate.

5.1.1. Wax Coating and Hot Melt

Wax may be used to coat the core particles, encapsulating drug by dissolution or dispersion in the molten wax. The waxy solution or suspension is dispersed by high-speed mixing into a cold solution, such as cold liquid paraffin. The mixture is agitated for at least 1 hr. The external phase (liquid paraffin) is then decanted and the microcapsules are suspended in a non-miscible solvent, and allowed to air-dry. Multiple emulsions may also be formed (93). For example, a heated aqueous drug solution can be dispersed in molten wax to form a water-in-oil emulsion, which is emulsified in a heated external aqueous phase to form a water-in-oil-in-water emulsion. The system is cooled and the microcapsules collected. For highly aqueous soluble drugs, a non-aqueous phase can be used to prevent loss of drug to the external phase (94). Another alternative is to rapidly reduce the temperature when the primary emulsion is placed in the external aqueous phase.

Wax-coated microcapsules, while inexpensive are often used, release drug more rapidly than polymeric microcapsules. Carnauba wax and beeswax can be used as the coating materials and these can be mixed in order to achieve desired characteristics (93). Wax-coated microcapsules have been successfully tableted. Small aerosol particles, 1–5 μm in diameter, have been condensation-coated from a vapor of a fatty acid or paraffin wax (95,96). These particles have been shown to exhibit reduced dissolution rates in vitro, corresponding to reduced absorption rates following deposition in the lungs of beagle dogs.

Polyanhydrides have been chosen for the preparation of microspheres because of their degradation by surface erosion into apparently non-toxic small molecules (97,98). The mixture of polymer and active ingredient is suspended in a miscible solvent, heated 5°C above the melting point of the polymer and stirred continuously. The emulsion is stabilized, by cooling below the melting point, until the particles solidify.

5.1.2. Spray Coating and Pan Coating

Spray coating and pan coating employ heat-jacketed coating pans in which the solid drug core particles are rotated and into which the coating material is sprayed. The core particles are in the size range of micrometers up to a few millimeters. The coating material is usually sprayed at an angle from the side into the pan. The process is continued until an even coating is completed. This is the process typically used to coat tablets and capsules.

Coating a large number of small particles may provide a safer and more consistent release pattern than coated tablets. In addition, several batches of microspheres can be prepared with different coating thicknesses and mixed to achieve specific controlled release patterns.

The Wurster process, a variation of the basic pan-coating method, is an adaptation of the fluid-bed granulator (99). The solid core particles are fluidized by air pressure and a spray of dissolved wall material is applied from the perforated bottom of the fluidization chamber parallel to the air stream and onto the solid core particles. Alternatively, the coating solution can be sprayed from the top or the sides into an upstream of fluidized particles. This adaptation allows the coating of small particles (100). The fluidized-bed technique produces a more uniform coating thickness than the pan-coating methodology. Problems can arise with inflammable organic solvents because of the high risk of explosion in the enclosed fluidizer chamber. Explosion proof units have been designed; however, over the past three decades, aqueous coating solutions are being used more and more.

Examples of aqueous coating solutions include water-soluble low molecular weight cellulose ethers (101,102), emulsion polymerization latexes of polymethacrylates, and dispersions of water-insoluble polymers such as ethylcellulose in the form of pseudolatex (103). These solvent-free coating solutions provide a range of different coatings from fast disintegrating isolating layers to enteric and sustained-release coatings. Lehmann (104) has reviewed different commercial methods, the conditions required for coating, and various coating formulas including illustrations of the types of equipment used.

5.1.3. Coacervation

Coacervation is the separation of a polymeric solution into two immiscible liquid phases, a dense coacervate phase, which is concentrated in the polymer, and a dilute equilibrium phase (105). Figure 2 illustrates the variables involved in a coacervation process (1). The concentrated polymer phase can be emulsified in the dilute phase and subsequently cross-linked to form microspheres. When only one polymer is present, this process is referred to as simple coacervation. When two or more polymers of opposite charge are present, it is referred to as complex coacervation (105). Simple coacervation

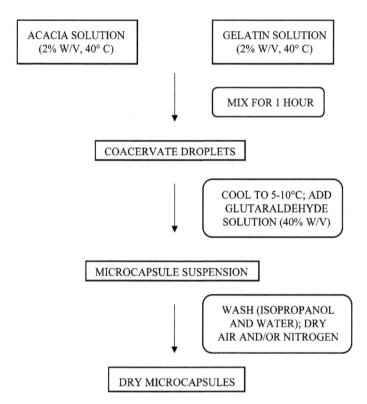

Figure 2 Flow diagram of the gelatin–acacia complex coacervation method.

is induced by a change in conditions that results in dehydration of the polymer, promoting polymer–polymer interactions over polymer–solvent interactions. This may be achieved by temperature change, the addition of a non-solvent or a change in the ionic strength. Electrostatic interactive forces drive complex coacervation between two or more macromolecules (106). Coacervates tend to form around any core material that may be present, such as hydrophobic drug particles or oil droplets. Subsequently, microcapsules can be formed by cross-linking. Cross-linking can be by heat or the use of chemical cross-linking agents such as glutaraldehyde (107–110). Coacervates may also be spray dried (refer Sec. 5.1.5).

There are a large number of variables involved in complex coacervation [pH, ionic strength, macromolecule concentration, macromolecule ratio, and macromolecular weight (105,111)] all of which affect microcapsule production. These variables can be manipulated to produce microcapsules with specific properties. Care must be taken since slight variations in these parameters can have a dramatic effect on coacervate yield and therefore microcapsule yield. This could cause problems during scale up and manufacturing. However, the range of conditions over which complex coacervation occurs can be extended by the addition of water-soluble non-ionic polymers, such as polyethylene oxide or polyethylene glycol (112,113). The presence of a small amount of these polymers allows microencapsulation to occur over an expanded pH range. For example, the pH range for coacervation of gelatin and acacia can be extended from pH 2.6–5.5 (111) to pH 2–9 (112). The pH range for simple coacervation can also be expanded in the presence of these water-soluble non-ionic polymers (112). Another factor to be aware of is that the use of chemical cross-linking agents and the application of heat may be harmful to the encapsulant materials, such as thermo- and chemically labile drugs and live cells. A stable coacervate system, formed without the use of chemical cross-linking agents or the application of heat, has been developed (114). This system is potentially useful for the delivery of protein and polypeptide drugs and other materials unable to withstand cross-linking procedures.

5.1.4. Divalent Ion Gelling

Gelling of alginate by dropping or spraying a solution of sodium alginate into a calcium chloride solution has been used to produce microcapsules. The divalent calcium ions cross-link the alginate, to form gelled droplets. This process is illustrated in Fig. 3 (1). These droplets can be permanently cross-linked by addition to a polylysine solution. This method was developed by Lim and Sun (115), for the encapsulation of live cells. Variations on this method with different polymers, such as chitosan, have been developed (116). For manufacturing

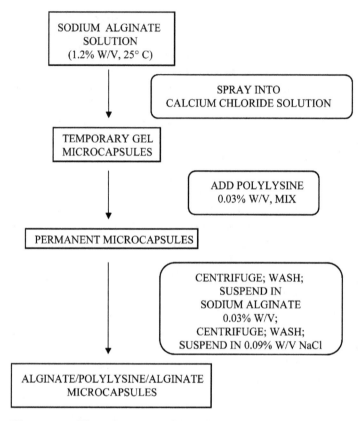

Figure 3   Flow diagram of the alginate–polylysine–alginate micro-encapsulation method.

purposes, an atomization method that produces a fine mist of droplets is used. An interaction chamber is suggested to confine the spray (25). Variations on this method include other methods of gelling such as temperature and solvent-induced gelling, where a polymer is sprayed into a nonsolvent or into a solvent held at reduced temperature.

5.1.5. Spray Drying

Spray drying is a fast one-step, closed-system process and is therefore ideal for the production of sterile materials for parenteral use. The process is illustrated in Fig. 4a. Figure 4b

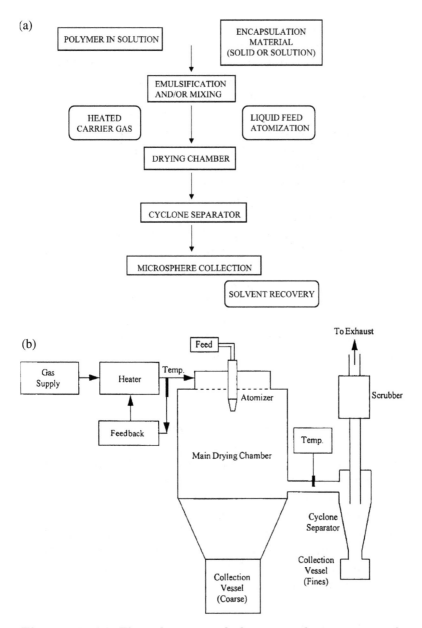

Figure 4 (a) Flow diagram of the spray-drying microsphere manufacturing method. (b) Schematic layout of open-cycle spray-dryer design.

depicts an open system spray drier (117). This method is easily scaled up to large batch sizes, is frequently used commercially for the production of microspheres and is applicable to a wide variety of materials. The polymer and drug are dissolved in a suitable solvent (aqueous or non-aqueous) or the drug may be present as a suspension or dissolved within an emulsion. For example, biodegradable PLGA microspheres can be prepared by dissolving the drug and polymer in methylene chloride. Microsphere size is controlled by the spray and feed rates of the polymer drug solution, the nozzle size, the temperature in the drying and collecting chambers, and the chamber sizes. The quality of spray-dried products can be improved by the addition of plasticizers that promote polymer coalescence and film formation and enhance the formation of spherical and smooth-surfaced microcapsules (118). Recently, a modified approach to spray drying has been developed which involves freeze drying the droplets after production (119–122). This method has the advantage of allowing an additional variable to be used to modify particle structure and the ability to prepare particles with greater stability during processing.

5.1.6. Solvent Evaporation

This is one of the earliest methods of microsphere manufacture. The polymer and drug must be soluble in an organic solvent, frequently methylene chloride. The solution containing the polymer and the drug may be dispersed in an aqueous phase to form droplets. Continuous mixing and elevated temperatures may be employed to evaporate the more volatile organic solvent and leave the solid polymer-drug particles suspended in an aqueous medium. The particles are finally filtered from the suspension. Figure 5 shows polylactic acid particles prepared in this manner (123).

5.1.7. Precipitation

Precipitation is a variation on the evaporation method. The emulsion consists of polar droplets dispersed in a non-polar medium (86). Solvent may be removed from the droplets by the use of

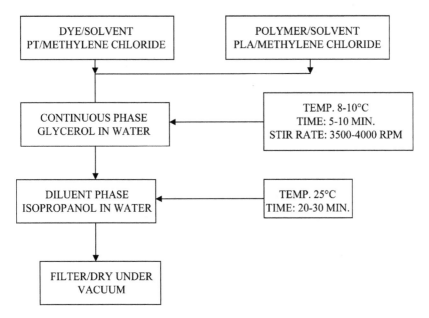

Figure 5 Flow diagram of the solvent evaporation method employed to manufacture the phenolphthalein (PT)-loaded polylactic acid (PLA) microspheres.

a cosolvent. The resulting increase in the polymer drug concentration causes precipitation forming a suspension of microspheres. In effect this is an enhanced evaporation process.

5.1.8. Freeze Drying

This technique involves the freezing of the emulsion (86); the relative freezing points of the continuous and dispersed phases are important. The continuous-phase solvent is usually organic and is removed by sublimation at low temperature and pressure. Finally, the dispersed-phase solvent of the droplets is removed by sublimation, leaving polymer-drug particles. Figure 6 illustrates the method of manufacture of the solvent removal precipitation method (124).

5.1.9. Supercritical Fluid Techniques

Supercritical fluid manufacture of particles involves using the capacity of certain substances to dissolve solutes in a unique

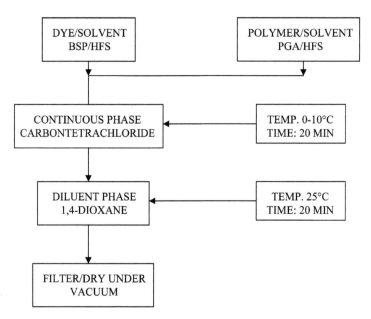

Figure 6 Flow diagram of aspects of the solvent precipitation method employed to manufacture the bromsulphthalein (BSP)-loaded polyglycolic acid (PGA) microspheres. HFS, hexafluoroacetone sesquihydrate.

region of their pressure–temperature phase diagram, shown in Fig. 7 (117,125,126). The capacity of a supercritical fluid can be manipulated in terms of its density and viscosity, which vary depending on the pressure and temperature under which they are being maintained. This principle has been employed using a number of materials, notably carbon dioxide, to prepare drug and excipient particles, with a high degree of crystal lattice and particle size uniformity (127). The most prominent methods of manufacture have involved: rapid expansion of supercritical fluids, bringing the solute to saturation solubility and then causing a homogeneous nucleation and; supercritical anti-solvent methods using the mixing of two supercritical fluids each exhibiting different capacities to dissolve drug to induce a controlled crystallization. More recently polymeric microspheres have been manufactured by an emulsion stabilization technique in supercritical fluids (128–131).

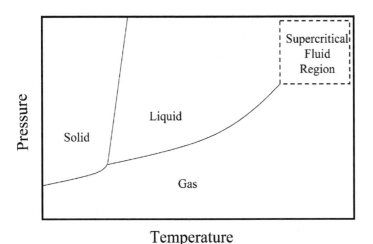

Figure 7 Pressure–temperature phase diagram for a pure component.

5.2. Scale-up Approaches

Unit operations have been discussed in the previous section. The principles and equipment used on the laboratory scale are the basis for pilot and ultimately production scale manufacturing. Basic process or chemical engineering approaches suggest that preserving vessel geometry, maintaining mixing conditions (Reynolds number), and considering heat and mass transfer will allow successful scale-up to occur. The reproducibility of unit doses prepared in bulk is usually very dependent upon subtle aspects of the production process. Hence it is rarely the case that scale-up occurs as a simple multiple of component masses or volumes, vessel geometries, mixing, or filtration conditions. Consequently, experiments must be conducted to evaluate such procedures. These experiments involve many variables and are suited to factorial design or response surface methodologies (132). While numerous pharmaceutical companies have developed scaled-up processes for microsphere manufacture, there have been rare publications on the commercial scale manufacture of microspheres (Floy, 1993, no. 1631). In addition, methods have been developed with consideration of the potential to scale-up the process (133). It should be remembered that sterility is of paramount importance in parenteral products. Consequently, the method of sterilization needs to be considered. The options are moist and dry heat, radiation, and ethylene oxide treatments (95). All of these have an impact on the structure of polymers and there is some concern over the toxicity of ethylene oxide. It appears that the least damaging approach for certain polymers is irradiation. However, this method requires validation for both the ultimate sterility and stability of the product. Irradiation can cause polymer degradation and affect release rates (140,141). Aseptic processing continues to be the method of choice although this is challenging to scale-up.

6. CONCLUSIONS

The use of microsphere delivery systems for parenteral administration has been the subject of an enormous research effort over the last quarter of the 20th century. It is surprising that there have not been more products based on this drug delivery approach. However, there are a number of plausible reasons for this poor success rate. The major drugs and disease states, according to the wisdom of the last 50 years, have been easily treated with orally delivered systemic therapies. These small molecular weight drugs were relatively inexpensive and the number of side effects was small, and acceptable, in light of the severity of disease. It is understandable that an early interest in the targeted delivery of anti-cancer agents arose, as these often exhibit severe systemic side effects and small doses are desirable. It should also be noted that the safety of polymeric delivery systems must be assessed, requiring large-scale and expensive toxicology studies. Nevertheless, there have been some successes notably Leupron Depot (Leuprolide acetate) for the treatment of prostate cancer.

In the last two decades, a significant change in attitude has occurred. The products of biotechnology which are extremely potent, costly and not readily administered orally require novel drug delivery approaches if they are to be successful therapeutically and commercially. The ability to treat subclasses of disease and to manage chronic diseases, with the implications that this has for the overall public health and the cost of care, is driving new developments in drug delivery. It is now cost-effective to perform the necessary toxicology studies on the components of new delivery systems since the targeting that can then be achieved may be the only successful approach to delivering these agents. In this light, peptides such as calcitonin, G-CSF, and growth factor have been studied for controlled delivery. Targeting of nucleic acids (oligonucleotides and genes), as therapy for congenital diseases, and vaccine delivery may offer the greatest promise for microsphere delivery systems as the potential outcome is a cure for the disease being treated. There is no question that the number of polymers and drugs and the breadth of applications being considered will increase as the 21st century unfolds.

REFERENCES

1. Burgess DJ, Hickey AJ. Microsphere technology and applications. In: Swarbrick J, Boylan J, eds. Encyclopedia of Pharmaceutical Technology. Vol. 10. New York, NY: Marcel Dekker, 1994:1–29.

2. Rohollah S, Bruskewitz RC, Gittleman MC, Graham SD, Hudson PB, Stein B. Leuprolide acetate 22.5 mg 12-week depot formulation in the treatment of patients with advanced prostate cancer. Clin Ther 1996; 18:647–657.

3. Okada H. One- and three-month release injectable microspheres of the LH–RH superagonist leuprorelin acetate. Adv Drug Deliv Rev 1997; 28:43–70.

4. Fujimoto S, Miyazaki M, Endoh F, Takahashi O, Shrestha RD, Okui K, Morimoto Y, Terao K. Effects of intra-arterially infused biodegradable microspheres containing Mitomycin C. Cancer 1985; 55(3):522–526.

5. Gupta PK, Hung CT, Perrier DG. Albumin microspheres. I. Release characteristics of adriamycin. Int J Pharm 1986; 33: 137–146.

6. Gupta PK, Hung CT, Perrier DG. Albumin microspheres. II. Effect of stabilization temperature on the release of adriamycin. Int J Pharm 1986; 33:147–153.

7. Gupta PK, Lam FC, Hung CT. Albumin microspheres. IV. Effect of protein concentration and stabilization time on the release rate of adriamycin. Int J. Pharm 1989; 51:253–258.

8. Jones C, Burton MA, Gray BN. Albumin microspheres as vehicles for the sustained and controlled release of doxorubicin. J Pharm Pharmacol 1989; 41:813–816.

9. Schwope AD, Wise DL, Howes JF. Lactic/glycolic acid polymers as narcotic antagonist delivery system. Life Sci 1968; 17:1877–1886.

10. Gardner DL, Patanus AJ, Fink DJ. Steroid release from microcapsules. In: Drug Delivery Systems. Gabelnick HL, ed. DHEW Publ. No. (NIH), 77–1238. Washington, DC: Department of Health, Education and Welfare, 1977: 265–278.

11. Tsung M, Burgess DJ. Preparation and stabilization of heparin/gelatin complex coacervate microcapsules. J Pharm Sci 1997; 86:603–607.

12. Tsung M, Burgess DJ. Preparation and characterization of gelatin surface modified PLGA microspheres. AAPS PharmSci 2001; 3[2]:E11.

13. Schmidt C, Wenz R, Nies B, Moll F. Antibiotic in vivo/in vitro release, histocompatibility and biodegradation of gentamicin implants based on lactic acid polymers and copolymers. J Control Release 1995; 37:83–94.

14. Burgess DJ, Davis SS. Potential use of albumin microspheres as a drug delivery system. II. In vivo deposition and release of steroids. Int J Pharm 1988; 46:69–76.

15. Pavanetto F, Genta I, Giunchedi P, Conti B, Conte U. Spray-dried albumin microspheres for the intra-articular delivery of dexamethasone. J Microencapsul 1994; 4:445–454.

16. Firsov AA, Nazarov AD, Fomina IP. Biodegradable implants containing gentamicin: drug release and pharmacokinetics. Drug Dev Ind Pharm 1987; 13:1651–1674.

17. Narayani R, Rao KP. Controlled release of anticancer drug methotrexate from biodegradable gelatin microspheres. J Microencapsul 1994; 11:69–77.

18. Slobbe L, Medlicott N, Lockhart E, Davies N, Tucker I, Razzak M, Buchan G. A prolonged immune response to antigen delivered in poly (e-caprolactone) microparticles. Immunol Cell Biol 2003; 81(3):185–191.

19. Hanes J, Dawson M, Har-el Y-el, Suh J, Fiegel J. Gene delivery to the lung. In: Hickey AJ, eds. Pharmaceutical Inhalation Aerosol Technology. New York: Marcel Dekker, Inc., 2003:489–539.

20. Kim TK, Burgess DJ. unpublished results.

21. Payne RG, McGonigle JS, Yaszemski MJ, Yasko AW, Mikos AG. Development of an injectable, in situ crosslinkable, degradable polymeric carrier for osteogenic cell populations. Part 3. Proliferation and differentiation of encapsulated marrow stromal osteoblasts cultured on crosslinking poly(-propylene fumarate). Biomaterials 2002; 23(22):4381–4387.

22. Payne RG, McGonigle JS, Yaszemski MJ, Yasko AW, Mikos AG. Development of an injectable, in situ crosslinkable, degradable polymeric carrier for osteogenic cell populations. Part 2. Viability of encapsulated marrow stromal osteoblasts cultured on crosslinking poly(propylene fumarate). Biomaterials 2002; 23(22):4373–4380.

23. Payne RG, Yaszemski MJ, Yasko AW, Mikos AG. Development of an injectable, in situ crosslinkable, degradable polymeric carrier for osteogenic cell populations. Part 1. Encapsulation of marrow stromal osteoblasts in surface crosslinked gelatin microparticles. Biomaterials 2002; 23(22): 4359–4371.

24. Kwok KK, Groves MJ, Burgess DJ. Production of 5–15 μm diameter alginate-polylysine microcapsules by an air-atomization technique. Pharm Res 1991; 8(3):341–344.

25. Abraham SM, Vieth RF, Burgess DJ. Novel technology for the preparation of sterile alginate-poly-l-lysine microcapsules in a bioreactor. Pharm Dev Technol 1996; 1(1):63–68.

26. Schroeder HG, Bivins BA, Sherman GP, DeLuca PP. Physiological effects of subvisible microspheres administered intravenously to beagle dogs. J Pharm Sci 1978; 67(4):C18–C23.

27. Martin A. Physical Pharmacy. 4th ed. Philadelphia: Lea and Febiger, 1993:324–361.

28. Fujiwara K, Hayakawa K, Nagata Y, Hiraoka M, Nakamura T, Shimizu Y, Ikada Y. Experimental embolization of rabbit renal arteries to compare the effects of poly L-lactic acid microspheres with and without epirubicin release against intraarterial injection of epirubicin. Cardiovasc Interv Radiol 2000; 23:218–223.

29. Hariharan M, Price JC. Solvent, emulsifier and drug concentration factors in poly(D,L-lactic acid) microspheres containing hexamethylmelamine. J Microencapsul 2002; 19(1):95–109.

30. Kobayashi D, Tsubuku S, Yamanaka H, Asano M, Miyajima M, Yoshida M. In vivo characteristics of injectable poly(DL-lactic acid) microspheres for long-acting drug delivery. Drug Dev Ind Pharm 1998; 24:819–825.

31. Coombes AG, Major D, Wood JM, Hockley DJ, Minor PD, Davis SS. Resorbable lamellar particles of polylactide as adjuvants for influenza virus vaccines. Biomaterials 1998; 19: 1073–1081.

32. Ciftci K, Hincal AA, Kas HS, Ercan TM, Sungur A, Guven O, Ruacan S. Solid tumor chemotherapy and in vivo distribution of fluorouracil following administration in poly(L-lactic acid) microspheres. Pharm Dev Technol 1997; 2:151–160.

33. Kuang L, Yang DJ, Inoue T, Liu WC, Wallace S, Wright KC. Percutaneous intratumoral injection of cisplatin microspheres in tumor-bearing rats to diminish acute nephrotoxicity. Anticancer Drugs 1996; 7:220–227.

34. Johansen P, Tamber H, Merkle HP, Gander B. Diphtheria and tetanus toxoid microencapsulation into conventional and end-group alkylated PLA/PLGAs. Eur J Pharm Biopharm 1999; 47(3):193–201.

35. Raghuvanshi RS, Katare YK, Lalwani K, Ali MM, Singh O, Panda AK. Improved immune response from biodegradable polymer particles entrapping tetanus toxoid by use of differ-

ent immunization protocol and adjuvants. Int J Pharm 2002; 245(1–2):109–121.

36. Miller JA, Oehler DD, Pound JM. Delivery of ivermectin by injectable microspheres. J Econ Entomol 1998; 91:655–659.

37. Chowdhury DK, Mitra AK. Kinetics of a model nucleoside (guanosine) release from biodegradable poly(DL-lactide-co-glycolide) microspheres: a delivery system for long-term intraocular delivery. Pharm Dev Technol 2000; 5:279–285.

38. Lunsford L, McKeever U, Eckstein V, Hedley ML. Tissue distribution and persistence in mice of plasmid DNA encapsulated in a PLGA-based microsphere delivery vehicle. J Drug Target 2000; 8:39–50.

39. Barrio GGd, Novo FJ, Irache JM. Loading of plasmid DNA into PLGA microparticles using TROMS (Total Recirculation One-Machine System): evaluation of its integrity and controlled release properties. J Control Release 2003; 86(1):123–130.

40. Chen W, Lu DR. Carboplatin-loaded PLGA microspheres for intracerebral injection: formulation and characterization. J Microencapsul 1999; 16:551–563.

41. Goldberg SN, Walovitch RC, Straub JA, Shore MT, Gazelle GS. Radiofrequency-induced coagulation necrosis in rabbits: immediate detection at US with a synthetic microsphere contrast agent. Radiology 1999; 213:438–444.

42. Hilbert AK, Fritzsche U, Kissel T. Biodegradable microspheres containing influenza A vaccine: immune response in mice. Vaccine 1999; 17:1065–1073.

43. Cleland JL, Lim A, Daugherty A, Barron L, Desjardin N, Duenas ET, Eastman DJ, Vennari JC, Wrin T, Berman P, Murthy KK, Powell MF. Development of single-shot subunit vaccine for HIV-1. 5. programmable in vivo autoboost and long lasting neutralizing response. J Pharm Sci 1998; 87:1489–1495.

44. Atkins TW. Fabrication of microspheres using blends of poly (ethylene adipate) and poly(ethylene adipate)/poly(hydroxy butyrate-hydroxyvalerate) with poly(caprolactone): incor-

poration and release of bovine serum albumin. J Biomater Sci Polym Ed 1997; 8:833–845.

45. Blanco MD, Bernardo MV, Sastre RL, Olmo R, Muniz E, Teijon JM. Preparation of bupivacaine-loaded poly(vepsiln.-caprolactone) microspheres by spray drying: drug release studies and biocompatibility. Eur J Pharm Biopharm 2003; 55(2):229–236.

46. Sinisterra RD, Shastri VP, Najjar R, Langer R. Encapsulation and release of rhodium (II) citrate and its association complex with hydroxypropyl-beta-cyclodextrin from biodegradable polymer microspheres. J Pharm Sci 1999; 88:574–576.

47. Chiba M, Hanes J, Langer R. Controlled protein delivery from biodegradable tyrosine-containing poly(anhydride-co-imide) microspheres. Biomaterials 1997; 18:893–901.

48. Tokumitsu H, Ichikawa H, Fokumori Y, Block LH. Preparation of gadopentetic acid-loaded chitosan microparticles for gadolinium neutron-capture therapy of cancer by a novel emulsion-droplet coalescence technique. Chem Pharm Bull 1999; 47:838–842.

49. Jameela SR, Kumary TV, Lal AV, Jayakrishnan A. Progesterone-loaded chitosan microspheres: a long acting biodegradable controlled delivery system. J Control Release 1998; 52:17–24.

50. Mi FL, Wong TB, Shyu SS. Sustained-release of oxytetracycline from chitosan microspheres prepared by interfacial acylation and spray hardening methods. J Microencapsul 1997; 14:577–591.

51. Patashnik S, Rabinovich L, Colomb G. Preparation and evaluation of chitosan microspheres containing bisphosphonates. J Drug Target 1997; 4:371–380.

52. Elcin YM, Dixit V, Gitnick G. Controlled release of endothelial cell growth factor from chitosan-albumin microspheres for localized angiogenesis: in vitro and in vivo studies. Artif Cells Blood Substit Immobil Biotechnol 1996; 24:257–271.

53. Tabata Y, Hijikata S, Muniruzzaman M, Ikada Y. Neovascularization effect of biodegradable gelatin microspheres incor-

porating basic fibroblast growth factor. J Biomater Sci 1999; 10:79–94.

54. Laurent A, Beaujeux R, Wassef M, Rufenacht D, Boschetti E, Merland JJ. Trisacryl gelatin microspheres for therapeutic embolization, I: development and in vitro evaluation. Am J Neuroradiol 1996; 17:533–540.

55. Nakaoka R, Tabata Y, Ikada Y. Potentiality of gelatin microsphere as immunological adjuvant. Vaccine 1995; 13:653–661.

56. Brown KE, Leong K, Huang CH, Dalal R, Green GD, Haimes HB, Jimenez P, Bathon J. Gelatin/chondroitin 6-sulfate microspheres for the delivery of therapeutic proteins to the joint. Arthritis Rheum 1998; 41:2185–2195.

57. Yoshikawa Y, Komuta Y, Nishihara T, Itoh Y, Yoshikawa H, Takada K. Preparation and evaluation of once-a-day injectable microspheres of interferon alpha in rats. J Drug Target 1999; 6:449–461.

58. Nehra A, Gettman MT, Nugent M, Bostwick DG, Barrett DM, Goldstein I, Krane RJ, Moreland RB. Transforming growth factor-beta 1 (TGF-beta 1) is sufficient to induce fibrosis of rabbit corpus cavernosum in vivo. J Urol 1999; 162:910–915.

59. Leblond FA, Simard G, Henley N, Rocheleau B, Huet PM, Halle JP. Studies on smaller (approximately 315 µM) microcapsules: IV Feasibility and safety of intrahepatic implantations of small alginate poly-L-lysine microcapsules. Cell Transplant 1999; 8:327–337.

60. Pulfer SK, Gallo JM. Enhanced brain tumor selectivity of cationic magnetic polysaccharide microspheres. J Drug Target 1998; 6:215–227.

61. Huang LK, Mehta RC, DeLuca PP. Evaluation of a statistical model for the formation of poly [acryloyl hydroxyethyl starch] microspheres. Pharm Res 1997; 14:475–482.

62. Jiang G, Qiu W, DeLuca PP. Preparation and in vitro/in vivo evaluation of insulin-loaded poly(acryloyl-hydroxyethyl starch)-PLGA composite microspheres. Pharm Res 2003; 20(3):452–459.

63. Habler O, Kleen M, Hutter J, Podtschaske A, Tiede M, Kemming G, Corso C, Batra S, Keipert P, Faithfull S, Mess-

mer K. Effects of hemodilution on splanchnic perfusion and hepatorenal function. II. Renal perfusion and hepatorenal function. Eur J Med Res 1997; 2:419–424.

64.   Conway BR, Eyles JE, Alpar HO. A comparative study on the immune responses to antigens in PLA and PHB microspheres. J Control Release 1997; 49:1–9.

65.   Jeffery H, Davis SS, O'Hagan DT. The preparation and characterization of poly(lactide-co-glycolide) microparticles. I: Oil-in-water emulsion solvent evaporation. Int J Pharm 1991; 77:169–175.

66.   Bodmer D, Kissel T, Traechslin E. Factors influencing the release of peptides and proteins from biodegradable parenteral depot systems. J Control Release 1992; 21:129–138.

67.   Zhang Y, Zale S, Sawyer L, Bernstein H. Effect of salts on poly(DL-lactide-co-glycolide) polymer hydrolysis. J Biomed Mater Res 1997; 34:531–538.

68.   Dilova V, Shishkova V. Albumin microspheres as a drug delivery system for dexamethasone: pharmaceutical and pharmacokinetic aspects. J Pharm Pharmacol 1993; 45:987–989.

69.   Kim HC, Burgess DJ. Effect of drug stability an analysis of release data from controlled release Microspheres. J Microencapsulation 2002; 19(5):631–640.

70.   Kim TK, Burgess DJ. Formulation and release characteristics of poly(lactic-co-glycolic acid) microspheres containing chemically modified protein. J Pharm Pharmacol 2001; 53(1):23–31.

71.   Benoit JP, Machais H, Rolland H, Velde VV. Biodegradable microspheres: advance in production technology. In: Benita S, ed. Microencapsulation: Methods and Industrial Applications. New York: Marcel Dekker, 1996:35–72.

72.   O'Donnell PB, McGinity JW. Preparation of microspheres by the solvent evaporation technique. Adv Drug Deliv Rev 1997; 28:25–42.

73.   Cao X, Shoichet MS. Delivering neuroactive molecules from biodegradable microspheres for application in central nervous system disorders. Biomaterials 1999; 20:329–339.

74. Takada S, Kurokawa T, Miyazaki K, Iwasa S, Oqawa Y. Utilization of an amorphous form water-soluble GPIIb/IIIa antagonist for controlled release from biodegradable microspheres. Pharm Res 1997; 14:1146–1150.

75. Nihant N, Schugens C, Grandfils C, Jerome R, Teyssie P. Polylactide microparticles prepared by double emulsion/evaporation technique. I. Effect of primary emulsion stability. Pharm Res 1994; 11:1479–1486.

76. Zambaux MF, Bonneaux F, Gref R, Miaincent P, Dellacherie E, Alonso MJ, Labrude P, Vigneron C. Influence of experimental parameters on the characteristics of poly(lactic acid) nanoparticles prepared by a double emulsion method. J Control Release 1998; 50:31–40.

77. Heya T, Okada H, Tanigawara Y, Ogawa Y, Toguchi H. Effects of counterion of TRH and loading amount on control of TRH release from copoly(DL-lactic/glycolic acid) microspheres prepared by an in-water drying method. Int J Pharm 1991; 69:69–75.

78. Ghaderi R, Sturesson C, Carlfors J. Effect of preparative parameters on the characteristics of poly(D,L-Lactide-co-glycolide) microspheres made by the double emulsion methods. Int J Pharm 1996; 141:205–216.

79. Cole ML, Whateley TL. Release rate profiles of theophylline and insulin from stable multiple w/o/w emulsions. J Control Release 1997; 49:51–58.

80. Sah H. A new strategy to determine the actual protein content of poly(lactide-co-glycolide) microspheres. J Pharm Sci 1997; 86:1315–1318.

81. Esposito E, Cortesi R, Cervellati F, Menegatti E, Nastruzzi C. Biodegradable microparticles for sustained delivery of tetracycline to the periodontal pocket: formulatory and drug release studies. J Microencapsul 1997; 14:175–187.

82. Allen T. Particle Size Measurement. 4th ed. London: Chapman and Hall, 1990:503–538.

83. Shaw DJ. Introduction to Colloid and Surface Chemistry. 3rd ed. London: Butterworths, 1986:108–126.

84. Redmon MP, Hickey AJ, DeLuca PP. Prednisolone-21-acetate poly(glycolic acid) microspheres: influence of matrix characteristics on release. J Control Release 1989; 9:99–109.

85. Kwok KK, Burgess DJ. A novel method of determination of sterility of microcapsules and measurement of viability of encapsulated organisms. Pharm Res 1992; 9(3):410–413.

86. DeLuca PP, Hickey AJ, Hazraty AM, Wedlund P, Rypacek F, Kanke M. Porous biodegradable microspheres for parenteral administration. In: Breimer DD, Speiser P, eds. Topics in Pharmaceutical Sciences. Amsterdam: Elsevier Science Publishers B.V (Biomedical Division), 1987:429–442.

87. Xie K, Fidler IJ. Therapy of cancer metastasis by activation of the inducible nitric oxide synthase. Cancer Metastasis Rev 1998; 17:55–75.

88. Illum L, Davies SS, Muller RH, Mak E, West P. The organ distribution and circulation time of intravenously injected colloidal carriers sterically stabilized with a block copolymer-poloxamine 908. Life Sci 1987; 40:367–374.

89. Ghassabian S, Ehteezazi T, Forutan SM, Mortazavi SA. Dexamethasone-loaded magnetic albumin microspheres: preparation and in vitro release. Int J Pharm 1996; 130:49–55.

90. Newman KD, Sosnowski DL, Kwon GS, Samuel J. Delivery of MUC1 mucin peptide by Poly(D,L-lactic-co-glycolic acid) microspheres induces type 1 T helper immune responses. J Pharm Sci 1998; 87:1421–1427.

91. McCabe WL, Smith JC, Harriott P. Unit Operations of Chemical Engineering. New York: McGraw Hill, 1993.

92. Hickey AJ, Ganderton D. Pharmaceutical Process Engineering. Vol. 112. New York, NY: Marcel Dekker, 2001.

93. Bodmeier R, Wang J, Bhagwatwar H. Process and formulation variables in the preparation of wax microparticles by a melt dispersion technique. II. W/O/W multiple emulsion technique for water-soluble drugs. J Microencapsul 1992; 9:9–107.

94. Benita S, Zonai O, Benoit J-P. 5-Fluorouracil:carnauba wax microspheres for chemoembolization: an in vitro evaluation. J Pharm Sci 1986; 75:847–851.

95. Hickey AJ, Fults KA, Pillai RS. Use of particle morphology to influence the delivery of drugs from dry powder aerosols. J Biopharm Sci 1992; 3(1/2):107–113.

96. Pillai RS, Yeates DB, Miller IF, Hickey AJ. Drug release from condensation coated inhalation aerosols. Proc Int Symp Control Release Bioact Mater 1992; 19:224–225.

97. Leong KW, Brott BC, Langer R. Bioerodible polyanhydrides as drug-carrier matrices. I: Characterization, degradation, and release characteristics. J Biomed Mater Res 1985; 19: 945–955.

98. Mathiowitz E, Langer R. Polyanhydride microspheres as drug carriers I. Hot-melt microencapsulation. J Control Release 1987; 5:13–22.

99. Wurster DE. Air-suspension techniques of coating drug particles: a preliminary report. J Am Pharm Assoc Sci Ed 1959; 48:451–454.

100. Robinson MJ, Grass GM, Lantz RJ. An apparatus and method for coating of solid particles. J Pharm Sci 1968; 57:1983–1988.

101. Porter SC. Aqueous film coating: an overview. Pharm Tech 1979; 3:54–59.

102. Porter SC. The effect of additives on the properties of an aqueous film coating. Pharm Tech 1980; 4:66–75.

103. Banker GS. The new, water-based colloidal dispersions. Pharm Tech 1981; 5:54–61.

104. Lehmann K. Fluid-bed spray coating. In: Donbrow M, ed. Microcapsules and Nanoparticles in Medicine and Pharmacy. London: CRC Press, 1991:73–97.

105. Bungenberg de Jong HG. Reversible systems. In: Kruyt HG, ed. Colloid Science. Vol. II. New York: Elsevier, 1949: 335–432.

106. Burgess DJ. Practical analysis of complex coacervate systems. J Colloid Interface Sci 1990; 140:227–238.

107. Burgess DJ, Davis SS, Tomlinson E. Potential use of albumin microspheres as a drug delivery system. I. Preparation and in vitro release of steroids. Int J Pharm 1987; 39:129–136.

108. Nixon JR, Microencapsulation. New York: MarcelDekker, 1976.

109. Nixon JR, Nouh A. The effect of microcapsule size on the oxidative decomposition of core materials. J Pharm Pharmacol 1978; 30:533–537.

110. Madan PL. Microencapsulation, 1. Phase separation or coacervation. Drug Dev Ind Pharm 1978; 4:95–116.

111. Burgess DJ, Carless JE. Microelectrophoretic studies of gelatin and acacia for the prediction of complex coacervation. J Colloid Interface Sci 1984; 98:1–8.

112. Jizomoto H. Phase separation induced in gelatin-base coacervation systems by addition of water-soluble nonionic polymers I: microencapsulation. J Pharm Sci 1984; 73:879–882.

113. Jizomoto H. Phase separation induced in gelatin-base coacervation systems by addition of water-soluble nonionic polymers II: Effect of molecular weight. J Pharm Sci 1985; 74:469–472.

114. Burgess DJ, Singh ON. Spontaneous formation of small sized albumin/acacia coacervate microcapsules. J Pharm Pharmacol 1993; 45:586–591.

115. Lim F, Sun AM. Microencapsulated islets as bioartificial endocrine pancreas. Science 1980; 210:908–910.

116. Rha CK, Rodrigues-Sanches D. U.S. Pat. 4,749,620, 1980.

117. Sacchetti M, Oort MMV. Spray-drying and supercritical fluid particle generation techniques. In: Hickey AJ, ed. Inhalation Aerosols: Physical and Biological Basis for Therapy. Vol. 94. New York: Marcel Dekker, 1996:337–384.

118. Wan LSC, Heng PWS, Chia CGH. Plasticizers and their effects on the microencapsulation process by spray-drying in an aqueous system. J Microencapsul 1992; 9:53–62.

119. Lam XM, Duenas ET, Cleland JL. Encapsulation and stabilization of nerve growth factor into poly(lactic-co-glycolic) acid microspheres. J Pharm Sci 2001; 90(9):1356–1365.

120. Lam XM, Duenas ET, Daugherty AL, Levin N, Cleland JL. Sustained release of recombinant human insulin-like growth

factor-I for treatment of diabetes. J Control Release 2000; 67(2–3):281–292.

121. Maa Y-F, Nguyen P-A. Method of spray freeze drying proteins for pharmaceutical administration. US 6284282, 2001.

122. Wang J, Chua KM, Wang C-H. Stabilization and encapsulation of human immunoglobulin G into biodegradable microspheres. J Colloid Interface Sci 2004; 271(1):92–101.

123. Hickey AJ, Tian Y, Parasrampuria D, Kanke M. Biliary elimination of bromsulphthalein, phenolphthalein and doxorubicin released from microspheres following intravenous administration. Biopharm Drug Dispos 1993; 14:181–186.

124. Sato T, Kanke M, Schroeder HG, DeLuca PP. Porous biodegradable microspheres for controlled drug delivery. I. Assessment of processing conditions and solvent removal techniques. Pharm Res 1988; 5:21–30.

125. Caliceti P, Elvassore N, Bertucco A. Preparation of insulin loaded polylactide microparticles by supercritical anti-solvent technique. Proc Int Symp Control Release Bioact Mater 2000; 1094–1095.

126. Ribeiro D, Santos I, Richard J, Thies C, Pech B, Benoit J-P. A supercritical fluid-based coating technology. 3: Preparation and characterization of bovine serum albumin particles coated with lipids. J Microencapsul 2003; 20(1):110–128.

127. York P, Hanna M, Shekunov BY, Humphreys GO. Microfine particle formation by SEDS (solution enhanced dispersion by supercritical fluids): scale up by design. Respir Drug Deliv 1998; VI:169–175.

128. Clark MR, DeSimone JM. Cationic polymerization of vinyl and cyclic ethers in supercritical and liquid carbon dioxide. Macromolecules 1995; 28(8):3002–3004.

129. Romack TJ, Combes JR, DeSimone JM. Free-radical telomerization of tetrafluoroethylene in supercritical carbon dioxide. Macromolecules 1995; 28(5):1724–1726.

130. Romack TJ, Maury EE, DeSimone JM. Precipitation polymerization of acrylic acid in supercritical carbon dioxide. Macromolecules 1995; 28(4):912–915.

131. McClain JB, Londono D, Combes JR, Romack TJ, Canelas DA, Betts DE, Wignall GD, Samulski ET, DeSimone JM. Solution properties of a CO2-soluble fluoropolymer via small angle neutron scattering. J Am Chem Soc 1996; 118(4):917–918.

132. Box GEP, Hunter WG, Hunter JS. Statistics for Experiments: An Introduction to Design, Data Analysis, and Model Building. New York: John Wiley and Sons, 1978.

133. Felder CB, Blanco-Prieto MJ, Heizmann J, Merkle HP, Gander B. Ultrasonic atomization and subsequent polymer desolvation for peptide and protein microencapsulation into biodegradable polyesters. J Microencapsul 2003; 20(5): 553–567.

134. Burgess DJ, Crommelin DJA, Hussain AS, Chen M-L. Assuring quality and performance of sustained and controlled release parenterals. Eur J Pharm Sci 2004; 21(5):679–690.

135. Kim T-K, Burgess DJ. Formulation and release characteristics of poly(lactic-co-glycolic acid) microspheres containing chemically modified protein. J Pharm Pharmacol 2001; 53(1):23–31.

136. Patil SD, Burgess DJ, Papadimitrikapoulos F. Dexamethasone-loaded PLGA microspheres/PVA hydrogel composites for inflammation control. Diabet Technol Ther 2004; 6(6): 887–897.

137. Hickey T, Kreutzer D, Burgess DJ, Moussy F. Dexamethasone/PLGA microspheres for continuous delivery of an anti-inflammatory drug for implantable medical devices. Biomaterials 2002; 23(7):1649–1656.

138. Hickey T, Kreutzer D, Burgess DJ, Moussy F. In vivo evaluation of a dexamethasone/PLGA microsphere system designed to suppress the inflammatory tissue response to implantable medical devices. J Biomed Mater Res 2002; 61(2):180–187.

139. Kim TK, Burgess DJ. Pharmacokinetic characterization of 14C-vascular endothelial growth factor controlled release microspheres using a rat model. J Pharm Pharmacol 2002; 54(7):897–905.

140. Montanari L, Cilurzo F, Selmin F, Conti B, Genta I, Poletti G, Orsini F, Valvo L. Poly(lactide-co-glycolide) microspheres

containing bupivacaine: comparison between gamma and beta irradiation effects. J Control Release 2003; 90(3): 281–290.

141. Bittner B, Mader K, Kroll C, Borchert H-H, Kissel T. Tetracycline-HCl-loaded poly(DL-lactide-co-glycolide) microspheres prepared by a spray drying technique: influence of γ-irradiation on radical formation and polymer degradation. J Control Release 1999; 59(1):23–32.

10

Case Study: Development and Scale-Up of NanoCrystal® Particles

ROBERT W. LEE

Elan Drug Delivery, Inc.,
King of Prussia, Pennsylvania, U.S.A.

1. INTRODUCTION

NanoCrystal® Technology was developed to meet a significant market need for delivery of poorly water-soluble drugs. This formulation problem is addressed using proprietary technology that presents drugs as extremely small—nanometer-sized

---

This chapter is based on two presentations that were given at the American Association of Pharmaceutical Scientists 37th Annual Pharmaceutical Technologies Conference at Arden House: Parenteral Products Integrating Science, Innovation, and Patient Needs, 15 January, 2002. The two presentations were: (i) Robert W. Lee, Product Development for Suspensions and Nanoparticulate Suspensions; and (ii) Robert W. Lee, Case Study II: Development and Scale-Up for a Nanoparticulate Suspension.

355

—particles. NanoCrystal particles are made by wet-milling active pharmaceutical ingredients (APIs), water, and a stabilizer to create a colloidal dispersion in the size range of 100–400 nm in diameter. These NanoCrystal particles do not aggregate due to the non-covalent adsorption of stabilizing polymers onto the particle surface, which decreases the surface-free energy. The hydrophilic polymers used to stabilize the colloidal dispersions can be found in marketed products and are generally recognized as safe* (GRAS) materials. These NanoCrystal particles can be further processed into all of the dosage forms traditionally used for administering drugs by oral, parenteral, inhalation, or topical routes. The applicability of this technology is defined solely by the drug candidate's aqueous solubility and is not constrained by therapeutic category or chemical structure. The NanoCrystal milling process is fairly gentle and has been applied to proteins.

Pharmaceutical companies estimate that approximately 60% of the new chemical entities that are synthesized each year have an aqueous solubility of less than 0.1 mg/mL. This low solubility is a significant cause of failure for discovery phase compounds. Poorly water-soluble compounds are both difficult to formulate and difficult to analyze in humans or animals and are often discarded. Additionally, the synthesis of water-soluble analogs often results in decreased bioactivity when compared to their insoluble counterparts. A significant number of the "Top 200" drugs exhibit clinical or pharmacoeconomic limitations that arise from their poor water solubility. The biological performance of these marketed products could likely be improved by the application of NanoCrystal Technology. Furthermore, NanoCrystal Technology is the only nanoparticulate technology that is fully scalable and validated. This technology is incorporated into Wyeth's Rapamune® (sirolimus) Tablets, an approved oral solid dosage form that is marketed in the US. The Rapamune

---

* Inactive Ingredient Guide, Division of Drug Information Resources, Food and Drug Administration, Center for Drug Evaluation and Research, Office of Management, January 1996.

Tablet formulation has two advantages over the solution formulation:

1. The tablet is marketed as a non-refrigerated unit dose, whereas the solution requires refrigeration due to chemical instability and needs to be mixed with juice prior to administration.
2. According to the package insert for Rapamune (sirolimus) Oral Solution and Tablets, the mean bioavailability of sirolimus after the administration of the tablet is about 27% higher relative to the oral solution.

NanoCrystal particles can be processed into a wide range of oral solid dosage forms using conventional, scalable unit operations that are well established in the pharmaceutical industry. NanoCrystal Technology can be incorporated into:

- Tablets—immediate release, film-coated, enteric-coated, rapidly disintegrating (waterless), chewable, sustained (extended) release, and bilayer (combination immediate- and extended-release).
- Capsules—immediate-release (dry-filled powder), immediate-release (bead-filled), and sustained (pulsatile) release.
- Powders for reconstitution.
- Sachets—immediate-release (dry-filled powder), and immediate-release (bead-filled).

Orally administered NanoCrystal formulations of poorly water-soluble drugs have been demonstrated to provide the following therapeutic benefits:

- Faster onset of action.
- Increased bioavailability.
- Reduced fed/fasted variable absorption (i.e., "food effect").
- Improved dose proportionality.

The potential benefits of NanoCrystal Technology relevant to the development and manufacture of solid dosage forms include:

- Improved chemical stability of the API.

- Improved physical stability and performance stability of the dosage form.
- Enhanced appearance (i.e., elegant, compact presentation).
- Improved processability and reduced impact of lot-to-lot variability in API.
- Rapid development.
- Predictable performance throughout product development, scale-up, and transfer to commercial manufacturing site.

NanoCrystal Technology provides improved performance characteristics for intravenous, subcutaneous, or intramuscular injection, including:

- Elimination of harsh vehicles by allowing for the use of an aqueous-based vehicle.
- Higher dose loading with smaller dose volume.
- Longer dose retention in blood and tumors.
- Low viscosity, enabling a quick push for injection.
- Capable of sterilizing by terminal heat, gamma irradiation, and filtration.

NanoCrystal Technology can be used to achieve optimal drug delivery to the lung. This is accomplished by formulating the drug into precisely controlled particles of the appropriate aerodynamic size, shape, and density. Drug can preferentially be delivered to the upper respiratory pathways, for the treatment of diseases such as asthma, or delivered to the deep lung for systemic availability or local effect in the deep lung. Delivery to the upper airways requires an aerodynamic particle size range of 3–5 μm in diameter. In contrast, deep lung deposition requires the particles to be less than 3 μm in diameter.

2. PHARMACEUTICS

A significant number of APIs possess low aqueous solubility, i.e., less than 0.1 mg/mL. There are limited options available for the formulation of these APIs for parenteral administration.

The formulator can use a solubilizing approach, i.e., organic cosolvents, solubilizing agents, and/or pH extremes. However, there are issues with solubilizing approaches. Organic, water-miscible cosolvents may be somewhat physiologically incompatible and thus may give rise to toxic effects. Solubilizing agents include cyclodextrins and surfactants. Cyclodextrins have issues with low loading capacity and their release profile may not be the desired one. Certain surfactants are known to give rise to anaphylaxis and thus have liabilities associated with their toxicity. The use of pH extremes to solubilize ionic APIs may be physiologically incompatible and may also cause the API to precipitate upon injection.

Another approach to formulating poorly water-soluble APIs for intramuscular administration is as traditional coarse (micron-sized) suspensions. There are commercially available coarse suspensions such as Bicillin® C-R 900/300 (penicillin G benzathine and penicillin G procaine suspension) and Depo-Provera® Contraceptive Injection. The use of coarse suspensions is limited to intramuscular injection and there are some limiting pharmaceutical issues associated with these types of formulations such as physical instability, problems with homogeneity/content uniformity, and challenges with producing a sterile product; besides, intravenous administration is also not possible. NanoCrystal Technology provides an attractive alternative for the formulation of poorly water-soluble APIs for parenteral administration, including the intravenous route. The advantages of NanoCrystal particles over coarse suspensions are as follows:

- Physical stability—the smaller particle size decreases the rate of settling and improves redispersibility.
- Homogeneity/content uniformity—NanoCrystal particles behave more like molecular solutions and are therefore much more homogeneous.
- Sterilization—there are more options available for sterilization of NanoCrystal particles compared to coarse suspensions. In select cases, NanoCrystal particles can be rendered small enough to allow for sterile filtration.

- Administration—the small particle size permits all routes of parenteral administration including intravenous.

NanoCrystal particles consist of API, water, and a stabilizer. A high-energy media-milling process is used to wet mill the slurry and create a colloidal dispersion with particles in the size range of 100–400 nm in diameter. These NanoCrystal particles do not aggregate due to the non-covalent adsorption of stabilizing polymers onto the surface of the particle, which decreases the surface-free energy. The polymers that are used to stabilize the colloidal dispersions are typically hydrophilic in nature and can be found in marketed products. These polymeric stabilizers are GRAS materials. The physical basis of NanoCrystal Technology is the dramatic increase in the surface area to mass ratio compared to unmilled API. This in turn leads to significantly enhanced rates of dissolution.

3. FORMULATION DEVELOPMENT

NanoCrystal particles are developed as simple formulations, with similar compositions compared to traditional parenteral formulations. Each NanoCrystal particle formulation contains, at a minimum, the API, a stabilizer (typically polymeric), and water. Additional excipients can be added, as required, to enhance product quality. The added excipients can include ingredients found in solution parenteral formulations such as anti-oxidants, anti-microbial preservatives, buffers/pH adjusters, reconstitution or bulking agents for lyophilized formulations, and tonicity modifiers. The key difference between solution formulations and NanoCrystal particle formulations is the stabilizer. The stabilizer is an essential component that provides physical stability by keeping the NanoCrystal particles discrete and separated. The stabilizer may also: (i) aid in redispersibility upon storage; (ii) in the case of lyophilized formulations aid in reconstitution; and (iii) in conjunction with particle size, influence pharmacokinetics. The stabilizer is typically GRAS, can be either non-ionic or ionic, and can be either polymeric or small molecule. Examples of stabilizers include polyvinylpyrrolidones (BASF Kollidon 12 PF and Kollidon 17 PF), Pluronic® F-108, Pluronic® F-68, and Tween-80.

For parenteral NanoCrystal particle products, there are two possible final dosage forms: ready-to-use NanoCrystal particles and lyophilized NanoCrystal particles. For the most part, NanoCrystal particle formulations require lyophilization. This is due to the lack of suitable stabilizers that are acceptable for parenteral administration and that are capable of producing stable NanoCrystal particle formulations (i.e., with a 2-year shelf life at room temperature). In every case investigated to date, a lyophilizable NanoCrystal particle formulation has been successfully developed. This refers to the ability to lyophilize and subsequently reconstitute the lyophilized NanoCrystal particle formulation to a particle size that is comparable to that of the pre-lyophilized formulation. An example of the redispersibility of a lyophilized NanoCrystal particle formulation is given in Table 1.

The most important aspect of any parenteral product is its sterility. This requirement imposes significant challenges in the development of sterile NanoCrystal particle formulations. However, these challenges are not insurmountable.

In the case of ready-to-use NanoCrystal particle formulations, there are three major methods of sterilization that are evaluated. These are terminal heat sterilization, sterile filtration, and aseptic processing. The use of terminal heat sterilization is desirable both from an ease of manufacturing and from a regulatory perspective. However, only a small

Table 1 Redispersed Mean Particle Size of a Lyophilized NanoCrystal Particle Formulation in Different Diluents

Reconstitution diluent	Mean particle size (nm)
Pre-lyophilized NanoCrystal particle formulation	99
Deionized water	104
5% dextrose	104
0.45% saline	107
6% Dextran 70 in 5% dextrose	102
10% Dextran 40 in 0.9% saline	109

fraction of NanoCrystal particle formulations are amenable to terminal heat sterilization. One example is N1177, an iodinated diagnostic-imaging agent. This particular NanoCrystal particle formulation has a pre-autoclaved mean particle size of about 170 nm. After autoclaving (116°C for 115 minutes) the N1177 NanoCrystal particle formulation, the mean particle size grew to about 270 nm. As in the case of terminal heat sterilization, only a small fraction of NanoCrystal particle formulations are capable of being milled to sufficiently small particle sizes (mean of less than 100 nm) to allow for sterile filtration. The ability to sterile filter NanoCrystal particle formulations is not only a function of particle size, but also particle morphology and concentration of API. These two parameters affect the extent of hydraulic packing and subsequent blinding of the filter. In the case of aseptic processing, sterile API is required or alternatively, the formulation can be sterilized by heating the pre-milled slurry. The stabilizer and any added excipients can be dissolved in water and sterile filtered as they are charged to the recirculation vessel. Once all of the sterile formulation components are in the recirculation vessel, the milling process can be conducted in a pre-sterilized, closed milling system. After the targeted particle size is achieved, the sterile NanoCrystal particle formulation can then be subjected to further aseptic processing, such as filling and lyophilization.

There are two principle methods for sterilization of lyophilized NanoCrystal particle formulations: aseptic processing and terminal sterilization using gamma-irradiation. In the case of aseptic processing, there are three possible routes:

1. Sterile filtration of the NanoCrystal particle formulation, followed by aseptic filling, and lyophilization.
2. Heat sterilization of the pre-milled slurry, followed by aseptic processing, and lyophilization.
3. Use of sterile API in combination with aseptic compounding, processing, filling, and lyophilization.

As mentioned previously, the bulk of the NanoCrystal particle formulations need to be lyophilized, due to physical instability. In the lyophilized state, the majority of the APIs that have been evaluated have withstood gamma-irradiation (typical sterilizing dose is 25 kGray).

4. PHARMACOKINETICS/ PHARMACODYNAMICS

The pharmacokinetics (pK) of NanoCrystal particles administered parenterally is a function of the formulation and physicochemical characteristics of the API. When Nano-Crystal particles are administered intramuscularly, the pK profiles may have the following attributes:

- Relatively quick onset times and sustained release plasma drug concentration vs. time profiles.
- Highly influenced by particle size (Fig. 1).
- Plasma drug levels may be as high for intramuscular administration as when the drug is administered intravenously (Fig. 2).

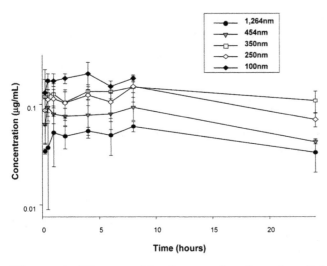

Figure 1 Pharmacokinetics as a function of particle size for intramuscular administration of a NanoCrystal particle formulation in rats.

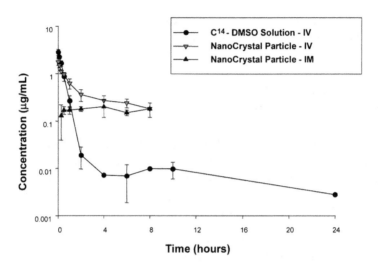

Figure 2 Pharmacokinetics of a solution formulation dosed intravenously compared to a NanoCrystal particle formulation dosed intravenously and intramuscularly in rats.

The pK of NanoCrystal particles administered intravenously may have the following attributes:

- sustained release pK profile;
- higher plasma levels and longer circulation times compared to solution formulation;
- similar pK profiles to cyclodextrin-based formulations (Fig. 3);
- good dose proportionality (Fig. 4).

In the case of intramuscular administration, NanoCrystal particle formulations can be formulated to be physiologically compatible and thus minimize myotoxicity (i.e., site of injection damage to the muscle) and pain on injection. This is due to the absence of irritating organic cosolvents, surfactants, and pH extremes. Additionally, isotonic formulations are possible. A phase I study was completed with naproxen NanoCrystal particles administered intramuscularly. The formulation contained a high concentration of naproxen, 400 mg/mL (the aqueous solubility of naproxen is 16 μg/ mL below the pK$_a$ of the carboxylic acid moiety in naproxen

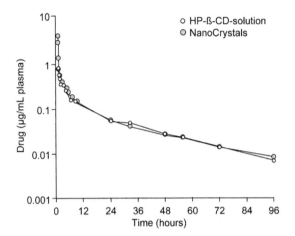

Figure 3 Pharmacokinetics of a NanoCrystal particle formulation compared to a cyclodextrin-solubilized formulation.

and ~3.5 mg/mL at pH 7), and a microbial preservative system. The two major goals for the clinical trial were no myotoxicity and no pain on injection. Both of these clinical endpoints were successfully achieved with this formulation.

When administering NanoCrystal particle formulations intravenously to dogs, the rate of infusion needs to be relatively slow due to an acute hemodynamic effect (1). It is

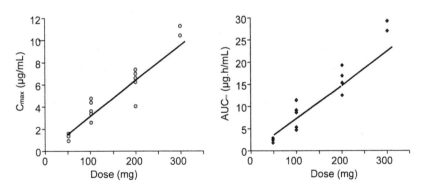

Figure 4 Plasma drug concentration as a function of dose for a NanoCrystal particle formulation.

Table 2 Hypotension in Dogs as a Function of Intravenous Infusion Rate of a NanoCrystal Particle Formulation

Concentration of solids (mg/mL)	Infusion rate (mL/min)	Dose rate (mg/min)[a]	Hypotension
50	1	50	Yes
50	0.1	5	No
10	5	50	Yes
10	1	10	Yes
10	0.5	5	No

[a]Based on an average dog weight of 10 kg.

based on the concentration of solids administered as a function of time. This effect is not unique to NanoCrystal particle formulations and has been observed with other particulate systems such as emulsions. To eliminate the hemodynamic effect, the infusion rate should be kept to 5 mg/min (Table 2).

5. MANUFACTURING PROCESS

NanoCrystal colloidal dispersions are manufactured using a proprietary high-energy media-milling process called the NanoMill™ process. The process consists of recirculating a slurry comprised of API, stabilizers, and water through the NanoMill (Fig. 5). This results in the comminution and dispersion of the API crystals from an initial size of about 10 to 100 μm to a final mean diameter of about 100–200 nm. Poly-Mill™ Polymeric Milling Media is used during the NanoMilling process to impart mechanical and hydraulic shearing plus impact forces within the NanoMill chamber. The API crystals are fractured and dispersed as they recirculate through the NanoMill and are sterically and/or electrostatically stabilized by non-covalent adsorption of the stabilizers onto their surface.

6. SCALE-UP

A wide range of NanoMill equipments have been designed and built to meet both internal and client requirements

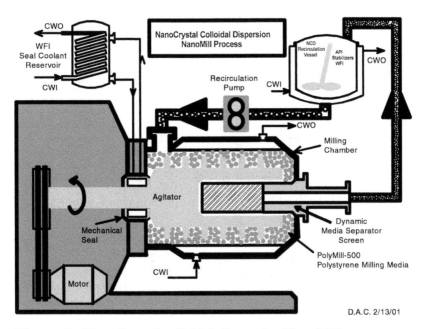

Figure 5 NanoCrystal colloidal dispersion NanoMill process.

(Table 3). The NanoMill process is directly scalable—from laboratory scale (100 mg API) to full commercial scale (500 kg API). This has been demonstrated for multiple compounds including both marketed products and new chemical entities. The primary parameter that measures scalability is replicating the particle size distribution of the NanoCrystal colloidal dispersions over the range of the NanoMills (Fig. 6). The excellent process scalability is obtained through precise control of the critical process parameters in the NanoMill process, including the consistency of the PolyMill Polymeric Milling Media.

An essential component of the NanoMill process is the PolyMill Polymeric Milling Media. PolyMill is a highly cross-linked copolymer of styrene and divinylbenzene. It was specifically designed to be chemically and biologically inert and possesses exceptional wear resistance. PolyMill is manufactured exclusively by the Dow Chemical Company for NanoCrystal applications. It is available in two size

Table 3 NanoMill® Specifications

	NanoMill-001	NanoMill-01	NanoMill-1 Mag-Drive	NanoMill-2	NanoMill-10	NanoMill-60
Chamber volume	10 mL	10 mL, 50 mL, 100 mL	500 mL; 1000 mL	2 L	10 L	60 L
Process	Batch	Batch	Batch	Recirculation	Recirculation	Recirculation
Minimum batch size[a]	100 mg	100 mg	10 g	1 kg	10 kg	100 kg
Maximum batch size[a]	1000 mg	10,000 mg	100 g	10 kg	100 kg	500 kg
Patent status	Patented	Patented	Patent pending	Patent pending	Patent pending	Patent pending
CE certification	No	Yes	Yes	Yes	Yes	Yes

[a]In terms of Active Pharmaceutical Ingredient.

Horiba Particle Size Distributions

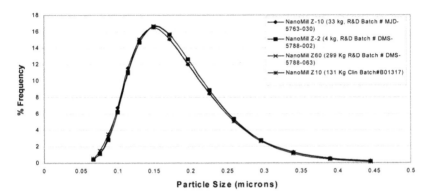

Figure 6 Particle size distributions produced on laboratory, pilot, and full commercial scale NanoMills.

classes—PolyMill-200 is nominally 200 μm in mean diameter and PolyMill-500 is nominally 500 μm.

7. SUMMARY

NanoCrystal Technology provides a platform for the rapid and efficient development and commercialization of poorly water-soluble drugs. It has been demonstrated to enhance the biological performance of drugs administered orally, parenterally, and by inhalation. NanoCrystal Technology has been fully developed, validated, and typically provides a more pharmaceutically elegant product. This proprietary technology is used in a commercial product and can be leveraged to reap the tremendous opportunities that exist in drug discovery and development, and the manufacturing of poorly water-soluble drugs.

ACKNOWLEDGMENTS

The author would like to thank William Bosch, David Czekai, Patricia Shreck, and Edward Tefft for providing technical material and/or reviewing this case study.

REFERENCE

1. Garavilla L, Peltier N, Merisko-Liversidge E. Controlling the acute hemodynamic effects associated with IV administration of particulate drug dispersions in dogs. Drug Develop Res 1996; 37:86–96.

11

Case Study: Formulation Development and Scale-Up Production of an Injectable Perfluorocarbon Emulsion

ROBERT T. LYONS

Allergan, Inc., Irvine, California, U.S.A.

1. INTRODUCTION

1.1. Biological Requirements for PFC Emulsions

Perfluorocarbon-based "blood substitutes" are more closely related to devices than to pharmaceuticals in the sense that they perform a function that is primarily physical. These blood substitutes are intended to promote a rapid and reversible exchange of blood gases without impairing the functions of

---

Studies being reported in this chapter were conducted solely at the former laboratories of Kabi Pharmacia in Alameda, California, U.S.A.

blood elements, the flow of fluids, or the performance of organs. To be acceptable for injection, a perfluorocarbon emulsion must meet specific requirements of hemocompatibility and biocompatibility (1). These requirements include the following:

- Be miscible with blood in all proportions without undergoing phase changes, precipitation, flocculation, or coalescence.
- Not activate or inhibit blood coagulation factors.
- Not impair the normal functioning of circulating blood elements.
- Neither disrupt the cell membranes of blood elements nor alter their permeability.
- Not cause clinically significant changes in the rheology of blood.
- Be free of toxic effects and systemic adverse reactions.
- Not physically occlude the microvasculature and cause embolisms.
- Be excreted at a predictable rate without being metabolized or stored in tissues.

The present case study will summarize key formulation and scale-up optimization studies performed during the development of a perfluorocarbon-in-water emulsion made with purified egg yolk phospholipids (EYP) and perfluoro-1, 3-dimethyladamantane (PFC). This structure is shown in Fig. 1. During development, the physiological requirements listed above served as a guide, and every significant change in either formulation or physical process was subjected to some type of biological screening, either *in vivo* or *in vitro*.

1.2. Clinical Applications for PFC Emulsions

At least six clinical indications have been proposed for this PFC emulsion (2). The first, cardioplegia during open-heart surgery, involves perfusing an admixture of emulsion with a cardioplegic salt solution. This fluid is designed to preserve the functional and structural integrity of the myocardium while subjected to hypothermic arrest during cardiopulmonary

Figure 1 Perfluorodimethyladamantane molecular structure. This highly stable caged molecule consists of four interlocking 6-membered rings.

bypass. A second proposed use, also related to heart surgery, involves *blood oxygenator priming* to minimize blood unit requirements. A third, m*yocardial ischemic rescue*, is designed to minimize reperfusion injury during restoration of blood flow following treatment for coronary artery occlusion. In this case, potential advantages of PFC emulsions compared to whole blood include lower viscosity, smaller drop size, and reduced oxygen radicals generated by inflammatory cells. A fourth indication is *radiation sensitizer* to improve oxygenation of solid tumors with anoxic cores prior to therapy with ionizing radiation. A fifth proposed use involves *organ preservation for transplantation*. Oxygenated PFC emulsions could extend the viability of organs that are now protected solely by electrolytic perfusion and hypothermia. A sixth and "ultimate" use for this product would be as a true *blood substitute for transfusion*, e.g., following carbon monoxide poisoning, decompression sickness, or hemorrhage. Such applications are especially valuable with rare blood types.

The latter indication forms the basis of a rigorous biological screen used during product development, namely, *total exchange perfusion* (1,3). Rats are anesthetized with an oxygen-halothane gas mixture; the test PFC emulsion is infused through a tail vein while removing a mixture of blood and test substance from the exterior jugular vein. This exchange perfusion is performed in an isovolemic manner to prevent a shock-like syndrome, down to hematocrits of 1–3% (v/v). Initially, the animal is allowed to breathe 90–100% oxygen. This phase is followed by a 3–5 day weaning schedule which gradually steps down to atmospheric oxygen levels as whole blood is regenerated. In a sense, survival of these blood-less animals represents the ultimate tolerance or "LD_{50} test" for any injectable product.

2. FORMULATION DEVELOPMENT

2.1. Overview

The initial formulation given to our development group was relatively simple: 50 g PFC plus 2.4 g purified EYP per 100 mL emulsion product. PFC emulsions are produced using a standard procedure (4) outlined in Fig. 2. Briefly, EYP is first dispersed in hot water for injection by means of a high shear overhead mixer (UltraTurrax); nitrogen-sparged PFC is then added at a controlled rate through a narrow orifice while continuing high shear mixing to form a coarse pre-emulsion or "premix." Next, this dispersion is transferred to a high-pressure homogenizer (e.g., APV Gaulin, Inc, Model M3) for emulsification at about 10,000 psig and 35–40°C under continuous nitrogen protection. The finished product is then filtered through a 10 μm stainless steel mesh into washed, quarter-liter, borosilicate glass bottles. These bottles are protected with nitrogen headspace gas, stoppered, secured with aluminum overseals, and then terminally heat sterilized in a rotating steam autoclave with air overpressure. Using this procedure, a sterile product typically exhibits a mean drop size of about 200–300 nm as measured by photon correlation spectroscopy (Coulter N4).

2.2. Biological Screening

As part of an initial biological screening for this experimental product, we conducted incubations with heparinized whole blood to evaluate emulsion effects on erythrocyte morphology *ex vivo*. Normal morphology is necessary for proper distribution of blood flow in the microcirculation. Deformation or *crenation* (5) is the result of membrane damage and can serve as a marker for hemo-incompatibility. Briefly, whole blood is diluted with fresh plasma (1:1) which is mixed with

PRODUCTION OF A PERFLUOROCARBON EMULSION

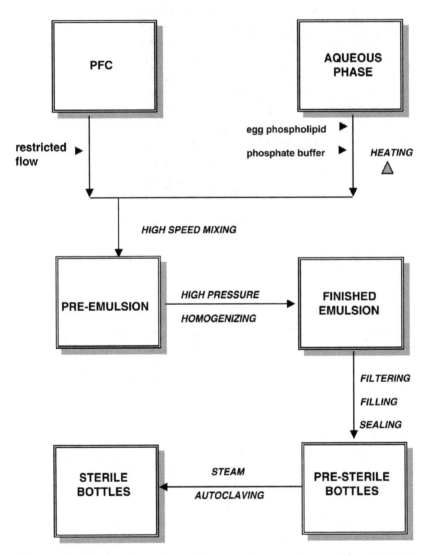

Figure 2 Production of a perfluorocarbon emulsion. Liquid PFC is added slowly to a hot aqueous dispersion of EYP with continuous high speed mixing to form a pre-emulsion (or premix). High-pressure homogenization is required to reduce the drop size distribution to the submicron range. During production, all steps are performed in a closed system under a protective blanket of nitrogen gas.

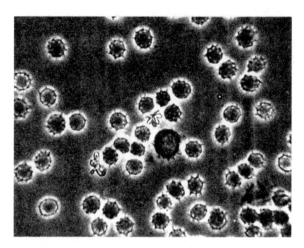

Figure 3 Erythrocyte crenation. Deformation of erythrocytes is a result of subtle membrane damage. Incubating test formulations with whole blood ex vivo serves as a sensitive biological screening tool to predict unacceptable cytotoxicity.

test emulsion (1:1) and then incubated for 10 min at 37°C. Erythrocyte morphology is evaluated by microscopy under phase-contrast optics (Zeiss ICM405, 410× magnification). As shown in Fig. 3, crenated erythrocytes exhibit prominent spoke-like projections called spicules. Addition of 3.5 mM oleic acid or 0.3 mM lysolecithin to these incubations will result in virtually 100% crenated red blood cells. At higher concentrations, these surface-active agents will cause total hemolysis.

While crenation testing evaluates effects of emulsion product on blood cells, a second useful biological screen *ex vivo* involves evaluating effects of blood plasma on emulsion integrity. Fresh or frozen plasma, anticoagulated with either citrate or EDTA, is incubated in varying proportions with test emulsion for 30 min at 37°C. Again using phase-contrast microscopy, we evaluate relative resistance of the emulsion to *flocculation* (6,7). Floccules appear as irregular-shaped, loose aggregates of emulsified oil droplets. These range in size from small "grape clusters" (2–10 µm) to massive "ice floes" of 100 µm or larger. An approved parenteral fat emulsion such as Intralipid® 20% (soybean oil emulsified with egg phospholipids) may be used as

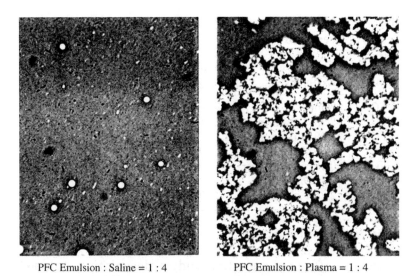

PFC Emulsion : Saline = 1 : 4   PFC Emulsion : Plasma = 1 : 4

Intralipid 20% : Plasma = 1 : 4

Figure 4 Emulsion flocculation. Serial dilutions of test emulsions are incubated ex vivo with human blood plasma. Microscopic evidence of flocculation predicts hemo-incompatibility for the new formulation.

a negative control for this test. A pronounced tendency to flocculate in plasma is predictive of poor rheological properties, undesirable product deposition in organs such as liver, spleen, and lungs, and elevated systemic toxicity. As shown in Fig. 4, initial samples of this PFC emulsion formulation showed a high tendency to flocculate when incubated as described.

Table 1   Effect of Excipients on Emulsion Flocculation and
Erythrocyte Crenation

Test emulsion	Excipient	Flocculation score	RBC crenation (% of total)	Relative flow rates[a]
50% PFC	None	+++	75%	0.48
50% PFC	2.25% glycerin	+++	75%	0.51
50% PFC	0.28 M L-alanine	++	25%	0.55
50% PFC	0.05 M phosphate	0	0	0.98
50% PFC	0.05 M phosphate (ADMIXED)	++++	75%	0.33
Intralipid 20%	2.25% glycerin	0	0	0.93

(0) = none, (+) = trace, (++) = moderate, (+++) = heavy, (++++) = severe.
[a]Normalized to blood:saline (3:1 v/v).

Relative effects of crenation and/or flocculation on blood rheology may be estimated by means of a glass flow viscometer (8). Flow times for blood:emulsion (3:1 v/v) mixtures are expressed as ratios to times for control blood–saline mixtures. Compared to Intralipid–blood mixtures, we observed significantly increased viscosities with PFC emulsion samples. In order to address this problem, a series of small-scale (400 mL) alternative emulsions were made. Test excipients were added prior to steam sterilization, and the sterile product was tested in blood mixtures for resistance to flocculation, ability to induce erythrocyte crenation, as well as for relative flow viscosity. A summary of some of these experiments is shown in Table 1.

We observed that added glycerin had no effect on either crenation or flocculation, while a neutral amino acid such as alanine had some beneficial action. However, addition of sodium phosphate (adjusted to pH 7.4) was very effective in preventing these effects and preserving low flow viscosity. Surprisingly, phosphate added (admixed) to previously sterilized emulsion was ineffective in this regard. Commercial Intralipid 20% (soybean oil emulsion) resisted both crenation and flocculation under these test conditions. As a result of a series of such studies, a modified formulation was adopted that includes 0.05 M sodium phosphate (pH 7.4) added prior to terminal steam sterilization (9).

Flocculation also occurs *in vivo*. In a typical experiment, Sprague–Dawley rats were infused with 20 mL/kg body weight (bw) of PFC emulsion via the tail vein. Blood samples were collected, anticoagulated with EDTA, and scored for flocculation by phase-contrast microscopy as described above. Under these conditions, large floccules were observed for at least 4 h post-infusion with the original formulation, while phosphate-containing emulsion was much more resistant over this time period.

3. PROCESS OPTIMIZATION

3.1. Premix Formation

Having established a viable formulation, developmental efforts for this PFC emulsion next focused on process optimization. As shown in Fig. 2, formation of a coarse oil-in-water dispersion or *premix* containing all excipients precedes high pressure homogenization. At the premix stage, relative size homogeneity of oil droplets is critical to producing a high quality finished product with a minimum number of large (i.e., >5 μm) droplets (10). We found several good methods to monitor premix formation and optimize high shear process time. The simplest is optical phase-contrast microscopy on in-process samples. Examples of photomicrographs (320×) are shown in Fig. 5. Under tested conditions, 10 min of processing time (at 20,500 rpm) appears sufficient to produce a very uniform dispersion. Since these parameters are highly correlated with the specific formulation, equipment type, and even batch size, process optimization must be an ongoing, project-specific activity.

Since microscopy is subjective and not very quantitative, we evaluated two other methods. The first is large particle counting using the Coulter ZM with a 100 μm orifice. Samples are diluted in 0.9% saline prior to counting. These data, summarized in Fig. 6, suggest a continuous formation of smaller particles (2–10 μm) with a parallel disappearance of larger droplets during premix processing. Since this analytical method is labor-intensive and failed to identify a process endpoint, we evaluated a third procedure using the same

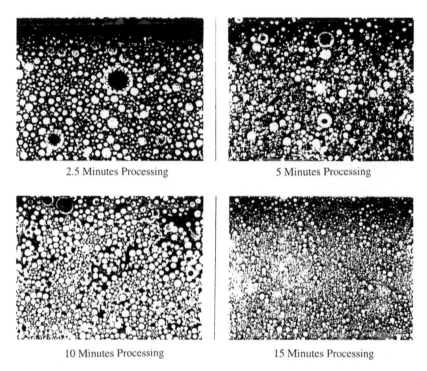

2.5 Minutes Processing   5 Minutes Processing

10 Minutes Processing   15 Minutes Processing

Figure 5 Evaluation of in-process premix samples by phase-contrast microscopy (1 mm = 2.3 µm). The preferred endpoint during premix formation is a relatively homogeneous drop size distribution.

capillary flow viscometer described above. As shown in Fig. 7, flow viscosity drops dramatically during the first 10 min of premix processing, with only small changes thereafter. From studies such as these, we were able to optimize the premix process time at each production scale. For other similar projects, our methods of choice for premix evaluation remain phase-contrast microscopy and flow viscosity.

3.2. High Pressure Homogenization

Earlier pilot studies had shown that better quality emulsions were prepared using the highest available homogenizer

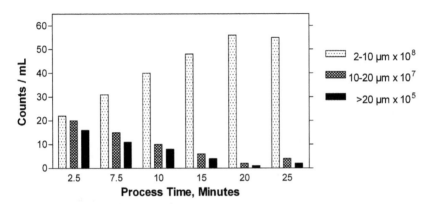

Figure 6 Evaluation of in-process premix samples by electrozone sensing. Shifts in particle size distribution during processing may be monitored by large particle counting using the Coulter ZM with a 100 μm orifice. This is a quantitative analysis of particle counts per milliliter, unlike microscopy or laser light scattering measurements.

pressure, 10,000 psig, with 15% of this value (1500 psig) chosen as the second stage back-pressure. Earlier studies also indicated that homogenization at lower temperatures, e.g., 5–10°C, resulted in poor quality emulsions while higher temperatures, e.g., above 60°C, resulted in extensive losses of volatile PFC. Intermediate temperatures, e.g., 35–40°C, produced the best results and this range was selected for future batches.

Our next challenge was to optimize the number of homogenizer passes through the spring-loaded valve system. Multipassing is known to narrow the drop size distribution, but to have relatively little effect on particle mean diameter. This phenomenon is shown in Fig. 8. However, an excessive number of passes may produce large PFC droplets due to over-processing and may promote more degradation of the unsaturated phospholipid emulsifier due to hydrolysis and oxidation (11). In addition, over-processing may result in significant loss of

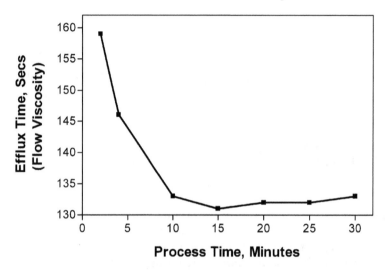

Figure 7 Evaluation of in-process premix samples by flow viscosity. A glass capillary viscometer was used to demonstrate a dramatic reduction in flow viscosity in premix samples as continued processing reduces drop size distribution.

volatile PFC raw material and will extend production time, adding to costs. For all these reasons, we sought a reliable method to optimize homogenizer passes.

During homogenization, in-process "pass samples" are removed and analyzed for *large particle counts* by the Coulter ZM as described above. Three size classes are monitored, and typical results are summarized in Fig. 9. We observe a rapid initial drop in count rate, followed by a slower but continuous reduction during further processing. However, based on these data, we were not able to establish an optimum process endpoint.

A second parameter used to evaluate the homogenization process is phospholipid binding (12). For this test, aliquots of non-sterile pass samples or sterile final product are centrifuged for 30 min at 36,000×*g* (Beckman J2-21 m/E centrifuge with JA-20 fixed angle rotor). Aliquots of resulting supernatant or of whole emulsion are vortex-mixed with ethanol (1:2 v/v).

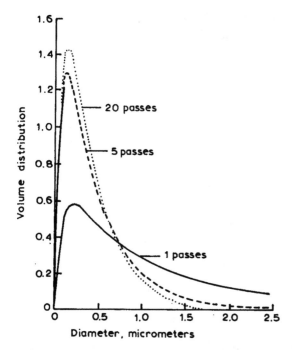

Figure 8 Evaluation of homogenizer multipassing by laser light scattering. Homogenizer multipassing narrows the emulsion drop size distribution but has relatively little effect on the mean diameter. This oil-in-water emulsion was prepared using a laboratory-size Gaulin homogenizer model 15MR. Particle size distributions were determined by laser light scattering measurements.

Samples are then evaporated in a vacuum oven (3 hr at 60°C) and phospholipid concentration determined gravimetrically. Phospholipid bound to the PFC fraction is estimated as the difference between total phospholipid in the whole emulsion minus the unbound fraction in the supernatant. Figure 10 shows that phospholipid binding to PFC droplets is essentially complete by pass 12, and a reciprocal drop in supernatant (free) phospholipid is observed during the process time-course. We also note some loss in binding during terminal heat sterilization (typically down to 40–50%). This loss parallels an observed shift in population size distribution towards larger sizes with less total surface area. For example, mean diameter for a

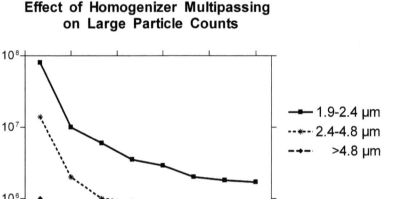

Figure 9 Evaluation of homogenizer multipassing by electrozone sensing. Large particle counts decline with processing time as submicron particles are being created but these submicron sizes are not detected by the Coulter ZM.

typical batch may shift from about 190 to 285 nm during autoclaving, with relatively small further changes occurring during prolonged storage. We believe that small amounts of phospholipid are sloughed from the surface of emulsified PFC droplets during autoclaving. This phospholipid then folds in upon itself in the aqueous phase. As a result, empty spherical vesicles (liposomes) are formed with phospholipids arranged in one or more bilayers (13).

A third in-process measurement is based on our observation that PFC emulsion turbidity, as measured by light transmittance at 410 nm, is closely correlated with unbound (supernatant) phospholipid concentration. This wavelength was chosen to give strong light scattering with minimum absorbance from the unsaturated phospholipids. Figure 11 shows that turbidity declines dramatically at 10,000 psig

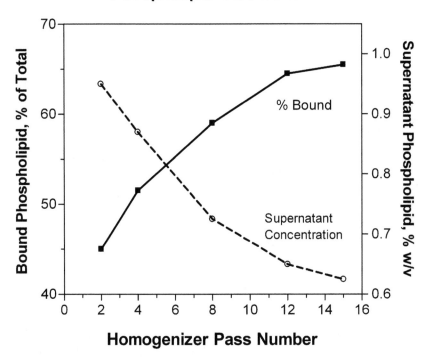

Figure 10 Evaluation of homogenizer multipassing by monitoring phospholipid binding. Homogenizer multipassing reduces the number of larger particles thereby increasing the total interfacial surface area per unit volume of emulsion. Phospholipids are recruited continuously from the aqueous phase and bind to newly created interfaces between PFC and water. When the ratio between bound and free (aqueous) phospholipid becomes constant, processing at the chosen conditions of temperature and pressure is complete.

during the first 4–6 passes at 35–40°C, but at 5–10°C, over 12 passes are required. We observed no evidence of over-processing under these conditions. Initially, measurements were made on pass samples using a UV/VIS spectrophotometer with sipper attachment (Beckman DU 640). Later, in-line spectrophotometric monitoring was accomplished

Effect of Multipassing on Emulsion Turbidity

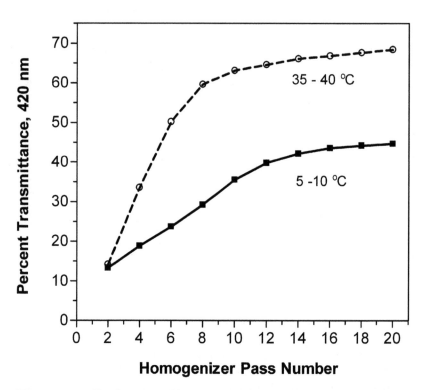

Figure 11 Evaluation of homogenizer multipassing by monitoring turbidity. Homogenizer multipassing reduces the concentration of unbound phospholipid that is most responsible for visible light scattering. Therefore, in-line measurement of light transmittance is a simple way to monitor the effect of continued processing on emulsion quality. This method is especially useful here since the refractive index of the dispersed phase (PFC) is close to that of the continuous phase (water).

by means of a probe colorimeter (Brinkmann Instruments; Westbury, NY; Model PC800) fitted with a 420 nm filter. A 1 cm path length fiber optic probe is inserted into the product stream with an in-line stainless steel T-fitting. A strip chart recorder gives us a permanent record of product turbidity during processing. Homogenization is continued until a pre-determined percentage of transmittance is attained (usually after 8–10 passes) (14,15).

4. PROCESS SCALE-UP

4.1. Nonemulsified Perfluorocarbon

Previous process development work was performed at the 4–8 L scale, filling 100 mL bottles. At the 50-L scale and in 500 mL bottles, process conditions such as temperature, pressure, and passes were revisited and re-optimized using the techniques described above. However, a new problem emerged. Figure 12 shows a photomicrograph (phase-contrast, 100×) of sediment taken from the bottom of a sterilized 500 mL bottle after standing for 1 week. In these containers, we observed large numbers of rapidly sedimenting, poorly emulsified PFC droplets ("fish eggs") with diameters in the 50–200 nm range. The presence of free or poorly emulsified PFC in this product was quantified by extraction with a lower molecular weight perfluorocarbon liquid, fluorodimethycyclohexane (F-DMCH),

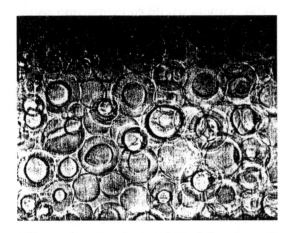

Figure 12 Poorly emulsified droplets of perfluorocarbon. These dense droplets sediment rapidly to form a layer at the bottom of the container. Phase-contrast microscopy (100×) of a sample from this layer reveals PFC droplets in the 50–200 nm range with a "fish egg" appearance.

followed by gas chromatographic analysis. Earlier tests demonstrated that F-DMCH could serve as an extraction solvent since it removed only non-emulsified (free) perfluorodimethyladamantane.

Briefly, an exact volume (e.g., 3–5 mL) of F-DMCH is injected by syringe directly into each bottle which is then shaken vigorously by hand for about 30 sec. Bottles are inverted and allowed to stand at room temperature for 1 hr, after which time a portion of F-DMCH is removed by syringe needle through the stopper. An aliquot of this extract is then injected into a gas chromatograph (Hewlett Packard 3388) equipped with an automatic sampler (7672A), a Supelcowax 10 column (30 m length, 0.32 mm ID, 0.25 μm film thickness), and a flame ionization detector. Operating conditions were: oven temperature 60°C, injector temperature 275°C, detector temperature 300°C, column head pressure 13 psi, with helium carrier gas at 0.70 mL/min. Retention time for F-DMCH is about 1.5 min compared to 2.1 min for extracted PFC, allowing simple quantitation of the latter.

We suspected that refluxing was occurring in the bottle during terminal heat sterilization at temperatures up to 121°C. During cooling, PFC vapors would condense and settle to the bottom as large, poorly emulsified droplets. When the mass of recovered (free) PFC was plotted vs. the square of headspace volume, we observed a linear relationship, shown in Fig. 13. We found that a minimal headspace volume (here, 7% of 500 mL) is essential to minimize the formation of non-emulsified PFC in sterile product.

4.2. Product Uniformity

The density of liquid perfluorodimethyladamantane is 2.025 g/mL at 20°C. For this reason, emulsified PFC droplets tend to sediment fairly rapidly, and the potential exists for non-uniformity in PFC content from bottle to bottle. This is especially true during scale-up, since the bottle filling operation time increases with the batch size, e.g., up to 2 hr for a 50 L batch. Any emulsion sedimentation in the holding tank would result in a gradual increase in product

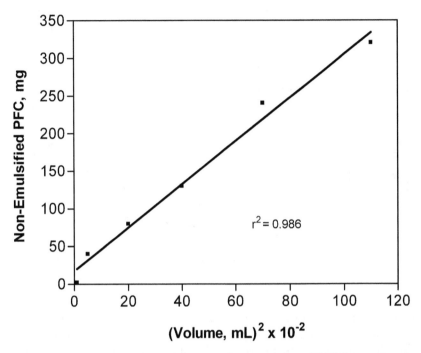

Figure 13 Effect of headspace volume on free PFC formation in autoclaved 100 mL bottles. During autoclaving at 121°C, PFC vapors form in the nitrogen headspace and then condense upon cooling to form small, non-emulsified oil droplets. This effect can almost be eliminated by minimizing headspace volume during filling.

PFC concentration across the fill series. In order to avoid this, we established a continuous recirculation of product in and out of the holding tank through a 1/4-in stainless steel tubing using a centrifugal pump (Eastern). Unfortunately, rapid recirculation caused a progressive increase in Coulter large particle counts. Typical data are shown in Fig. 14. We now believe that shear forces generated by high flow velocity through the narrow tubing was responsible for continuous de-emulsification and particle coalescence (16). Eventually, this problem was solved by means of a low shear,

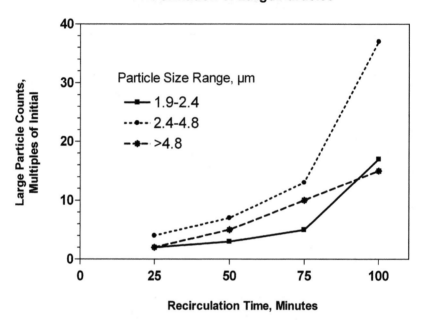

**Effect of Recirculation on
Formation of Large Particles**

Figure 14 Evaluation of product recirculation during filling by electrozone sensing. Large particle counts increase steadily when the PFC emulsion is recirculated too rapidly during the filling operation. Data were generated using the Coulter ZM instrument with a 100 μm aperture and 0.9% saline as diluent.

positive displacement pump (Waukesha Pumps; Waukesha, WI) coupled to larger ID tubing to reduce flow velocity. Bottle-to-bottle uniformity was verified by analyzing every 25th bottle in an entire 50 L batch for both PFC concentration and drop size distribution.

5. CONCLUSIONS

The development of an injectable PFC emulsion is complicated by formidable challenges of both a biological and physicochemical nature. We found that the early establishment of several reproducible and relevant biological screens is essential during formulation optimization and process development.

One inherent problem relates to the immiscibility of PFC liquid and phospholipid emulsifier. Raising the ionic strength in the aqueous phase may help by forcing more phospholipid to the oil–water interface. In addition, process conditions must be carefully optimized to avoid both over-processing and high shear conditions after homogenization, e.g., during final filtration. These adverse treatments can strip phospholipid from the interface and induce de-emulsification.

PFCs are selected for high vapor pressure to facilitate excretion via the lungs. However, this high vapor pressure means that careful temperature control during processing is critical. It also means conditions leading to vaporization and condensation must be avoided, e.g., autoclaving bottles with large headspace volumes.

During scale-up production of perfluorocarbon emulsions, extra care must be taken to avoid sedimentation and stratification of heavy emulsified droplets. For this reason, bottle-to-bottle uniformity must be verified in the finished product, especially with regard to PFC content and drop size distribution.

An injectable product very similar to the one described in this review was manufactured in multiple 50 L batches at our Clayton, North Carolina facility. This product exhibited satisfactory storage stability and was well-tolerated in an animal model. Unfortunately, this project was halted for non-technical, business reasons by our collaborating partner. We can only hope that "lessons learned" will benefit and expedite future projects of this type, both at our facility and perhaps at yours.

REFERENCES

1. Biro GP, Blais P. Perfluorocarbon blood substitutes. CRC Crit Rev Oncology/Hematology 1987; 6(4):310–371.

2. Tremper KK, ed. Perfluorochemical Oxygen Transport. Vol. 23. Boston: Little, Brown and Company, 1985.

3. Geyer RP. Review of perfluorochemical-type blood substitutes. In: Proceedings of the Tenth International Congress for Nutrition, Kyoto, Japan, 1975:3–19.

4. Collins-Gold LC, Lyons RT, Bartholow LC. Parenteral emulsions for drug delivery. Adv Drug Delivery Rev 1990; 5:189.

5. Brecher G, Bessis M. Present status of spiculed red cells and their relationship to the discocyte-echinocyte transformation: a critical review. Blood 1972; 40:333.

6. Hulman G, et al. Agglutination of Intralipid by sera of acutely ill patients. Lancet 1982; 2:1426.

7. Lindh A, et al. Agglutinate formation in serum samples mixed with intravenous fat emulsions. Crit Care Med 1985, 13:151.

8. Hamai H, et al. Viscosimetric study of fluorocarbon emulsions and of their mixtures with blood. J Fluorine Chem 1987; 35:259.

9. Lyons R. Inhibiting aggregation in fluorocarbon emulsions. US Patent No. 5,073,383 (1991).

10. Pandolfe WD. Effect of premix condition, surfactant concentration, and oil level on the formation of oil-in-water emulsions by homogenization. J Dispersion Sci Tech 1995; 16:633.

11. Herman CJ, Groves MJ. The influence of free fatty acid formation on the pH of phospholipid-stabilized triglyceride emulsions. Pharmaceut Res 1993; 10:774.

12. Groves MJ, et al. The presence of liposomal material in phosphatide stabilized emulsions. J Dispersion Sci Tech 1985; 6:237.

13. Lasic DD. The mechanism of vesicle formation. Biochem J 1988; 256:1.

14. Deackoff LP, Rees LH. Testing homogenization efficiency by light transmission. APV Gaulin Technical Bulletin No. 63, 1981.

15. Lashmar UT, Richardson JP, Erbod A. Correlation of physical parameters of an oil-in-water emulsion with manufacturing procedures and stability. Int'l J Pharmaceutic 1995:315–325.

16. Han J, Washington C, Melia CD. A concentric cylinder shear device for the study of stability in intravenous emulsions. Eur J Pharm Sciences 2004; 23:253–260.

12

Case Study: A Lipid Emulsion—Sterilization

THOMAS BERGER

Pharmaceutical Research & Development,
Hospira, Inc., Lake Forest, Illinois, U.S.A.

1. OUTLINE

The following information outlines a case study pertaining to the engineering and microbiological activities required for an emulsion product in order to gain FDA approval prior to manufacture. Minimum details for each research, development, and production activity are discussed to demonstrate the supporting documentation that may be required in order to file a New Drug Application with the FDA.

An organized sequential flow of activities must occur as a new parenteral formulation is developed in an industrial research and development (R&D) environment, and subsequently processed in a manufacturing facility. Sterilization of pharmaceutical emulsions must be established and verified through a series of activities that confirm the product has been rendered free of any living microorganisms. In the case of moist heat sterilization, which is discussed here, the R&D phase activities must include sterilization developmental engineering consisting of sterilization cycle development; container thermal mapping; microbial closure studies; parenteral formulation microbial growth; D-value analysis; container maintenance of sterility (mos) studies; and final formulation stability studies.

Production phase activities for moist heat sterilization must include an initial sterilization vessel certification, which demonstrates that the vessel will deliver the defined sterilization process in a consistent and reproducible manner. Emulsion and container closure microbial validation studies must be conducted at subprocess production sterilization conditions employing heat resistant microorganisms. Equipment validation, filtration studies and assessment of the bioburden on component parts, and in the environment, must also be ascertained.

The developmental and production phase sterilization technology activities must be included in the documentation submitted as part of a New Drug Application. The procedures must follow the FDA guideline requirements for products that are either terminally sterilized or aseptically processed. These studies allow the establishment with a high level of sterilization assurance, the correct sterilization cycle F_0 (equivalent sterilization time related to the temperature of 121°C and a z-value of 10°C), temperature, and product time above 100°C to be used for the sterilization of a specific parenteral formulation in a particular container/closure system.

2. INTRODUCTION

This section will address sterilization and associated activities that occur in the R&D and production areas.

2.1. R&D Area

1. Sterilization engineering.
2. Thermal mapping studies.

3. Emulsion: microbial moist heat resistance analysis.
4. Closure microbial inactivation studies.
5. Lipid emulsion predicted spore logarithmic reduction (PSLR) values.
6. Accumulated F(Bio) (biologically derived sterilization value) for lipid emulsions.
7. R&D emulsion oil phase studies.
8. Maintenance of sterility studies.
9. Bacterial endotoxin.

2.2. Production Environment

1. Engineering penetration and distribution (P&D) studies.
2. Sterilization cycles.
3. Sterilizer microbial emulsion subprocess validations.
4. Sterilizer microbial closure subprocess validation.
5. Production environment bioburden screening program.

Refer to Young's detailed discussion of sterilization by moist heat processing (1).

2.3. Regulatory Submission

Checklist for aseptically processed and terminally sterilized products.

3. R&D AREA

3.1. Sterilization Engineering

Sterilization engineering personnel primarily focus their efforts on determining whether a new parenteral formulation packaged in a particular container configuration can be sterilized in a current sterilization cycle or whether a new cycle must be developed. Sterilization feasibility studies usually occur preliminarily to ascertain the physical effects of the cycle on a container and its emulsion. Once the basic engineering parameters (temperature, time, F_0) are established, engineering thermal container mapping studies are performed (2). F_0 is the integrated lethality or equivalent minutes at 121.1°C for the hottest and coldest thermocouple containers.

Product attributes that can be affected by a steam sterilization cycle include:

- product sterility;
- closure integrity;
- emulsion potency;
- pH, color;
- shelf-life stability (potency of emulsion);
- visible and subvisible particulates.

3.2. Thermal Mapping Studies

An R&D vessel is smaller than a production vessel but can simulate the sterilization cycles conducted in the larger production vessels. Container thermal mapping studies are performed in an R&D vessel.

- To locate the coldest zone or area inside a container.
- To determine the cold zone in an R&D vessel and the relationship to the location monitored by the production thermocouple.
- The data generated are used during the setting of production sterilization control parameters.

When conducting thermal mapping studies, there are various factors to be considered, and these are dependent upon the

- type of container (flexible or rigid);
- shaking/static solutions;
- viscosity;
- autoclave trays/design/surface contact;
- autoclave spray patterns/water flow.

Engineering map data obtained for lipid emulsions contained within a 1000 mL glass container are shown in Tables 1 and 2.

Table 1 Heat Input (F_0 units)

| Tc Number | Run CLHK00.049 | | Run CLHK01.050 | | Avg (Std Dev) |
	bt1 1	bt1 2	bt1 1	bt1 2	
1, 12	7.91	C7.28	8.13	C7.36	7.67 (0.415)
2, 13(PC)	7.79	7.49	8.02	7.64	7.74 (0.226)
3, 14	C7.46	7.40	C7.71	7.47	C7.51 (0.137)
4, 15	7.64	7.80	7.87	7.96	7.82 (0.135)
5, 16	12.66	12.90	12.95	12.91	12.86 (0.132)
6, 17	12.73	12.46	12.77	12.68	12.66 (0.138)
7, 18	12.78	12.69	12.95	12.91	12.83 (0.120)
8, 19	13.32	13.33	13.42	13.78	13.46 (0.223)
9, 20	14.21	14.33	14.03	14.56	14.28 (0.222)
10, 21	H15.87	H17.24	H15.18	H16.09	H16.10 (0.856)
11, 22	15.47	16.56	14.77	16.07	15.72 (0.773)
H-C	8.41	9.96	7.47	8.73	8.64 (1.028)
PC-C	0.33	0.21	0.31	0.28	0.28 (0.053)

Note: H denotes hottest TC location; C denotes coldest TC location; PC denotes approximate location of the Production Profile TC; Data from TC#9 used with a post-calibration variance +0.25°C at 100°C; All heat input values are calibration corrected.

The following summarizes typical heat map data obtained from a 1000 mL glass Abbovac intravenous container filled with 1035 mL of lipid emulsion:

1. Heat input in F_0 units.
2. Emulsion heat rates in minutes.
3. Figure 1 depicts the thermocouple locations
4. Figure 2 contains the average heat input (F_0) at various locations.

Thermocouple probes (Copper Constantan, type T, 0.005 in diameter) were used to monitor 11 emulsion locations within the 1000 mL container. The thermocouple probes were positioned at various distances (in inches) as depicted in Fig. 1. Each container was filled with 1035 mL of the lipid emulsion, evacuated to 20 in of mercury, and sealed with an aluminum overseal.

A flat perforated rack on a reciprocating shaker cart was used in the autoclave. The cycle's target temperature was

Table 2 Solution Heat Rates (minutes)

	Run CLHK00.049		Run CLHK01.050		
	btl 1	btl 2	btl 1	btl 2	Avg (Std Dev)
Coldest location					
Thermocouple no	3	12	3	12	—
Time to 100°C	19.0	19.0	19.0	19.0	19.00 (0.000)
Time ≥ 100°C	21.0	21.0	21.0	21.0	21.00 (0.000)
Time ≥ 120°C	4.0	3.0	4.0	3.0	3.50 (0.577)
Time 120–100°C	4.0	5.0	4.0	5.0	4.50 (0.577)
Max temp (°C)	120.82	120.77	120.92	120.77	120.82 (0.071)
Heat input (F_0)	7.46	7.28	7.71	7.36	7.45 (0.187)
Production profile TC location					
Thermocouple no	2	13	2	13	—
Time to 100°C	19.0	19.0	19.0	19.0	19.00 (0.000)
Time ≥ 100°C	22.0	21.0	22.0	21.0	21.50 (0.577)
Time ≥ 120°C	4.0	3.0	4.0	3.0	3.50 (0.577)
Time 120–100°C	5.0	5.0	5.0	5.0	5.00 (0.000)
Max temp (°C)	120.91	120.82	120.91	120.92	120.89 (0.047)
Heat input (F_0)	7.79	7.49	8.02	7.64	7.74 (0.226)

Note: H denotes hottest TC location; C denotes coldest TC location; PC denotes approximate location of the Production Profile TC; Data from TC#9 used with a postcalibration variance +0.25°C at 100°C; All heat input values are calibration corrected.

1000 mL ABBOVAC GLASS I.V. CONTAINER
- HEAT MAPPING STUDY

THERMOCOUPLE LOCATIONS

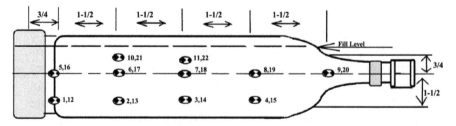

Figure 1 The numbers inside the 1000 mL glass emulsion bottle are the thermocouple locations for duplicate bottles from two separate runs. The numbers outside the bottle are distances in inches.

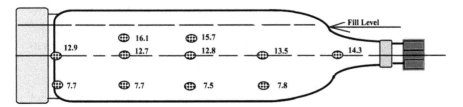

HEAT MAPPING STUDY

AVERAGE HEAT INPUT (F_0) AT VARIOUS LOCATIONS

Figure 2 The numbers inside the 1000 mL glass emulsion bottle are the average heat input (F_0) at the various thermocouple locations.

123°C, recirculating water spray cycle with 70 rpm of axial agitation, 30 psig (pounds per square inch) of air over-pressure and a minimum requirement of $6.0F_0$ units in the coldest location.

When the sterilization cycle was controlled to give a heat input of approximately $7.5F_0$ units in the coldest emulsion area, the average coldest emulsion area was measured by thermocouple number (TC#) 3,14. The average hottest emulsion area was measured by TC# 10,21. The difference between the hottest and coldest emulsion areas ranged from 7.5 to $10.0F_0$ units with an average of $8.6F_0$ units. Therefore, when the coldest emulsion area registered $7.5F_0$ units, the hottest emulsion area would average $16.1F_0$ units.

The emulsion area approximating the production profile thermocouple location was measured by TC#2,13 and averaged $7.7F_0$ units when the coldest emulsion was approximately $7.5F_0$ units. (Refer to Figs. 1 and 2.)

3.3 Emulsion: Moist Heat Resistance Analysis

A BIER vessel is an acronym for a biological indicator (BI) evaluator resistometer vessel that meets specific performance requirements for the assessment of BIs as per American National Standards developed and published by AAMI (Association for the Advancement of Medical Instrumentation) (3).

One important requirement for a BIER steam vessel as used in our studies is the capability of monitoring a square wave heating profile.

Figure 3 is a schematic of the steam BIER vessel used to generate the D and z-value data. D-value is the time in minutes required for a one log or 90% reduction in microbial population. The z-value is the number of degrees of temperature required for a 10-fold change in the D-value.

The family category of lipid emulsions and their respective $D_{121°C}$ and z-values as well as classification in terms of microbial resistance are shown in Table 3. A categorization of parenteral formulations with associated $D_{121°C}$ and z-values and their potential impact on microbial resistance using the BI, *Clostridium sporogenes,* were previously reported (4). In addition, the methodologies used for D and z-value analyses were likewise cited (4). The data in Table 3 indicate that 20% emulsion (List 4336) is at the top of the list, since it affords the most microbial moist heat resistance. It is therefore the emulsion that should be microchallenged

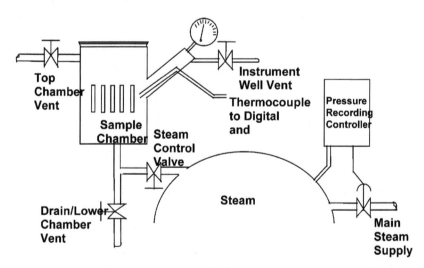

Figure 3   This is a schematic of the steam BIER unit used to generate spore crop (BI) and product D and z-values. Nine 5-mL glass ampules filled with various emulsion formulations can be tested at one time in the represented sample chamber.

Table 3 IV Lipid Emulsion Ranking

List #	Solution	$D_{121°c}$	z-Value	Predicted spore log reduction
4336	20% Emulsion	0.7	10.6	7.1
0720	10% Emulsion w/increased linolenate	0.7	11.4	7.5
9790	10% Emulsion w/100% soybean oil	0.6	10.1	8.0
9791	20% Emulsion w/100% soybean oil	0.7	12.8	8.2
0566	20% Emulsion w/increased linolenate	0.6	10.6	8.3
9786	10% Emulsion w/50% safflower & 50% soybean oil	0.6	10.7	8.4
9789	20% Emulsion w/50% safflower & 50% soybean oil	0.6	12.7	9.5

The columns represent the list number of the product, the emulsion or product name, its average $D_{121°C}$ value and z-value and finally the PSLR value.

(inoculated with spores) as part of the emulsion validation scheme. D and z-value data have been reported for other BIs such as *Bacillus stearothermophilus* (5,6) and *Bacillus subtilis* 5230 (7). There are many factors that can affect moist heat resistance including a BI's age, the sporulation media used, as well as the particular spore strain employed (8).

3.4. Closure Microbial Inactivation Studies

In lieu of using the large type steam sterilizers in the production environment, microbial inactivation at the closure/bottle interface of an emulsion container can be assessed in a developmental R&D sterilizer. The closure microbial inactivation (kinetic) studies can address how the size of the container, type of closure compound used as well as closure preparatory processes (leaching, washing, siliconing, autoclaving, etc.) influence microbial inactivation. Kinetic studies are conducted at various time intervals in a given sterilization cycle. Surviving organisms are ascertained by direct plate

(DP) count or fraction-negative (F/N) methodologies. Various BIs can be used and test data have been generated demonstrating the value of using both a moist heat organism (*C. sporogenes*) and a dry heat organism (*B. subtilis*) BIs for the sterilization validation of closure systems (9).

3.5. Lipid Emulsion PSLR Values

Lipid emulsion moist heat resistance values ($D_{121°C}$ and z-values) were generated in the steam BIER vessel using the BI *C. sporogenes* as shown in Table 3. The columns in Table 3 represent the list number of the product, the emulsion or product name, its average $D_{121°C}$ value and z-value and finally the PSLR value. Those parenteral formulations with the lowest PSLR value(s) are those that should be microbially validated at subprocess conditions, since these provide the most microbial resistance.

3.6. Accumulated F(Bio) for Lipid Emulsions

Accumulated F(Bio) by list number and z-values (Table 4) was used to construct the PSLR ranking for lipid emulsions as previously discussed for Table 3. The F(Bio) is the heat input for the biological solution based on the emulsion's moist heat D and z-values. By inputting the sterilizer temperatures from the coldest thermocouple of an engineering run for a particular container/sterilization cycle, the emulsion can be ranked according to PSLR values. The combined $D_{121°C}$ and z-value allows comparison of moist heat rankings between emulsions.

The data in Table 4 demonstrate that L. 4336, a 20% emulsion, has the lowest PSLR (7.105), thereby affording the highest moist heat resistance upon inoculation. Generation of this table allows prediction of which emulsion to microbiologically challenge as part of validation in the production sterilizer.

IV emulsions were inoculated in the oil phase after emulsification and filtration (Tables 5 and 6).

Table 4 Accumulated F(BIO) by List Number and z

| Solution | | F (PHY) | L 4336 | L 720 | L 9790 | L 9791 | L 566 | L 9786 | L 9789 | L 4335 |
Temp (°C)	Time (min.)	$z = 10.0$	$z = 10.6$	$z = 11.4$	$z = 10.1$	$z = 12.8$	$z = 10.6$	$z = 10.7$	$z = 12.7$	$z = 11.1$
105.4	1	0.0269	0.0330	0.0419	0.0278	0.0592	0.0330	0.0340	0.0579	0.0384
110.1	1	0.0793	0.0915	0.1082	0.0813	0.1380	0.0915	0.0935	0.1359	0.1019
114.1	1	0.1991	0.2181	0.2427	0.2023	0.2834	0.2181	0.2212	0.2806	0.2336
116.2	1	0.3228	0.3442	0.3709	0.3265	0.4134	0.3442	0.3476	0.4106	0.3611
118.1	1	0.5000	0.5200	0.5445	0.5035	0.5819	0.5200	0.5232	0.5794	0.5356
119.1	1	0.6295	0.6462	0.6663	0.6324	0.6966	0.6462	0.6489	0.6946	0.6591
119.4	1	0.6745	0.6897	0.7079	0.6772	0.7352	0.6897	0.6921	0.7334	0.7014
119.2	1	0.6442	0.6604	0.6799	0.6470	0.7092	0.6604	0.6630	0.7073	0.6729
118.5	1	0.5483	0.5672	0.5903	0.5515	0.6253	0.5672	0.5703	0.6230	0.5819
117.8	1	0.4667	0.4872	0.5124	0.4702	0.5513	0.4872	0.4905	0.5487	0.5033
116.2	1	0.3228	0.3442	0.3709	0.3265	0.4134	0.3442	0.3476	0.4106	0.3611
114.1	1	0.1991	0.2181	0.2427	0.2023	0.2834	0.2181	0.2212	0.2806	0.2336
110.6	1	0.0889	0.1020	0.1197	0.0911	0.1510	0.1020	0.1042	0.1487	0.1130
105.9	1	0.0301	0.0367	0.0463	0.0312	0.0648	0.0367	0.0379	0.0634	0.0426
101.7	1	0.0115	0.0148	0.0198	0.0120	0.0305	0.0148	0.0153	0.0296	0.0178
Total F		4.7436	4.9734	5.2646	4.7826	5.7366	4.9734	5.0107	5.7044	5.1573
D value			0.70	0.70	0.60	0.70	0.60	0.60	0.60	0.40
PSLR			7.105	7.521	7.971	8.195	8.289	8.351	9.507	12.893

The data demonstrate that L. 4336, 20% emulsion, is the emulsion that has the lowest PSLR (7.105) thereby affording the highest moist heat resistance upon inoculation.

Table 5 Plate Count Results of the IV Lipid Emulsion Inoculated with the BI, *C. sporogenes*, ATCC 7955, and Spores in the Oil Phase

5 pass emulsion	Initial count of emulsion	5.0×10^3 mL
	After pass #5	6.2×10^3 mL
	After 0.8 UM filtration	6.0×10^1 mL
15 pass emulsion	Initial count of emulsion	6.9×10^3 mL
	After pass #15	1.1×10^4 mL
	After 0.8 UM filtration	<10/mL

Based on the data, one would not have to routinely inoculate the BI in the oil phase prior to performing an emulsion microbial validation since the bacterial population count does not change significantly upon multiple emulsion processing steps.

3.7. R&D Sterilization Validation of IV Emulsion Inoculated in the Oil Phase after Emulsification and Filtration

Since the bacterial population count does not change significantly upon multiple processing steps (5 pass vs. 15 pass), it is not necessary to routinely inoculate the BI in the oil phase prior to performing a microbial validation.

3.8. Maintenance of Sterility Studies

The maintenance of sterility (MOS) studies are run on all moist heat terminally sterilized products with closure or componentry systems in order to demonstrate that the

Table 6 Plate Count Results of the IV Lipid Emulsion Inoculated with BI, *B. stearothermophilus*, ATCC 7953, and Spores in the Oil Phase

5 pass emulsion (lipid emulsion with emulphor)	Initial count of emulsion	4.3×10^4 mL
	After pass #5	4.3×10^4 mL
	After 0.8 UM filtration	<10/mL
15 pass emulsion (lipid emulsion with emulphor)	Initial count of emulsion	4.4×10^4 mL
	After pass #15	9.0×10^4 mL
	After 0.8 UM filtration	50/mL

Based on the data, one would not have to routinely inoculate the BI in the oil phase prior to performing an emulsion microbial validation since the bacterial population count does not change significantly upon multiple emulsion processing steps.

closure or componentry system is capable of maintaining the emulsion and fluid path in a sterile condition throughout the shelf life of the product

In an MOS study, the product container is sterilized at a temperature which is higher than the upper temperature limit of the chosen sterilization cycle and for a time that is greater than the maximum time limit for the cycle or producing an F subzero level greater than the maximum F subzero level for the cycle. The rationale for the selection of the maximum temperature and heat input level for the pre-challenge sterilization is that rubber and plastic closures are subjected to thermal stresses during sterilization and those stresses are maximized at the highest temperature and the longest time allowed.

3.9. Endotoxin Studies

Endotoxins are lipopolysaccharides from the outer cell membrane of Gram-negative bacteria. Endotoxins can be detected by the manual gel-clot method known as the limulus amebocyte lysate test (LAL). There are also various quantitative methods (turbidimetric and chromogenic) which use more rapid automated methodologies. All final product formulations are required to be tested for endotoxins and the method must be validated using three different lots of the final product. Emulsion formulations, if colored or opaque, cannot be tested by the turbidimetric method and therefore must use the LAL test.

4. PRODUCTION ENVIRONMENT

4.1. Engineering Penetration and Distribution Studies

Perform triplicate studies with thermocouples penetrating the product containers as well as thermocouples distributed outside the product containers in a production sterilizer at nominal operating process parameters.

4.2. Sterilization Cycles

There are two types of batch sterilizers that can be used for the sterilization of lipid emulsions, the shaking cycle and the rotary cycle. The filled and sealed containers are sterilized by exposure to circulated hot water spray or saturated steam at a specific temperature for a specified F_0. Air overpressure is used during heating, sterilizing, and cooling. The time, temperature, and pressure requirements are set to predetermined values to assure that the product will receive a thermal input equivalent to an F_0 minimum of, e.g., 8.0. Critical and key process parameters specified are to be controlled, monitored, and recorded. The target temperature of the circulating sterilizing medium during the peak dwell portion of the cycle must be maintained in a specific range, e.g., 121–123. For the shaking cycle, the product must be agitated lying on its side providing oscillatory movement along the container centroidal axis. The agitation frequency maintained throughout the cycle must be, e.g., 70 ± 3 rpm. For the rotary cycle, the product must be agitated throughout the sterilization process. The containers shall be horizontally positioned in rack(s) in a fixed manner and the rack(s) rotated at, e.g., 10 ± 2 rpm.

4.3. Sterilizer Microbial Emulsion SubProcess Validation

Table 7 shows the microbial emulsion validation conducted at subprocess conditions in a fully bulked load in a production sterilizer. The acceptance criteria of 6 spore logarithmic reduction (SLR) must be achieved for the BI, *C. sporogenes* and a 3 SLR for the more moist heat resistant BI, *B. stearothermophilus*. Each emulsion (20 containers) is inoculated with the appropriate BI at a target level of 1.0×10^6 and 1.0×10^2 for *C. sporogenes* and *B. stearothermophilus*, respectively. The 20 inoculated containers are distributed throughout the production sterilizer for sterilization at subprocess conditions.

4.4. Sterilizer Microbial Closure SubProcess Validation

Table 8 shows the microbial closure validation at subprocess conditions in a fully bulked production sterilizer. The BIs used were *C. sporogenes* and *B. subtilis*. Acceptance criteria of 3 SLR must be achieved for the moist heat BI (*C. sporogenes*) and the dry heat BI (*B. subtilis*). The R&D sterilization

Table 7 Microbial Fraction Negative (F/N) Analysis Following Exposure in a Moist Heat Cycle with Agitation, C. sporogenes vs. B. stearothermophilus. Lipid Emulsion Microbial Solution Validation. Sterilization Validation of IV Lipid Emulsion (Inoculated in Oil Phase) in the 200 mL Abbovac Bottle with Fraction Negative Method

Organism	Code	Avg. # spores/ bottle	Positive		Positive	
			#Positive controls	#Negative controls	#Test samples	SLR
C. sporogenes	5C6	4.8×10^5	2/2	0/4	0/20	>7.0
C. sporogenes	15C6	6.4×10^5	2/2	0/4	0/20	>7.1
B. stearo.	5B2	7.6×10^1	2/2	0/4	0/20	>3.2
B. stearo.	15B2	7.7×10^1	2/2	0/4	0/20	>3.2

F_0 Range: (c) 5.8–7.6
Agitation: 67–73.
Temperature range: 120–125°C.

Table 8 Microbial Fraction Negative (F/N) Analysis Following Exposure in a Moist Heat Cycle with Agitation, *C. sporogenes* vs. *B. stearothermophilus*. Lipid Emulsion Lipid Emulsion Microbial Closure Validation Sterilization Validation of 200 mL Abbovac Bottle Inoculated Closure Surface Coated with IV Fat Emulsion in Cycle with Agitation

Microorganism	Initial population/ stopper	#Positive controls	#Negative controls	Test samples	SLR
			#Positive		
C. sporogenes	8.4×10^3	2/2	0/4	0/20	> 5.2
B. subtilis	3.0×10^4	2/2	0/4	0/20	> 5.8

F_0 range: 6.3–6.4.
Agitation: 67–73 cpm.
Temperature range: 120–125°C.

validation of the inoculated closure system coated with the IV fat emulsion in a 200 mL Abbovac container is shown in Table 8. The surface of the stopper that comes into direct contact with the sidewall of the bottle was inoculated with the appropriate BI, dried and then a few drops of emulsion were placed over the inoculum to simulate manufacturing conditions.

The Halvorson and Ziegler equation is used to calculate the SLKR value as follows (1):

a. Positive for the indicator microorganism.
b. $SLR = \log a - \log b$, where $a =$ initial population of spores; $b = 2.303 \log(N/q) = in(N/q)$, where $N =$ total number of units tested, $q =$ number of sterile units. When $N = q$, assume one (1) positive for the purpose of calculating an SLR.
c. $F_0 =$ integrated lethality or equivalent minutes at 121.1°C for the hottest and coldest thermocouple containers.

4.5. Production Environment Bioburden Screening Program

Refer to the flow diagram in Fig. 4. A negative heat shock at 10 min exposure would indicate the bioburden's resistance as a $D_{121°C}$ of less than 0.079 min (a $0.0079 F_0$/min is accumulated at 100°C). A positive heat shock at 30 min exposure signifies that the surviving organisms should have a more detailed moist heat analysis conducted (e.g. 121, 118, and 112°C exposures).

5. REGULATORY SUBMISSION

The following checklists pertain to the sterilization portion of documents required in support of an FDA submission for aseptically processed and terminally sterilized products (10).

If a parenteral formulation can tolerate heat, then the moist heat sterilization process is the method of choice when compared to aseptic processing (11).

PRODUCTION ENVIRONMENT BIOBURDEN
SCREENING PROGRAM

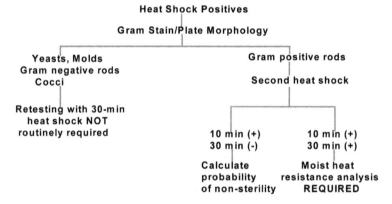

Figure 4 Heat shock is a method used for screening thermally resistant microorganisms. The application of a known amount of moist heat (approximately 10 or 30 min if required at less than 100°C) allows the isolation of bioburden microorganisms that potentially have moist heat resistance from microorganisms that have no moist heat resistance.

The following summarize the documents required to support a New Drug Application for product formulations that are aseptically processed or terminally sterilized.

5.1. Aseptic Processing

Microbiological sterilization and depyrogenation:

- Depyrogenation validation of glass/stopper.
- Microbiological sterilization validation of the stopper(s) (steam sterilizer).
- Microbiological sterilization validation of representative items for the family category of setups (filling line items, e.g., filling line needles, stoppers, etc.) in the steam sterilizer.
- Microbiological sterilization validation of representative items for the family category of filters in the steam sterilizer.

Stability of BIs:

- Engineering information:

  - Performance qualification for equipment, e.g., sterilizers, ovens, etc.
  - Thermocouple diagram during PQs.

- Procedures and specifications for media and environmental data.
- Sterility testing methods and release criteria:

  - Bacteriostasis, fungistasis.
  - Sterility testing.

- Bacterial endotoxin test product validation data.

5.2. Terminal Sterilization

- The sterilization process:

  - the operation and control of the production autoclave;
  - the autoclave process and performance specifications;
  - specification of the sterilization cycle.

- Autoclave loading patterns:

  - description/diagram of *representative* autoclave loading patterns.

- Thermal qualification of the cycle:

  - heat distribution in the production autoclave;
  - heat penetration in the production autoclave.

- Depyrogenation validation of container/closure prior to sterilization.
- Microbiological efficacy of the cycle:

  - identification;
  - resistance;
  - stability of BIs.

- Information and data concerning the identification, resistance, and stability of BIs used in the biological validation of the cycle should be provided. Include ATCC number, stock, resistance value and test date in each microbial validation report.

- The resistance of the BI relative to that of the bio-burden microbiological challenge studies:

  - microbial challenge of emulsion in a production vessel;
  - microbial challenge of closure in a production vessel.

- Demonstrate container integrity following maximum processing exposure:

  - Maintenance of sterility

- BET validation data:

  - LAL compatibility worksheet;
  - bulk drug and final product inhibition/enhancement data.

- Sterility testing methods and release criteria:

  - bacteriostasis, fungistasis sterility testing.

- Preservative efficacy at time zero, 3 months accelerated stability and time expire.

REFERENCES

1. Young RF. In: Morrissey RF, et al., eds. Sterilization with Steam Under Pressure, Sterilization Technology. New York: Van Nostrand Reinhold, 1993:120.

2. Owens JE. In: Morrissey RF, et al., eds. Sterilization of LVP's and SVP's, Sterilization Technology. New York: Van Nostrand Reinhold, 1993:254.

3. Association for the Advancement of Medical Instrumentation. Resistometers used for characterizing the performance

of biological and chemical indicators, Vol. 1.2. Arlington, VA: Association for the Advancement of Medical Instrumentation, 2003:1.

4. Berger TJ, Nelson PA. The effect of formulation of parenteral emulsions on microbial growth-measurement of D- and z-values. PDA J Pharm Sci Tech 1995; 49:32.

5. Feldsine PT, Shechtman AJ, Korczzynski MS. Survivor kinetics of bacterial spores in various steam-heated parenteral emulsions. Develop Industr Microbiol 1970; 18.

6. Caputo RA, Odlaug TE, Wilkinson RL, Mascoli CC. J Parent Drug Assoc 1979; 33:214.

7. Berger TJ, Chavez C, Tew RD, Navasca FT, Ostrow DH. Biological indicator comparative analysis in various product formulations and closure sites. PDA J Pharm Sci Tech 2000; 54:101.

8. Pflug IJ, Holcomb RG. In: Block SS. ed. Principles of Thermal Destruction of Microorganisms Disinfection, Sterilization and Preservation, 3rd ed. Philadelphia: Lea and Febiger, 1983:759.

9. Berger TJ, May TB, Nelson PA, Rogers GB, Korczynski MS. The effect of closure processing on the microbial inactivation of biological indicators at the closure-container interface. PDA J Pharm Sci Tech 1998; 52:70.

10. FDA guideline for submitting documentation for sterilization process validation in applications for human and veterinary drug products. Federal Register, 58, No. 231, Friday, December 3, 1993, Notices, p. 63996.

11. Cooney PH. Aseptic fill vs. terminal sterilization. Presented at the Pharmaceutical Technology Conference, Cherry Hill, NJ, September 16–18, 1986.

13

Case Study: Formulation of an Intravenous Fat Emulsion

BERNIE MIKRUT

Pharmaceutical Research
& Development, Hospira, Inc., Lake Forest,
Illinois, U.S.A.

1. INTRODUCTION

The history of i.v. fat emulsions can be traced as far back as 1873 when Holder infused milk in cholera patients. In the 1920s, Yamakawa (1) in Japan produced a product called "Janol" from caster oil, butter, fish oil, and lecithin which had many side effects. It was not until 1945 that Stare et al. (2) produced the first relatively non-toxic emulsion using purified soy phospholipids. This product was further refined by Schuberth and Wretlind (3) in 1961, who made 1506 infusions in 422 patients using a soybean oil emulsion made with purified egg phospholipids with no untoward reactions in humans. This led to the product, Intralipid®, which was approved in Sweden in 1961. Intralipid was approved in the US in 1975 and Liposyn® was approved in the US in 1979.

2. FORMULATION

i.v. fat emulsions are oil-in-water emulsions of soybean or a 1:1 soybean/safflower oil mixture emulsified using purified egg phosphatide. Tonicity is adjusted with glycerin and pH is adjusted with sodium hydroxide. i.v. fat emulsions with fat contents of 10%, 20%, and 30% are commercially available.

2.1. Oil

The current products available in the US use either a 1:1 combination of safflower oil and soybean oil or soybean oil exclusively. Worldwide, other products are available which contain medium chain triglyceride (MCT) oil in combination with soybean oil. Both safflower and soybean oils are listed in USP 23 and their respective fatty acid profiles are summarized in Table 1.

Safflower and soybean oils are highly unsaturated and prone to oxidation through initial peroxide formation. Therefore, they must be maintained under nitrogen gas protection during storage and handling. Both oils contain some saturated waxes and sterols which must be removed by the standard oil industry practice of "winterization." In this process, the oils are refrigerated for a length of time during which the waxes and sterols crystallize out and settle to the bottom of the drum. The oils are then quickly cold filtered

Table 1 Fatty Acid Composition

Safflower oil, USP	Soybean oil, USP
Palmitic acid 2–10%	Palmitic acid 7–14%
Stearic acid 1–10%	Stearic acid 1–6%
Oleic acid 7–42%	Oleic acid 19–30%
Linoleic acid 72–84%	Linoleic acid 44–62%
	Linolenic acid 4–11%

to remove these unwanted components and stored under nitrogen gas protection prior to their use in emulsion manufacture. The oils must be food-grade oils and of high chemical purity, pyrogen-free and free of herbicides and pesticides. The FDA requires these oils to be tested to show the absence of herbicides and pesticides (4).

2.2. Emulsifier

Highly purified egg lecithin is used as the emulsifier in all commercial i.v. fat emulsions. Historically, soy phosphatides were used; however, they were rejected due to untoward clinical effects. Pluronic F68 was investigated, but was discarded because of toxic effects (5,6). The main components of egg phospholipid are phosphatidylcholine (PC) and phosphatidylethanolamine (PE), along with minor components. Pure PC and PE make poor emulsions. The minor components of lecithin are necessary to produce a stable emulsion (7). The components of a typical egg phospholipid used in the manufacture of i.v. fat emulsions are presented in Table 2.

2.3. Tonicity Adjuster

The tonicity adjuster of choice is glycerin at a concentration of 2.25% (Intralipid) or 2.5% (Liposyn II/III). Glycerin was used by Schuberth and Wretlind (3) in their classic work in 1961 and is still used today. Dextrose is not used since it has been reported to interact with egg phospholipid to produce brown discoloration upon autoclaving and storage (8,9).

Table 2 Typical Egg Phospholipid Composition

PC	73.0%
Lysophosphatidylcholine (LPC)	5.8%
PE	15.0%
Lysophosphatidylethanolamine (LPE)	2.1%
Phosphatidylinositol (PI)	0.6%
Sphingomylin (SP)	2.5%

(Adapted from Ref. 10.)

2.4. Others

2.4.1. pH

Small amounts of sodium hydroxide are used to adjust the pH to approximately 9.5 during manufacture. This pH level has two effects: it causes the ionization of the acidic phospholipids present in the egg phospholipid mixture, creating a net negative charge for droplet repulsion; it also forms some free fatty acids. These fatty acids form sodium soaps and further stabilize the emulsion by acting as auxiliary emulsifiers. The ionization characteristics of the individual phospholipids are summarized in Table 3 (10).

2.4.2. Preservatives

No preservatives are used in i.v. fat emulsions. Small quantities of vitamin E and BHA/BHT are present since these occur in the original soybean or safflower oils. i.v. fat emulsions have been shown to be a good growth medium for microbial growth (11) and therefore are designed as a single-dose product.

Table 3 Ionization Characteristics of Phospholipids

Phosphatide	Ionic species	Ionization characteristics
Phosphatidic acid (PA)	Primary phosphate, PO_4^{2-}	Strong acid (pK_a 3.8, 8.6)
PC, LPC	Secondary phosphate-choline, $PO_4^{-}-NMe_3^{+}$	Isoelectric over a wide range of pH
PE	Secondary phosphate-amine, $PO_4^{-}-NH_3^{+}$	Negative at pH 7 (pK_a 4.1, 7.8)
Phosphatidylserine (PS)	Secondary phosphate-carboxyl-amine, $PO_4^{-}-COO^{-}-NH_3^{+}$	Negative at pH 7.5 (pK_a 4.2, 9.4)
Phosphatidylinositol (PI)	Secondary phosphate-sugar, $PO_4^{-}-sugar$	Negative above pH 4 (pK_a 4.1)

(From Ref. 10.)

3. PROCESSING

Several methods of emulsion manufacture can and have been used, but the equipment of choice for i.v. fat emulsions is the standard high-pressure homogenizer. The egg phospholipid is first dispersed in a portion of the water for injection (WFI) or dissolved in the oil. Both of these methods have been used successfully to manufacture acceptable emulsions. Abbott currently disperses egg phosphatide in WFI, the glycerin is added and the mixture homogenized to a fine dispersion. This dispersion is filtered through a 0.45 μm membrane and more WFI added. Oil is then added with agitation to form a crude emulsion, which is homogenized further to form the finished emulsion. pH is adjusted with sodium hydroxide at several points during the process so that the final emulsion has a pH of approximately 9.0 prior to autoclaving. Two different methods of homogenization have been used. The tank-to-tank method homogenizes the emulsion alternately from one tank to another until the desired globule size is attained. The other, the recirculation method, uses only one tank and continuously recirculates the emulsion through the homogenizer and back to the tank until the desired globule size is attained. The graphical representation (Fig. 1) from the homogenizer manufacturer correlates the efficiency of the two methods. The entire process must be oxygen-free as much as possible. All WFI is degassed and nitrogen gas purging/flushing is used throughout the process. A typical manufacturing scheme is outlined in Fig. 2. Since the final emulsion product has a mean globule size very close to the usual 0.45 μm filtration of intravenous fluids, emulsion stability would be compromised if the final product was filtered through a 0.45 μm membrane. All i.v. fluids are filtered through at least a 0.45 μm filter to reduce particulates. In this case, it was decided to filter each of the ingredients through a 0.45 μm filter prior to homogenization. The oil is also filtered through a 0.45 μm filter prior to homogenization. The 0.8 μm filtration is after the final homogenization to reduce the particulates introduced during the manufacturing procedure. Any breakdown of the emulsion which may have been caused

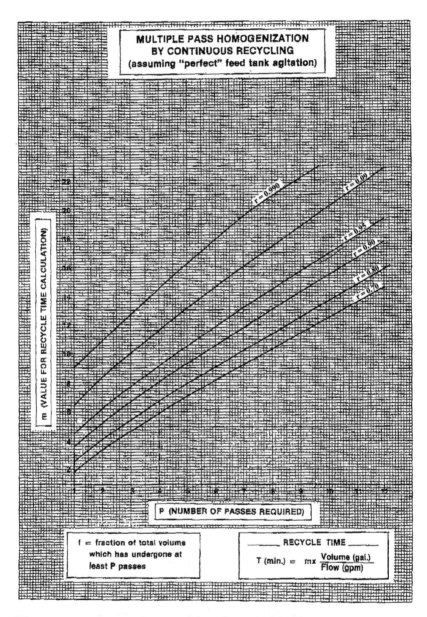

Figure 1 Comparison of tank-to-tank homogenization vs. continuous recirculation. (From APV Gaulin bulletin TB-71.)

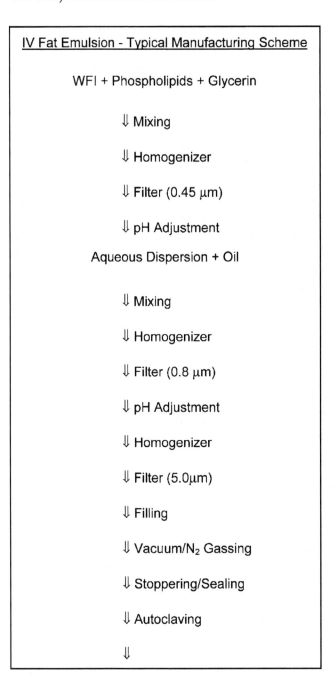

Figure 2   Typical manufacturing scheme for i.v. fat emulsions.

by the filtration at this point is repaired by the final homo-genization pass. The final 5.0 μm filtration is a regimesh stain-less steel filter just prior to the emulsion going to the filling line.

4. FILLING/PACKAGING

The currently marketed i.v. fat emulsions are available in 50, 100/200, 200, 500 and 1000 mL glass containers which are evacuated and flushed with nitrogen gas. The use of nitrogen gas reduces the degradation of the emulsion by oxidation as well as the formation of toxic by-products. Unpublished work in Abbott Laboratories to evaluate the packaging of i.v. fat emulsions in flexible containers using a full aluminum foil over-wrap has shown that a crack/break or pin hole in the aluminum foil layer will let in oxygen and result in by-products which have been shown to be toxic to mice. Both Type I and Type II glass containers are approved for packaging of i.v. fat emulsions.

5. STABILITY EVALUATION

The main stability-indicating parameters used in the stability evaluation of i.v. fat emulsions are pH, free fatty acids, extraneous particulates, visual evaluation, and globule size distribution.

5.1. pH and Free Fatty Acids

The pH of i.v. fat emulsions before autoclaving is approxi-mately 9.0. Subsequent to terminal heat sterilization, the pH drops to approximately 8.5. This drop in pH is normal and is the result of hydrolysis of the PC and PE to their respective lyso compounds and the subsequent formation of free fatty acids. Since the emulsions have no buffering capacity, the pH drops upon autoclaving and also on aging. The Abbott pH specifications are 6.0–9.0 over the 24-month shelf life. i.v. fat emulsions with a pH of less than 5.0 have significantly decreased zeta potential values and are subject to coalescence and globule size growth as noted by Davis (10).

5.2. Particulate

Particulate evaluation of i.v. fat emulsions is determined by the conductance of a microscopic particle counting method, wherein the sample is filtered through a 0.8 μm gray, gridded membrane to isolate solid extraneous particles. The light-obscuration method is unsuitable because of the composition of the sample, i.e., liquid micro-droplets of oil in a water-based vehicle. These droplets are counted as particles by the light-obscuration sensor and erroneous data are produced. When the microscopic method is attempted, emulsion droplets pass through or are absorbed by the filter and are not observed on the filter membrane. Particulate contamination can be created by stainless steel wear of the homogenizer or mixer parts and is observed as fine black particulates. Viton rubber or teflon wear particles can occur if the homogenizer plunger packings use these materials. Also, Ca^{++} or Mg^{++} contamination may result in particles originating from the precipitation of salts of the free fatty acids which are formed by hydrolysis during processing, autoclaving, and storage.

5.3. Visual Evaluation

Visual evaluation is very important since no instrumental globule-sizing technique can detect free oil droplets floating on the emulsion surface. One technique of visualizing these droplets is to view the emulsion surface at an angle using an inspection lamp in a darkened room. The oil droplets show up as shiny specks on the dull emulsion surface. This technique requires practice in order to see beyond the glass surface and focus on the emulsion surface.

Creaming of i.v. fat emulsions is normal and the emulsion is easily re-dispersed with gentle agitation. Creaming is observed in the bottle as a whitish layer in the top portion of the emulsion and a darker, less opaque layer toward the bottom. This is a result of the low-density oil droplets slowly rising to the surface as a result of gravitational forces. Upon inversion of the container, the emulsion should be uniform in color and opacity.

5.4. Globule Size

Globule size and distribution are the most important factors in i.v. fat emulsion stability. Measurement of globule size and distribution must be evaluated using several instrumental techniques since no one technique can measure the entire size range. In addition to the instrumental techniques described here, visual evaluation should always be conducted.

5.5. Accelerated Stability Testing

5.5.1. Temperature

i.v. fat emulsions have been shown to be stable for up to 6 months at 40°C. This corresponds to approximately 24 months at 25–30°C. Longer storage at 40°C results in a significant increase in free fatty acid formation due to phosphlipid hydrolysis and a concomitant decrease in pH. Abbott normally tests i.v. fat emulsions at 25, 30, and 40°C.

5.5.2. Freeze–Thaw Cycling

Cycles that comprise slow freezing at –20°C, followed by undisturbed thawing at room temperature can be used to evaluate i.v. fat emulsion stability. Unpublished data at Abbott have shown that one cannot correlate this to shelf life at room temperature, but it can be useful for rank-order stability evaluation of various emulsion formulations.

5.5.3. Stress Shake

Horizontal shaking at approximately 200 cycles per minute can also cause emulsion breakdown and is useful for rank-order stability evaluation of various emulsion formulations.

REFERENCES

1. Yamakawa S. J Jpn Soc Intern Med 1920; 17:1.
2. Stare FJ, et al. Studies on fat emulsions for intravenous alimentation. J Lab Clin Med 1945; 30:488.

3. Schuberth O, Wretlind A. Intravenous infusion of fat emulsions, phosphatides and emulsifying agents. Acta Chirgurgica Scandinavica 1961; (suppl 278).

4. FDA Deficiency Letter, dated November 29, 1991, for Liposyn® III 30%.

5. Pelham D. Rational use of intravenous fat emulsions. Am J Hosp Pharm 1981; 38:198–208.

6. Geyer RP, Mann GV, Young J, et al. J Lab Clin Med 1948; 33: 163.

7. Scharr PE. Cancer Res 1969; 29:258.

8. Hansrani PK, Davis SS, Groves MJ. The preparation and properties of sterile intravenous emulsions. J Paren Sci Tech 1983; 37(4):145.

9. Tayeau F, Neuzil E. Bordeaus Med 1972; 10:1117.

10. Davis SS. The stability of fat emulsions for intravenous administration. In: Proceedings of the Second International Symposium on Advanced Clinical Nutrition, 1983, pp. 213–239.

11. Unpublished data from Abbott Laboratories.

14

Case Study: DOXIL, the Development of Pegylated Liposomal Doxorubicin

FRANK J. MARTIN

ALZA Corporation, Mountain View,
California, U.S.A.

1. INTRODUCTION

Over the past 30 years, liposomes have been proposed as a vehicle for improving the delivery (and thereby the therapeutic utility) of dozens of drugs. The cytotoxic anthracycline antibiotics doxorubicin and daunorubicin and the polyene antibiotics amphotericin B and nystatin are perhaps the most often cited examples. The vast majority of these publications originated from academic laboratories and thus do not generally address the pharmaceutical attributes required for the regulatory approval of a commercially viable product.

Development of a liposomal product in most respects parallels that of any other ethical pharmaceutical product. A medical need must be identified and the product must be shown in well-controlled clinical trials to meet that need and to do so safely. Moreover, it must have at least comparable activity to other drug products approved for the same clinical indication. Implicit in the ability to conduct clinical trials is the availability of the drug product in sufficient quantity to supply clinical investigators. It must also be of proper quality to meet regulatory requirements. A commercially successful product must also be cost-effective, reproducibly made in large scale and stable enough to be supplied through the normal channels of distribution.

Four liposomal products meet these requirements and are approved in the US and/or Europe, DOXIL^{TM} (pegylated liposomal doxorubicin, also known as Caelyx^{TM} in Europe), DaunoXome^{TM} (liposomal daunorubicin), Ambisome^{TM} (liposomal amphotericin B), and Myocet^{TM} (liposomal doxorubicin). The case study reported here examines the formulation design, clinical evaluation, and regulatory strategy used in the development and registration of DOXIL. The case study following this one (Chapter 15 of this book) focuses on the formulation design, clinical evaluation, and regulatory strategy used in the development and registration of AmBisome.

2. BACKGROUND

Early liposome formulations of doxorubicin were shown to significantly reduce cardiotoxicity and acute lethality in animals but on a dose-equivalent basis, were no more active than the unencapsulated drug (1). Why was there no improvement of anti-tumor activity? In retrospect, two related biological responses provoked by the injection of liposomes appear to be to blame. Firstly, liposomes released a proportion (up to 50%) of their encapsulated doxorubicin as a consequence of opsonization by components of blood (lipoproteins, albumin, complement components, formed elements) (2). Obviously, drug that is lost in this way is not available to be delivered to a tumor in *encapsulated* form. Secondly, the liposomes that survive destabilization in blood are rapidly removed from circulation by fixed macrophages in the liver and spleen (the mononuclear phagocyte system or MPS; also known by an older designation as the reticuloendothelial system or RES) (3). Once internalized by macrophages, the liposome matrix is digested by lysosomal lipases and the drug is released intracellularly. This combination of leakage and MPS uptake virtually eliminates any opportunity for "true" targeting, as drug loaded liposomes never reach the tumor.

3. DEFINE PROBLEM

3.1. Improve Anti-tumor Activity of Doxorubicin by "Passive" Liposome Targeting

To successfully deliver an encapsulated drug to tumors, the liposome carrier must retain the drug while in blood, the medium through which the liposomes must pass to reach the target. Moreover, the liposomes must recirculate for the period of time needed to access the tumor and possess the physical characteristics that allow them to actually enter the tumor.

The liposome literature of the late 1970s and early 1980s is replete with reports from the laboratories of liposome scientists who attempted to engineer liposomes to circulate longer in blood and remain intact while doing so. Bona fide structure/function relationships emerged from this work (4). For example, small ($<50\,$nm) liposomes composed of high phase transition lipids and cholesterol were found to resist degradation in blood and to circulate at least for a few hours in rodents (5). These results were later reproduced in human cancer patients (6). In the mid-1980s surface modification of liposomes was explored as a strategy to improve recirculation times further. The rationale driving this approach was to create a liposome that behaved like a tiny-formed element in blood (i.e., an erythrocyte or platelet). Indeed, circulation times were significantly improved when specific glycolipids such as a brain-derived ganglioside (GM_1) or a plant phospholipid (hydrogenated phosphatidyl inositol) were included in the formulation (7,8). Moreover, prolonged circulation times were highly correlated with improved distribution

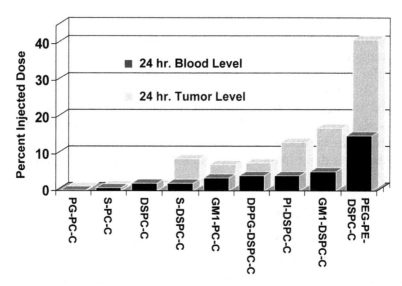

Figure 1 Correlation between liposome circulation time and tumor uptake.

of liposomes to implanted tumors in mice (Fig. 1) (9). This finding confirmed the belief that reducing the rate of MPS uptake (increasing circulation time) would allow i.v. injected liposomes to access systemic tumors.

Following these hopeful developments with carbohydrate-coated liposomes, other surface modification approaches were pursued. The most promising results were achieved by grafting polymer groups to the liposome surface (10). Circulation half-lives in excess of 12 hr in rats were found with polyethylene glycol-coated liposome formulations (Fig. 2) (11). A comparison of the pharmacokinetics in cancer patients among various liposome formulation is shown in Fig. 2 (6,12,13).

3.2. Provide Required Pharmaceutical Attributes

Adequate shelf-life stability, a scalable, reproducible production method and validation of sterility assurance methodology are required for any injectable pharmaceutical product.

With respect to stability, liposome products represent a special case. The safety and efficacy of the system is critically

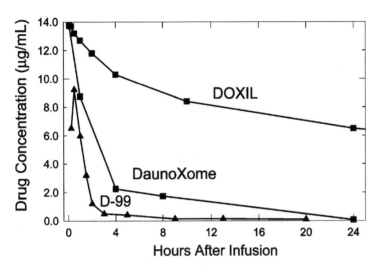

Figure 2 Human plasma pharmacokinetics of liposomes.

related to the *encapsulated* form of the drug. Indeed, toxicology studies required to qualify the product for clinical testing are performed on the encapsulated drug. Therefore, for the claimed shelf-life of a particular product, the drug must remain encapsulated (at least within predetermined limits). Any leakage during storage could, and very likely would, change the safety and efficacy profiles of the product, which is unacceptable from a regulatory perspective.

The reproducibility of production is also critical. Simple chemical entities must meet strict purity and potency standards and all excipients must be of suitable quality. Physical characteristics of liposomes profoundly influence their pharmacology. So, in addition to chemical standards related to the active ingredient and excipients (including lipids), liposome products must reproducibly conform to equally strict physical specifications. These include percent encapsulation (i.e., amount of "free" or unencapsulated drug in the product—which regulators may rightly regard as a contaminant), amount of drug carried in each liposome (drug "loading," usually expressed as mass or moles of drug per unit lipid), size, and the rate of release of the drug.

Sterility assurance is another requirement that presents interesting challenges to the developers of injectable liposome products. In general, terminal sterilization with heat or ionizing radiation is not possible in the case of liposome encapsulated drugs due to the sensitivity of the drug and/or the lipids to degradation under these conditions. Aseptic processing is possible but validation is costly and burdensome. Terminal filter sterilization using reliable, industry-validated filtration systems is the method of choice.

3.3. Craft Regulatory Approval Strategy

Any improvement of the anti-tumor activity of a drug provided by liposome encapsulation could be offset by an increase in any one of the side effects which limit the dose patients are able to tolerate. Well-designed and executed preclinical toxicology/pharmacology studies are absolutely necessary to provide reassurance that the therapeutic index of the encapsulated drug is demonstrably superior to that of the unencapsulated drug. Without such information, it would be foolhardy to embark on an expensive product development program.

Clinical development is generally the most expensive and time consuming element of the product development cycle. In the case of cytotoxic cancer drugs, registration is usually based on results obtained from the typical sequence of clinical trials. During Phase I, the safety profile and preferred dose and dosing schedule for the agent are established. Phase II trials fine tune the dose and confirm clinical activity in well-defined patient populations. The primary source of data required for approval is usually derived from "well-controlled" Phase III trials which are designed to demonstrate both safety and efficacy, usually relative to some established therapy.

For regular marketing approval of oncology drugs substantial evidence of efficacy from "adequate and well-controlled" trials is necessary. Pivotal registration trials must have a valid control (i.e., a population to which the results of the product under testing can be compared) and provide an objective, quantitative measurement of the drug's effect.

Endpoints typically are disease-free survival, overall survival or a surrogate for one of these. Optimally, from the perspective of the regulatory authorities, the comparison would be to standard therapy in a blinded study (i.e., patients are prospectively randomized to the new product or the comparator without either the patient nor the caregiver knowing which one is actually administered). In some instances, standard therapy for a specific type of cancer is not officially approved, but support for the therapy as a standard of care has been established in the peer-reviewed literature.

At times, there are no proven therapeutic options that would be suitable to be used as a control arm of a comparative study. In non-life-threatening diseases, placebo controls are often used. In the case of cancer, placebos are understandably not acceptable to study participants and their physicians. In this case less optimal controls can be relied upon. Two different doses of the product could be compared with the prospect that one may provide greater benefit than the other. Historically controlled trails rely upon a comparison of the benefit of the new drug in a specific tumor type to a series of retrospectively collected cases of the same cancer type treated with standard therapy. Although historical controls are appropriate at times, regulators generally regard them as a poor substitute for prospectively randomized trials. For example, the standard of care may have changed between the time the control patients were treated at the time the new drug was tested and there is no way of telling whether this influenced the outcome of the comparison. Nevertheless, several cancer drugs including paclitaxel have been approved on the basis of historically controlled trials (14).

Clearly the tumor type and patient population for clinical trails will be selected based on the sensitivity of the tumor to the encapsulated drug, patients' tolerance of the product, and the therapeutic benefit provided by the product. If a meaningful clinical benefit can be established in a population of patients afflicted with a life-threatening tumor and who have exhausted all other treatment options, accelerated review by regulatory authorities and more rapid that normal approval may be an option.

New Drug Application (NDA) regulations in the United States were modified in 1992 to include a provision (CFR Title 21, Part 314, Subpart H) which allows for accelerated approval of drugs intended to treat life-threatening diseases in situations when the drug appears to provide benefit over available therapy, but does not meet the standards required of regular approval. With respect to cancer therapy, the example that is often cited is accelerated approval based on a surrogate end-point (e.g. partial response rate or time to tumor progression) which is likely to predict clinical benefit (e.g. complete response rate, survival) but not yet established to the degree that would be required to support regular approval. As a condition, approvals based on Subpart H require the sponsor to conduct post-marketing trials to validate that the surrogate marker used actually does predict objective clinical benefit.

Accelerated marketing approval is an attractive option, for both small and large pharmaceutical companies. But there are many attendant risks. In the first place, if no proven treatment options exist in the selected clinical indication, there may not be an opportunity to compare the new liposomal drug product with an existing therapy. That is, randomized, comparative clinical trails, which represent the "gold standard" for pivotal registration trials, are not possible because there is no proper comparator. In this case, so-called "open-label" non-comparative trials must be relied upon for approval. In a real sense the comparator in this case is a historical understanding of the typical course of the disease process. For example, patients with advanced non-small cell lung cancer who have failed all standard chemotherapy do not typically improve spontaneously, but rather their disease progresses with a median survival time of only a few months. In such a population, if intervention with a liposomal anti-cancer drug (or any other drug for that matter) provides objective responses or demonstrable benefit to a reasonable number of patients, regulatory approval could be sought without the benefit of comparative data. Following this accelerated strategy, the burden of proving without question that the patients are truly refractory to existing therapy and that the benefit is meaningful to the patients falls squarely on the drug sponsor. But, if the liposomal drug product performs well, the design and execution of a clinical trials program aimed at accelerated approval is a real possibility.

Another potential shortcoming for an accelerated approval approach is that the approved use (label claim) will be limited to a small number of patients that fall into the chemo-refractory or salvage therapy categories. In this case, post-marketing studies are typically conducted to expand the label claims, and thus the market potential, for the product.

4. SOLUTIONS TO PROBLEM: DOXIL (PEGYLATED LIPOSOMAL DOXORUBICIN)

4.1. Liposome Design and Scientific Rationale

4.1.1. Conventional Liposomes

The active drug substance encapsulated in DOXIL liposomes is doxorubicin hydrochloride, a cytotoxic anthracycline antibiotic isolated from cultures of *Streptomyces peucetius* var. *caesius*. Doxorubicin interacts strongly with nucleic acids, presumably by specific intercalation of the planar anthracycline nucleus with the DNA double helix, and inhibits DNA and RNA metabolism in vitro and in vivo. Cell culture studies have demonstrated that the drug exerts its cytotoxic action on rapidly proliferating tumor cells (IC_{50} values are generally $<1\,\mu g/mL$) with rapid cell penetration, perinuclear chromatin binding, and rapid inhibition of mitotic activity and nucleic acid biosynthesis.

Doxorubicin HCl (which is often referred to by its trade name Adriamycin®) is an approved anti-neoplastic agent and has been in clinical use for over 20 years. Human tumors shown to be responsive to doxorubicin HCl include acute leukemia, resistant Hodgkin's and non-Hodgkin's lymphomas, sarcoma, neuroblastoma, ovarian and endometrial carcinoma, breast carcinoma, bronchogenic carcinoma, lung cancer and thyroid and bladder carcinoma (15). AIDS-Related Kaposi's sarcoma (KS) is somewhat responsive to Adriamycin as a single agent and in combination regimens (16). Dose-dependent toxicities, including stomatitis/mucositis, nausea/vomiting, bone marrow suppression, and cardiomyopathy, limit the amount of Adriamycin patients are able to tolerate.

Tumors, including the cutaneous and visceral lesions characteristic of KS, depend on blood vessels for exchange of gases, nutrients, and metabolic waste products. Neovascularization is necessary to support tumors larger than a few millimeters in diameter. The permeability of vessels in tumors is significantly higher than those residing in normal tissues (17). Vessels supplying KS lesions are particularly permeable as evidenced by edema and extensive extravasation of formed blood elements (perivascular streams of extravasated red blood cells are typically seen in KS lesions) (18). This increased vascular permeability has been attributed to several factors: the existence of fenestrated and discontinuous capillaries, the existence of blood channels without an endothelial lining (19) increased occurrence of trans-endothelial channels and higher trans-endothelial pinocytotic transport (20).

Conventional liposomal formulations of doxorubicin have been proposed as a means to reduce doxorubicin HCl-related toxicities and thereby improve the drug's therapeutic index. The scientific rationale for the use of liposomal formulations of doxorubicin HCl is discussed below, first for conventional liposomes and then for Stealth liposomes.

Conventional liposomes used for drug delivery purposes are generally small in size (<300 nm) and composed of naturally occurring or synthetic phospholipids, with or without cholesterol. The exposed outer surfaces of such liposomes are susceptible to attack and destabilization by components present in biological fluids. Following intravenous injection, a liposome of this type is rapidly recognized as a foreign body and cleared from the circulation in a dose-dependent fashion by elements of the immune system: primarily by specialized phagocytic cells residing in the liver and spleen, the MPS. It is believed that binding of plasma proteins (lipoproteins, immunoglobulins, complement) to the liposome surface triggers such macrophage uptake.

Internalization of liposome-encapsulated anti-tumor agents by MPS cells has the potential to diminish exposure of other body tissues to the irritating effects of such drugs. Liposomal encapsulation of doxorubicin has been proposed as a means of reducing the side effects of this highly active anti-tumor agent. By taking advantage of MPS clearance of encapsulated drug, exposure of other healthy tissues to high plasma concentrations of doxorubicin is reduced. Doxorubicin-related nausea/vomiting and cardiomyopathy are related to the drug's peak levels in plasma. By using liposome encapsulation to sequester the majority of an injected dose in the MPS, in theory, plasma levels of free drug are attenuated and safety improved. The drug is eventually released from MPS organs and distributes to peripheral tissues in the free form. In this case, the pharmacokinetic pattern is intended to mimic that seen following administration of doxorubicin as a divided-dose or prolonged infusion, regimens known to reduce drug-related side effects (21–25). Indeed, it has been shown that administration of liposome-encapsulated doxorubicin reduces the drug's acute and chronic toxicities in preclinical animal models (26). Moreover, results from animal models indicate that doxorubicin delivered in this fashion retains its activity against non-hepatic tumors (27). The pharmacokinetics and safety of various clinical formulations of conventional liposomal doxorubicin have been reported in the scientific literature (13,28–30). Clinical pharmacokinetic measurements confirm that conventional liposome formulations are cleared rapidly from plasma. These data also suggest that a considerable amount of encapsulated doxorubicin HCl is released into the plasma *prior* to MPS uptake (2,31).

4.1.2. Long Circulating Liposomes

Design Features

Recognizing that rapid liposome clearance, coupled with release of encapsulated drug, severely limits the potential of liposomes to transport *encapsulated* drugs to systemic tumors, strategies have been sought to stabilize liposomes in plasma and prolong their circulation following administration. Similarly, efforts have been made to optimize liposome size (32–35).

DOXIL (pegylated liposomal doxorubicin) is a long-circulating "Stealth®" liposomal formulation of doxorubicin HCl. This new type of liposome contains surface-grafted segments of the hydrophilic polymer methoxypolyethylene glycol (MPEG). These linear MPEG groups extend from the liposome surface creating a protective coating that reduces interactions between the lipid bilayer membrane and plasma components. A schematic representative of a Stealth liposome, not drawn to scale, is presented in Fig. 3.

The critical design features of the Stealth liposome include:

- Polyethylene glycol ("Stealth" polymer) coating: reduces MPS uptake and provides long plasma residence times.
- Average diameter of approximately 100 nm: balances drug carrying capacity and circulation time, and allows extravasation through endothelial defects/gaps in tumors.
- Low permeability lipid matrix and internal aqueous buffer system: provide high drug loading and stable encapsulation, i.e., drug retention during residence in plasma.

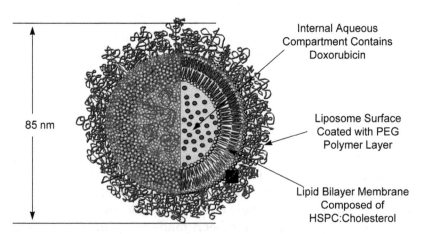

Figure 3 Diagram of PEG-stabilized Stealth liposome.

The "steric stabilization" effect provided by MPEG is believed to be responsible for the remarkable stability of DOXIL in plasma (36). The MPEG coating also inhibits the interaction (close approach) of liposomes with macrophage cells, thus reducing hepatic uptake and prolonging liposome residence time in the circulation (37). Comparative pharmacokinetic measurements in rodents and dogs indicate that doxorubicin HCl has a prolonged plasma residence time when administered as DOXIL relative to Adriamycin (15–30 hr, compared to a distribution half-life of 10 min for Adriamycin) (11). The long residence time of DOXIL was confirmed in studies conducted in cancer patients and AIDS patients with KS (13,38,39). In these studies, DOXIL remained in circulation with a distribution half-life of 40–50 hr while the distribution half-life of doxorubicin HCl is reported to be less than 10 min.

Stability

The internal buffer system used to load and retain doxorubicin into DOXIL liposomes is critical to achieve optimal targeting of the drug to tumors. Any drug release while in the circulation (i.e., in route to the tumor) would detract from the total amount delivered to the tumor in encapsulated from.

Weak bases like doxorubicin can be loaded into preformed liposomes under the influence of an ammonium ion gradient. This approach which is illustrated in Fig. 4 is used to load doxorubicin into DOXIL liposomes (40). Liposomes are formed and sized by extrusion in the presence of 250 mM ammonium sulfate (41). The external ammonium sulfate is subsequently replaced with a non-electrolyte solution of sucrose by cross-flow filtration. A solution of the hydrochloride salt of doxorubicin is then added to the external phase. An equilibrium between the protonated and deprotonated form of the primary amine group of doxorubicin is established. Although at neutral pH the protonated from of the drug is highly favored, a small proportion is in the deprotonated or neutral form. Owing to its greater hydrophobicity relative to the protonated form, this neutral form of doxorubicin is free to move rapidly through the bilayer membranes of the liposomes. Once a molecule of doxorubicin enters the liposome and complexes with sulfate

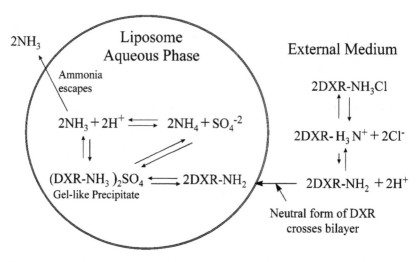

Figure 4   Ammonium sulfate-loading of doxorubicin.

ions, an insoluble sulfate salt forms and precipitates as a gel-
like structure. For every mole of doxorubicin that enters the
liposome, a mole of ammonia is released. This process
continues until virtually all of the doxorubicin is loaded as a
gel-like precipitate within the liposomes (42). This method
provides remarkable stability against leakage of the drug.
Indeed, DOXIL is stable at refrigerator temperatures as an
aqueous suspension for over 2 years. Remarkably, the stability-
determining factor is not drug leakage, but drug potency (see
discussion below and Table 1).

Extravasation in Tumors

Light and electron microscopic examination of C-26 colon
carcinoma and KS-like lesions show high concentrations of
liposomes in interstitial areas surrounding capillaries in mice
treated with Stealth liposomes containing colloidal gold particles as a liposome marker (43,44). These findings suggest that
such Stealth liposomes circulate for a sufficient period of time
and are small enough to extravasate through the capillaries
supplying tumors.

Following treatment of tumor-bearing mice with DOXIL,
doxorubicin concentrations achieved in tumors are higher

Table 1 Stability of DOXlL Batch 4-DOX-03

Month	DOX potency (mg DOX/mL)	Encapsulation (%)	Particle size (nm)	LPC (mg LPC/mL)
0	2.05	97.7	83	< 0.12
1	2.04	97.4	87	< 0.12
2	2.07	97.8	83	< 0.12
3	2.01	98.2	82	0.14
6	1.99	97.9	83	< 0.12
9	1.99	97.1	83	0.13
12	1.97	98.5	84	< 0.12
18	2.02	98.7	83	0.33
24	1.94	89.7	83	0.32
36	1.93	98.4	83	0.44

(Fig. 5) and anti-tumor activity is greater compared to animals receiving comparable doses of unencapsulated drug (45–50). These findings suggest that DOXIL, by virtue of its plasma stability and slow clearance, might have a higher therapeutic ratio than earlier liposome formulations of doxorubicin.

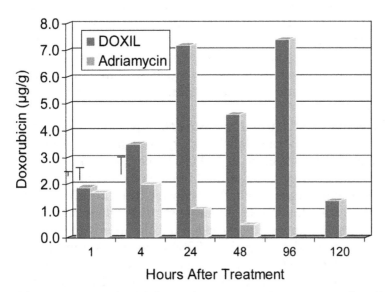

Figure 5 Uptake of doxorubicin in murine tumors after DOXIL and Adriamycin.

As in the animal models, there is evidence from clinical studies that DOXIL selectively distributes to tumor relative to normal tissue. This was first shown in biopsies of KS lesions and adjacent normal skin (51). Forty-eight hours after an injection of DOXIL, the average doxorubicin concentration in tumors was 19 times (range 3–53) that in skin. In another set of experiments, doxorubicin levels in KS lesions were measured after treatment with either DOXIL or doxorubicin (52–63). Clinical experience suggests DOXIL is active against advanced AIDS-related KS (55–63).

At 72 hr, concentrations of doxorubicin in the KS lesions were 5–11 times greater following DOXIL administration than after dosing with conventional doxorubicin. In two patients with metastatic breast cancer, doxorubicin levels were measured in bone fragments and in tumor-free adjacent muscle during surgical repair of a pathological fracture 6 and 12 days after a dose of DOXIL (64). Doxorubicin levels in bone were 10-fold those seen in the muscle (Table 7). Doxorubicin fluorescence and specific nuclear staining showed good co-localization. This suggests that the observed differences in doxorubicin levels between tumorous and normal tissue are a result, at least in part, of differences in intracellular drug levels. Thus, while DOXIL appeared to be distributed to normal tissues, there is evidence that it selectively concentrates in tumors and that the encapsulated drug may offer some protection to normal tissues by lessening overall exposure.

Selective distribution of Stealth liposomes was also demonstrated with radiolabeled pegylated liposomes of the same size and composition of DOXIL in patients with several common malignancies, including breast cancer, squamous cell cancer of the head and neck, lung cancer, cancer of the cervix, and high grade glioma (65). In these studies, tumor uptake of the radiolabeled liposomes exceeded uptake in any of the normal tissues with the exception of RES tissues. Accumulation of radiolabeled DOXIL was studied in patients with locally advanced sarcomas undergoing radiotherapy, with significant accumulation of the drug observed in each of the seven patients studied (66). Similarly, relative to surrounding normal tissues, biopsy-proven higher intratumoral concentrations of doxorubicin were reported following administration of DOXIL to 10 patients with breast cancer, one patient each with ovarian cancer and hepatoma, and 3 patients with liver metastases (67). In addition, preferential tumor accumulation of radiolabeled pegylated liposomal doxorubicin was demonstrated by planar and SPECT scintography following treatment of 15 patients with brain metastases or glioblastoma (68).

In addition to increasing doxorubicin localization in tumor tissues, the encapsulation of doxorubicin in Stealth liposomes could also result in a reduction of some of the adverse reactions associated with doxorubicin HCl administration. For example, the cardiotoxicity caused by high cumulative doses of doxorubicin is believed to be related to the high peak plasma concentration of doxorubicin HCl after its administration using the standard 3-week schedule. It is well established in the literature that the incidence of cardiomyopathy is significantly reduced when the drug is administered using a 1-week or a prolonged infusion schedule (21–25). The encapsulation of doxorubicin HCl will effectively reduce the peak drug concentration in plasma, therefore mimicking the prolonged infusion regimen.

4.2. Large Scale Production and Stability

4.2.1. Production

A flow diagram illustrating the key step of the DOXIL production process is presented in Fig. 6. An ethanolic solution of lipids (cholesterol, soy bean derived phosphatidylcholine, and MPEG-phosphatidylethanolamine) is introduced into a warm solution of 250 mM ammonium sulfate. During this process, a crude suspension of multilamellar liposomes spontaneously forms as the lipid solvent power of the ethanol is lost by dilution with the aqueous phase. The suspension is then passed under pressure through a series of capillary pore filters of defined pore size, ultimately yielding a mean liposome diameter of 100 nm (69). Residual ethanol and the external ammonium sulfate are removed by a diafiltration process during which the external phase is replaced with isotonic

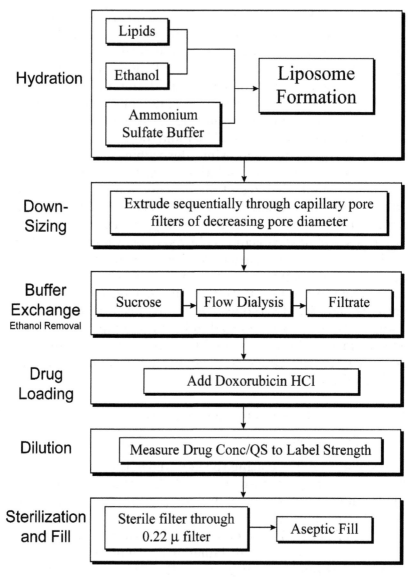

Figure 6 Flow diagram of DOXIL production process.

sucrose. A solution of doxorubicin hydrochloride is added and loading of the liposomes is driven by the process described in greater detail in Fig. 4. There is no need for a free drug removal step since >95% of the doxorubicin is loaded by the method. The drug potency is measured and the bulk suspension diluted until the label-strength concentration (2 mg/mL) is achieved. The suspension is then passed through a sterile-grade filter into a holding tank in a class-100 clean room. Filling of vials makes use of a standard fill line.

Commercial batches of DOXIL are at the 400 L scale, yielding about 32,000 20-mg vials. As is the case with most pharmaceuticals, scale-up of the production process began at a bench-top volume and moved to the commercial scale in discrete stages. Formulation development work relied on 10–50 mL batches. At these volumes, small glass equipment is used under the type of laminar flow hood found typically in a cell culture laboratory. At the time toxicology and Phase I clinical supplies are needed, the scale is increased to 10 L and stainless steel tanks replace glass vessels and the process is moved into a clean room. Down-sizing of the liposomes continues to be done using flat stock polycarbonate membranes housed in a high pressure cell. The next step up, to 100 L, comes as larger Phase III clinical trials become imminent. At the 100 L scale, the process was validated using conventional pharmaceutical processing tanks, pumps, filters, sterilization and filling equipment. At this stage, sizing by extrusion was done using polycarbonate membranes housed in cartridges. In this way, it was possible to connect a parallel series of such cartridges to accommodate the increased volume.

4.2.2. Stability

Physical and chemical stability are critically important for regulatory and commercial reasons. DOXIL is currently labeled with an 18 month shelf-life, based on real-time stability studies. It is likely that this time can be extended as more real-time stability data are generated. Several of the design features of DOXIL are responsible for this robust stability profile (Table 1).

By virtue of the PEG coating, the colloidal stability of DOXIL is remarkable. No size growth or precipitation is seen during storage at 2–8°C for at least 2 years. Similarly, the ammonium sulfate loading method provides for stable drug retention for long periods. In fact, no measurable drug leakage occurs over the shelf-life of DOXIL. The stability-limiting process during storage of DOXIL is a slow but steady decline of drug potency. Lysophosphatidylcholine appears over time, but after 36 months of storage, it remains well below 1.0 mg/mL, the level at which the product would fall out of regulatory specifications.

4.3. Clinical and Regulatory Strategy

The DOXIL overall registration strategy is illustrated in Fig. 7. The first indication sought for DOXIL was treatment of patients with refractory AIDS-related KS. In the principal clinical trial supporting this indication, patients were considered "refractory" by an independent panel of three physicians expert in the treatment of KS if they met both of the following criteria:

1. Prior treatment with at least two systemic chemotherapy drugs for treatment of AIDS-related KS for at least two cycles of therapy. (One, but not both, of the drugs could have been a vinca alkaloid; alpha-interferon was not considered a systemic chemotherapeutic drug.)
2. Had progressive disease or could not tolerate continued standard therapy due to drug toxicities.

Given the seriousness of the indication, the original NDA for DOXIL was based on the interim efficacy results from two studies: Study 30-12, the principal study supporting efficacy in treatment-refractory patients, and Study 30-03, a supportive study. An overview of DOXIL clinical trials contained in the NDA is presented in Table 2.

During the course of development, the DOXIL formulation evolved. The initial formulation, DOXIL 1, was unbuffered and stored frozen. Subsequently, a liquid formulation was developed. The early liquid formulation, designated DOXIL 2, had limited use in the clinic. The final liquid formulation, DOXIL 3, differs from DOXIL 2 solely in the buffer used. The final formulation was used in the primary AIDS-related KS

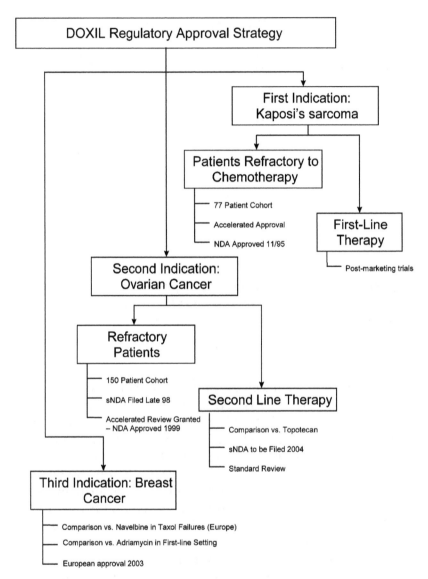

Figure 7 Flow diagram of DOXIL regulatory strategy.

clinical study, Study 30-12, in part of the supportive clinical study, Study 30-03, and in a pharmacokinetic study, Study 30-14. DOXIL 3 (hereafter referred to as DOXIL) has been used in a complete non-clinical pharmacology and toxicology

Table 2   Overview of DOXIL Clinical Studies Contained in DOXIL NDA

Study number	Location	Formulation(s)
Clinical pharmacology studies		
30-02	Israel	DOXIL 1
30-05	US	DOXIL 1
		DOXIL 3
30-14	US	DOXIL 3
Phase 2/3 efficacy studies—AIDS-related KS		
30-03	Australia/Europe	DOXIL 1
		DOXIL 2
		DOXIL 2
30-12	US/Europe	DOXIL 3

program and was the formulation proposed for commercial marketing. A summary of the DOXIL formulations used in the clinical trials is presented in Table 3.

4.3.1. Clinical Pharmacology

Three clinical pharmacology/pharmacokinetic studies were conducted for the DOXIL NDA. One study, Study 30-14, was an evaluation of the pharmacokinetics and KS tumor levels of DOXIL, the proposed commercial formulation. The previous frozen formulation, DOXIL 1, has been studied in two earlier clinical pharmacology trials, Study 30-02 and Study 30-05. Comparisons to conventional doxorubicin HCl (Adriamycin) were included in Study 30-05 and Study 30-02.

The pharmacokinetics of DOXIL are characterized by a biexponential plasma concentration–time curve, with a short first phase and a prolonged second phase that accounts for the majority of the AUC. Plasma concentration and AUC are dose-dependent, given as 10 or 20 mg/m^2, but disposition kinetics are independent of dose. Volume of distribution is relatively small, just several-fold the plasma volume and clearance is low.

Several lines of evidence support the conclusion that the majority of the doxorubicin remains encapsulated within the liposome after i.v. administration of DOXIL. In a pharmacoki-

Table 3 Summary of DOXIL Formulations used in Clinical Trials

Component	Frozen formulation DOXIL 1	Liquid formulations DOXIL 2	DOXIL 3
Doxorubicin HCl	2.0 mg/mL	2.0 mg/mL	2.0 mg/mL
HSPC	9.58 mg/mL	9.58 mg/mL	9.58 mg/mL
MPEG-DSPE	3.19 mg/mL	3.19 mg/mL	3.19 mg/mL
Cholesterol	3.19 mg/mL	3.19 mg/mL	3.19 mg/mL
Sucrose[a]	94 mg/mL	94 mg/mL	94 mg/mL
Tromethamine	—	1.21 mg/mL	—
Histidine	—	—	1.55 mg/mL
Ammonium sulfate[a]	1 mg/mL	2 mg/mL	2 mg/mL
α-Tocopherol	0.0195 mg/mL	—	—
Deferoxaminemesylate	0.132 mg/mL	—	—
Water for injection	qs 1mL	qs 1mL	qs 1mL
Formulation pH	5.5	6.5	6.5

[a]Estimated based on theoretically calculated liposome encapsulated volume.

netic study reported by Gabizon et al. (13), the fraction of the liposome-encapsulated and total drug in circulation after DOXIL treatment was actually measured. Essentially all the doxorubicin measured in plasma was liposome-associated (Fig. 8). Similar results were found in non-clinical pharmacokinetic studies in rats, i.e., at least 90–95% of the doxorubicin measured in plasma, and possibly more, is liposome-encapsulated (11). Moreover, low blood levels of doxorubicinol suggest that 99% of the drug remains liposome-encapsulated after DOXIL treatment.

In Study 30-14, the doxorubicin concentration of KS lesions ranged from approximately 3- to 16-fold higher than doxorubicin levels in normal skin biopsies collected from the same patients at the same time points. In Study 30–05, doxorubicin levels in KS lesions 72 hr after treatment with DOXIL 1 were found to range from 5- to 11-fold higher than after the same dose of Adriamycin. Animal studies have shown that tumor levels of doxorubicin are higher after treatment with DOXIL even at 1 hr post-treatment (70).

The pharmacokinetics of DOXIL appears to be significantly different from those reported for Adriamycin (Fig. 9).

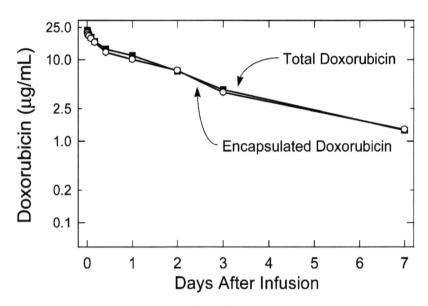

Figure 8   Human pharmacokinetics of DOXIL.

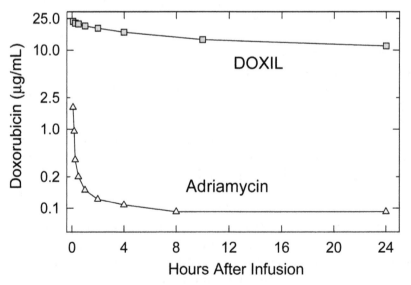

Figure 9   Comparative Human pharmacokinetics of DOXIL and Adriamycin.

DOXIL has a significantly higher AUC, lower rate of clearance (approximately 0.1 L/hr) and smaller volume of distribution (5–7 L) than those parameters reported for Adriamycin. The first phase of the biexponential plasma concentration-time curve after DOXIL administration is relatively short (half-life 1–3 hr), and the second phase, which represents the majority of the AUC, is prolonged (half-life 50–55 hr). Doxorubicin C_{max} after DOXIL administration is 15- to 40-fold higher than after the same dose of Adriamycin, and the ratio quickly increases as Adriamycin is rapidly cleared from circulation; however, the majority of plasma doxorubicin remains liposome-encapsulated after DOXIL treatment. Because of the high percentage of liposome encapsulation in DOXIL, initial free drug levels in the plasma appear to be significantly lower than those measured after administration of an equal dose of Adriamycin (11,13,71,72).

No formal bioequivalence studies of DOXIL, the proposed commercial formulation, and the previous formulation, DOXIL 1, were conducted. However, using the data from Study 30-14 and Study 30-05, the pharmacokinetics and KS lesion localization of DOXIL can be compared to those of DOXIL 1 (Table 4). These two studies were conducted in patients with similar characteristics by the same investigator. No statistical tests were utilized to compare the pharmacokinetics of the two formulations, since the studies were not prospectively intended to be compared. Instead, a direct visual

Table 4 Summary of Median (Range) DOXIL and DOXIL 1 Pharmacokinetic Parameters

Parameter	DOXIL ($N = 42$)	DOXIL 1 ($N = 9$)
$AUC_{0 \to \infty}$[a]$(\mu g/mL/mg\ hr)$	13.2 (1.50–28.3)	8.09 (4.87–14.7)
V_c (L/m^2)	2.34 (1.12–3.79)	2.27 (1.60–2.75)
V_p (L/m^2)	0.452 (0.206–1.29)	1.21 (0.463–1.65)
V_{ss} (L/m^2)	2.79 (1.44–4.51)	3.25 (2.16–4.39)
CLt $(L/hr/m^2)$	0.0413 (0.0171–0.358)	0.0670 (0.0340–0.108)
$t_{1/2\lambda1}$ (hr)	3.13 (0.542–16.5)	3.76 (2.01–4.90)
$t_{1/2\lambda2}$ (hr)	48.7 (6.00–98.9)	41.3 (19.8–54.0)

[a]$AUC_{0-\infty}$ is normalized to dose.

comparison of median pharmacokinetic parameters was applied. Pharmacokinetic parameters are summarized in Table 4. By inspection, it appears that only V_p, $t_{1/2\lambda2}$, and $AUC_{0\rightarrow\infty}$ may be different between formulations. These observations are consistent with what is known about these formulations. The decreased peripheral volume of the DOXIL formulation may reflect its decreased leakage rate of doxorubicin. In contrast, the DOXIL 1 formulation, which has a higher leakage rate of free doxorubicin, expresses a larger V_p, reflecting the large volume of distribution of free doxorubicin. The increase in terminal half-life and $AUC_{0\rightarrow\infty}$ of DOXIL are both the result of decreased release rate of doxorubicin from DOXIL liposomes. It appears that DOXIL may provide a larger and more sustained exposure to doxorubicin than does DOXIL 1.

4.3.2. Pivotal Clinical Studies

Five clinical trials were included in the DOXIL NDA (Table 5). The three clinical pharmacology studies (Studies 30-02, 30-05, and 30-14) have been summarized above. The efficacy of DOXIL for its proposed indication is supported by two studies, Studies 30-03 and 30-12, which are described below.

It is customary for the sponsor of a new drug product to consult with the FDA prior to conducting pivotal clinical studies to obtain regulatory advice and clinical guidance. Based on such interactions with the Division of Oncology and Pulmonary Drug Products of the FDA, the DOXIL NDA sought approval for DOXIL in a limited indication, namely, treatment of AIDS patients with advanced KS which has failed standard first-line combination systemic chemotherapy due to disease progression or unacceptable toxicity. This patient population had no clearly established treatment options. The key points discussed at meetings with FDA focused on the patient population being studied and the methods used to document anti-tumor response and clinical benefit.

- There was general agreement that an NDA could be filed based on Phase 2 open-label data provided that the sponsor documented a clinical response of mean-

Table 5 Clinical Studies Submitted in Support of Clinical Efficacy/Benefit

Study number	Study sites	N (M/F)[a]	Diagnosis	Mean age (range)	Dose	Formulation
30-03	22 sites in Australia and Europe	242/5	Kaposi's sarcoma	39.1 years (16–70 yrs)	10–40 mg/m^2 every 2 weeks	DOXIL 1 DOXIL
30-12	18 sites in US and Europe	137/0	Kaposi's sarcoma	38.3 years (24–68 yrs)	20 mg/m^2 every 3 weeks	DOXIL

ingful duration in a population of about 50 KS patients who had failed standard systemic chemotherapy.

- The FDA recommend that unanimous (three out of three) vote by a panel of AIDS-KS clinical experts be used to identify treatment failure patients among those enrolled in clinical studies.
- Only those patients who had failed on prior combination therapy with Adriamycin, bleomycin, and vincristine (or vinblastine) (ABV) or bleomycin and vincristine (or vinblastine) (BV) were to be considered as treatment failure candidates.
- Supportive efficacy data were to be presented in the NDA from Study 30-03.
- Clinical benefit would be documented in narrative patient summaries based on case records for patients who achieve a tumor response based on indicator lesion assessment.

In accordance with the FDA's recommendations, a cohort of 77 treatment failure patients was identified among those enrolled in the primary efficacy study, Study 30-12.

Efficacy Studies Supporting the Indication

Therapeutic response and clinical benefit of DOXIL were assessed using each patient as his own control. Efficacy data were submitted in 383 patients but the primary efficacy analysis of response to DOXIL was based on objective documentation of changes in the characteristics of five indicator lesions in a cohort of 77 patients retrospectively identified as having disease progression on prior systemic combination chemotherapy as being intolerant to such therapy.

The results of the analyses were presented for the 77 patients selected by a panel of three AIDS-KS experts as having met the definition of failure on prior systemic combination chemotherapy either due to disease progression or unacceptable toxicity. Table 6 lists the efficacy results presented as: (i) the cohort of 41 patients submitted in the original NDA, (ii) the additional 36 patients submitted as a clinical amendment to the NDA, and (iii) the 77 patients combined.

Table 6 Comparison of Response by Indicator Lesion Assessment-
Conservative vs. Best Response

Method of analysis	41 Patients	36 Patients	77 Patients
Conservative method (best response)			
Partial	15 (36.6%)	13 (36.1%)	26 (33.8%)
Stable	8 (19.5%)	9 (25.0%)	19 (24.7%)
Progression	18 (43.9%)	14 (38.9%)	32 (41.6%)
Primary efficacy analysis (best response)			
Partial	27 (65.9%)	25 (69.4%)	52 (67.5%)
Stable	10 (24.4%)	10 (27.8%)	20 (26.0%)
Progression	4 (9.8%)	1 (2.8%)	5 (6.5%)

Demographics: The 77 patients were white, homosexual males and the median CD4 count for this group was 10 cells/mm^3. Their age ranged from 24 to 54 years, with a mean age of 38 years. Using the ATCG staging criteria I at baseline 78% of the patients were at poor risk for tumor burden (D, 96% at poor risk for immune system (1) and 58% at poor risk for systemic illness (S). Their mean Kamofsky status score was 74%. All 77 patients had cutaneous or subcutaneous lesions; 40% also had oral lesions, 26% reported pulmonary lesions; and 14% of patients reported lesions of the stomach/ intestine. The median duration of exposure to DOXIL for these 77 patients was 155 days and ranged from 1 to 456 days. The median cumulative dose was 154 mg/m^2 and ranged from 20 to 620 mg/m^2. Sixty-six of the 77 patients had progressive disease and 11 had toxicity with their prior first line therapies.

Response to treatment: The best response to treatment was partial response for 52 patients (68%), stable disease for 20 (26%), and progressive disease for five (6%). The median duration of PR was 64 days (mean, 74).

Forty-three of the 77 patients experienced progression of their KS during therapy with Adriamycin (in combination with I or 2 other cytotoxic agents) prior to receiving DOXIL. Of these 43 patients, 26 (60.5%) had a partial response as their best response on DOXIL therapy while 4 patients (9.3%) had progression as the best response after progressing on an Adriamycin-containing regimen.

Comparison of primary and secondary tumor response:
Table 6 presents a comparison of the primary and secondary measures of response for the 77 patients. Investigator assessment of response (56%) was slightly lower than the response seen using the indicator lesion assessment (68%). However, according to the investigator assessment criteria, one patient achieved a clinical complete response. Of the 77 patients, there are 24 (31%) patients where the indicator lesion assessment yields a more favorable assessment of response and in nine cases (11.7%), the investigator assessment is more favorable.

The median duration of partial response is 113 and 64 days for the investigator assessment and indicator lesion assessment, respectively. This duration of response comprises an important length of time in the life expectancy of a patient population with advanced AIDS.

Response by conservative analysis: In addition to the efficacy analyses described above, data were analyzed by a more conservative method, which limits responders to those who did not progress prior to having achieved a response, and whose response lasted for at least 21 days and occurred for at least two cycles. Table 6 shows that using this conservative definition 26 (34%) patients achieved a partial response, 19 (25%) achieved stable disease, and 32 (42%) had progressive disease. The median duration of the best partial response was 65 days.

Of the 77 cases, for 41 (53%), the primary indicator lesion assessment and the conservative assessment are in agreement, five (7%) patients had their original PR response reduced because a PD occurred before they experienced a PR, and 31 (40%) remaining patients had their response reduced because the original response had a duration of less than 21 days. Seven of these 31 patients have been conservatively classified as a PD, even though they were not categorized as having had progressive disease by the other analyses.

Clinical benefits: Table 7 shows the results of assessment of clinical benefit by best response category. Overall, 51% of patients experienced pain reduction, 79% edema reduction, 50% complete flattening, and 55% color improvement.

The results of these analyses show that clinical benefit is associated with response. These clinical benefits were seen

Table 7   Clinical Benefits by Best Response (77 Patients)

Best response	N	Complete flattening	Color improvement	Pain reduction	Edema reduction
PR	52	36	34	13	19
SD	20	0	4	6	2
PD	5	0	0	0	1
Total[a]	77	36/72 (50%)	38/69 (55%)	19/37 (51%)	22/28 (79%)
Median duration of benefit (days)		>149	>164	>216	>176

[a]Denominator represents number of patients with potential for changes, i.e., baseline edema and baseline existence of moderate/severe pain, red/purple color of at least one indicator lesion, and at least one raised indicator lesion.

mostly in those patients achieving a partial response and 48% of patients showed improvement in more than one category.

Efficacy conclusions: Treatment with DOXIL is efficacious and provides a reasonable tumor response and a clear and meaningful clinical benefit, in the treatment of patients with AIDS-related KS who have failed or are intolerant to conventional combination chemotherapy. The primary analysis of efficacy based on assessment of indicator lesions and the secondary analysis based on investigator assessment provide nearly identical results. Documented clinical benefits include flattening of all indicator lesions, improvement in the color of indicator lesions from purple or red to a brown or more neutral color, a reduction in KS-associated pain and a reduction in KS lesions-associated edema and nodularity.

DOXIL used in the treatment of the broader population of KS patients was also seen to be efficacious. Results from all patients treated in Studies 30-03 and 30-12 were consistent with those achieved in the refractory patient subset.

As demonstrated in Study 30-05, there was a prolonged doxorubicin plasma circulation time after the administration of DOXIL relative to the administration of Adriamycin. This long circulation time was associated with higher doxorubicin concentrations in KS lesions of patients who received DOXIL.

These concentrations exceed those found in normal skin as demonstrated in Study 30-14.

Taken together, these four clinical trials establish that doxorubicin encapsulated in long-circulating liposomes remains circulating in the blood stream for extended periods of time, allowing for accumulation of doxorubicin in KS lesions which translates into a reduced tumor bulk and clinical benefit for a population of patients with no other therapeutic options.

Safety Conclusions

Leukopenia: Most DOXIL-treated patients experienced leukopenia, an adverse reaction known to be associated with Adriamycin therapy and common among AIDS patients not receiving chemotherapy. Of the KS patients in the NDA safety database, 60% experienced leukopenia possibly or probably related to DOXIL therapy. The mean minimum value of ANC was 1250 cells/mm^3, with 13% of patients experiencing at least one episode of ANC <500 cells/mm^3. Despite the frequency of neutropenia observed, bacterial or fungal septicemia attributable to DOXIL were uncommon across all studies (0.7%) and only 11 patients discontinued DOXIL therapy due to bone marrow suppression.

Opportunistic infections: The relationship between DOXIL and the incidence of opportunistic infections is not known. Opportunistic infections occurred in 355 of 706 (50%) patients. The most commonly reported infections were Candida (24%), CMV (20%), herpes simplex (10%), and PCP (9%).

Acute infusion reactions: One type of adverse event may be related to the liposomal dosage form—an idiosyncratic acute reaction to the infusion. Acute reactions occurring during or soon after DOXIL infusion, characterized by sudden onset of facial flushing and in some cases shortness of breath, chest/back pressure/pain, and hypotension were experienced by 7% of patients (48 of 706 patients) in the NDA safety database. In most cases, the reaction occurred in the first cycle of treatment and the patient was retreated without complication. These reactions were generally self-limited and ceased within a few minutes after the infusion was temporarily stopped, or when the rate of infusion was decreased. However, several patients were treated with medication for the reaction and for six patients the event was severe enough to warrant discontinuation from study.

Hand-foot syndrome: Of 706 AIDS-KS patients who received DOXIL in any clinical trial, 24 (3%) developed skin eruptions consistent with a previously recognized palmar-plantar erythrodysesthesia. This syndrome encompasses swelling, pain, erythema and for some patients desquamation of the skin on the hands and feet. Some patients experienced a rash other than on the hands and feet and are included here. In general, this reaction was mild and resolved with interruption or discontinuation of therapy. However, three patients were terminated due to the event.

In preclinical testing, skin lesions developed in dogs and rats receiving multiple doses of DOXIL. A similar pattern of skin toxicity has been reported for standard Adriamycin and other chemotherapeutic agents including 5-flurouracil administered via prolonged infusions.

Alopecia, nausea, and vomiting: Other common toxicities associated with standard Adriamycin therapy, such as alopecia, nausea and vomiting, considered related to DOXIL administration were relatively infrequent (9%, 13%, and 4%, respectively). The lack of significant nausea/vomiting attributable to DOXIL therapy is a particular benefit for these patients who are already generally cachectic as a result of their FHV disease. The low incidence of drug-induced alopecia is also an advantage for patients who remain active, continue to work, and interact socially.

Risk vs. Benefit Discussion. Patients with advanced AIDS-KS who have failed standard combination cytotoxic chemotherapy have no clearly established treatment options. In this population, the benefits provided by DOXIL therapy outweigh the risks of DOXIL-related toxicities. DOXIL-related leukopenia is manageable. The incidences of cardiac adverse events and opportunistic infections do not appear to be significantly greater than those expected for similar patients receiving conventional cytotoxic chemotherapy. Occasional hypersensitivity reactions and skin eruptions occur but are self-limited and not life threatening. DOXIL-related alopecia, nausea, and vomiting are infrequent, so these toxicities, which are commonly associated with chemotherapy, would not be expected to significantly detract from these patients' quality of life. DOXIL therapy stabilizes KS for over 3 months in patients who would almost certainly progress without treatment. Partial responses are associated with clinical benefits including pain reduction, flattening, and discoloring of lesions and resolution of lesion-associated edema. Overall, the use of single-agent DOXIL provides meaningful clinical benefit for KS patients with manageable risk.

For patients in whom standard multi-agent systemic chemotherapy has failed, there are no clearly proven alternative treatments available. Alpha-interferon, an agent approved for the treatment of KS, is not expected to provide meaningful benefit to KS patients with advanced HIV disease. New combinations of the chemotherapeutic agents proven effective for the treatment of AIDS-related KS (Adriamycin, bleomycin, etoposide, vincristine, and vinblastine) could be attempted. However, combinations including bleomycin and a vinca alkaloid with or without Adriamycin had already failed the patients in our primary efficacy study.

For those patients for whom toxicity was the cause of treatment failure, single- or multiagent chemotherapy could be reintroduced after a recovery period. However, untreated KS probably would progress in the interval and the patients who resumed chemotherapy would be at high risk for the same toxicities.

Neutropenia, although common, is manageable with growth factor support (C-CSF or GM-CSF). Of KS patients treated with DOXIL 1.5% discontinued therapy because of bone marrow suppression. The risk of developing DOXIL-induced cardiotoxicity in a KS population is not known, but based on animal studies, it is probably not greater than, and may be less than, with standard Adriamycin. We have no indication that DOXIL produces an incidence of cardiac events higher than what one expects in a population of patients with advanced HIV infection. Acute infusion reactions are self-limited and symptoms resolve quickly after stopping or slowing the infusion. Skin eruptions are infrequent, reversible, and not dose-limiting in almost all patients. Since DOXIL-related alopecia, nausea, and vomiting are relatively infrequent, these toxicities, which are common with Adriamycin chemotherapy, are unlikely to detract significantly from the quality of life for these patients.

DOXIL therapy improves or stabilizes KS for a median of 4 months in patients who would almost surely progress without treatment. Partial responses are associated with important clinical benefits including pain reduction, flattening, and color improvement of lesions and resolution of lesion-associated edema.

Given that patients in these studies have advanced AIDS-related KS, have failed standard systemic cytotoxic chemotherapy, have no clearly proven treatment options, and whose KS would progress if untreated, the benefits provided by DOXIL therapy outweigh the risks of DOXIL-related toxicity.

4.4. NDA Contents, FDA Review, and Approval

The DOXIL NDA was formatted in the customary manner and consisted of over 500 volumes. One section was the proposed labeling (actual text of the product insert) annotated with references to other sections of the NDA which contained data to support each label claim. Other sections contained the scientific rationale and the marketing history for the product. The bulk of the submission consisted of the chemistry manufacturing and controls, non-clinical pharmacology, and clinical data (clinical data alone accounted for 155 volumes). The clinical sections included study reports from five trials including pharmacokinetics, efficacy, and safety data. Additionally, as required by the FDA, integrated summaries of efficacy and safety as well as an overall summary of the NDA were included.

The original NDA was submitted in September, 1994. The core of the submission was response data for a cohort of 41 "refractory" KS patients treated with DOXIL. Supportive efficacy data from several large open-label trials and safety data derived from five clinical trials totaling 753 patients were also submitted. Since the FDA expressed concern that the number of patients in the key refractory cohort was too small, a clinical amendment consisting of efficacy data for a cohort of 36 additional refractory patients was submitted in November, 1994. The submission with a combined refractory cohort of 77 KS patients as detailed above was reviewed by FDA and a hearing was held by the Oncologic Drugs Advisory Committee (ODAC) of the FDA on February 14, 1995. The purpose of the ODAC hearing was to address the clinical questions unresolved during the agency's review. Paraphrasing, the FDA posed the following questions to the ODAC members.

1. Are the historically controlled clinical trials that represent the basis for approval in this submission "adequate and well-controlled" (i.e., do they meet the statutory requirements for approval)?
2. Is DOXIL active in KS patients who have failed standard first-line therapy?
3. Given its safety and efficacy profiles, does DOXIL provide benefit to these patients?

Although the committee members voted positively on the questions of efficacy and safety of DOXIL, they evenly split on the question regarding the statutory adequacy of the historically controlled trial design (vis-a-vis the preference for comparative clinical trials). ODAC members agreed that the natural history of KS was reasonably well documented and that no agents were officially approved for second-line therapy, so historically controlled trials were reasonable. However, one KS expert on the panel believed that a variety of existing cytotoxic agents might be active in the second-line setting. Moreover, response to DOXIL therapy was based primarily on shrinkage (flattening) of a subset of "indicator" lesions followed in the 77 refractory patients. Although the ODAC members and the FDA medical reviewers were convinced that DOXIL therapy provided remission of KS tumors, it was not obvious that such a response provided meaningful clinical benefit to the patients. (With other solid tumors such as lung or ovarian, documentation of tumor shrinkage is considered sufficient in itself as evidence of clinical benefit. But, since AIDS-related KS was a fairly new disease, the underlying immune dysfunction was progressive, and KS was only rarely the cause of death in these patients, the benefit of tumor shrinkage in the overall context of AIDS was and is not clear-cut.) Thus the ODAC was reluctant to base its recommendation for approval of a full NDA for DOXIL on tumor response data alone. But, since a positive vote on the adequacy of the clinical trials is required for approval, FDA representatives at the hearing suggested that DOXIL might be eligible for "accelerated approval." Accelerated approval can be based on a surrogate marker of clinical benefit. (Examples of drugs receiving accelerated approval include AIDS anti-virals where the surrogate endpoint for efficacy was improvement of CD4 cell count (73).) In the case of the DOXIL submission, the FDA representatives proposed that tumor response could be considered a surrogate endpoint of benefit. In this context, the committee then voted unanimously to recommend DOXIL for accelerated approval (74,75). Based on the committee's recommendation and further review by the FDA, DOXIL was approved for marketing in November, 1995 under FDA accelerated approval regulations.

4.5. Post-Marketing Trials

4.5.1. Blinded Comparative Trial in KS

As a condition for accelerated approval, the sponsor must agree to conduct additional clinical trails (so called Phase IV studies) to confirm that the surrogate endpoint relied upon for accelerated approval actually reflects meaningful benefits to patients. In the case of DOXIL, the sponsor agreed to conduct a blinded comparative trial of DOXIL vs. DaunoXome, another liposome product known to be active against KS (6,27). This trial began in 1996 and was completed in 2001. In this trial, refractory KS patients were randomized to either DOXIL or DaunoXome (the randomization is 1 to 4 in favor of DOXIL). Both drugs were administered "blinded" (that is, neither the patient nor the caregiver knows which drug is being given). The endpoints of the trail included safety, tumor response assessed by careful serial measurement of all KS lesions, and clinical benefits such as reduced need for transfusions, resolution of edema, improvement in the range of motion of limbs, etc. An sNDA was submitted in 2001 seeking full approval but the FDA concluded that in the 5-year time period during which the trial had been conducted changes in anti-retroviral therapy confounded the efficacy assessment. The sponsor is currently discussing a follow on trial to confirm benefit in a population of patients in which anti-viral therapy is standardized.

4.5.2. Cardiotoxicity Assessment

Doxorubicin is known to cause irreversible cardiotoxicity in patients as cumulative exposure reaches or exceeds about 400–500 mg/m^2. It was important to know whether DOXIL provided any reduction in this dose-limiting toxicity. Endomyocardial biopsy has previously been shown to be a safe, sensitive, and specific method of demonstrating anthracycline-induced cardiac lesions. DOXIL has been shown in preclinical models to induce less cardiotoxicity than non-liposomal doxorubicin. An endomyocardial biopsy study was undertaken to assess histopathological changes in the hearts of KS patients receiving cumulative doses of DOXIL >400 mg/m^2.

Myocardial tissue from 10 KS patients who had received cumulative DOXIL (20 mg/m^2/biweekly) of 440–840 mg/m^2 was evaluated for evidence of anthracycline-induced cardiac damage (76). These results were compared to those of patients who had received cumulative doxorubicin doses of 410–671 mg/m^2 in two earlier cardiac biopsy protocols. Two control groups of patients who had not received cardiac irradiation were selected on the basis of cumulative doxorubicin dose ($\pm$10 mg/m^2) and peak dose (60 or 20 mg/m^2, group 1), or peak dose alone (20 mg/m^2, group 2). Electron micrographs of biopsies from the DOXIL and doxorubicin treated patients in group 1 were read blinded by two cardiac pathologists. Adjustments were made to some scores in group 1 to account for differences in peak dose.

DOXIL patients had significantly lower cardiac biopsy scores compared with those of matched doxorubicin controls.

In all three analyses, the average cumulative dose of doxorubicin given as DOXIL was higher than in the control patients. The mean biopsy scores for the DOXIL and doxorubicin groups, respectively, were 0.5 ± 0.6 vs. 2.5 ± 0.7 ($p < 0.001$) for group 1, 0.5 ± 0.6 vs. 1.4 ± 0.7 ($p < 0.001$) for group 2, and 0.5 ± 0.6 vs. 2.1 ± 0.7 ($p < 0.001$) for group 1 after adjustment. These results suggested that DOXIL is less cardiotoxic than doxorubicin. Studies in larger groups of patients receiving higher doses of DOXIL are underway to confirm these findings.

4.5.3. Investigator-Sponsored Trails in Solid Tumors

Since doxorubicin is active against a wide range of tumor types, it is not unexpected that investigators would experiment with DOXIL in diseases other than KS. Indeed, small Phase II trials of single-agent DOXIL are completed or underway in a wide range of tumors including soft tissue sarcoma, non-Hodgkin's lymphoma, carcinoma of the head and neck, renal cell carcinoma, multiple myeloma, and ovarian and breast carcinoma. In addition combinations of DOXIL and other agents including Navelbine, cyclophosphamide, paclitaxel, taxotere, and cisplatin have been investigated. Many of these trials are sponsored by individual investigators.

4.5.4. Post-Marketing Trials

First-Line AIDS KS

Two prospectively randomized clinical trails have compared the activity of DOXIL to combinations of Adriamycin (doxorubicin)/bleomycin/vincristine (ABV) and to bleomycin/vincristine (BV) as first-line treatment of AIDS-KS. In the DOXIL vs. ABV study the dosing interval was 2 weeks whereas in the DOXIL vs. BV study the interval was 3 weeks (63,77,78). A total of 254 patients were treated with DOXIL, 125 with ABV, and 120 with BV. Demographics and baseline disease status were well balanced among the groups. Response was 52% in the combined DOXIL group, 25% in the ABV group, and 23% among BV patients. A greater number of DOXIL patients were able to complete the trials (68% for DOXIL vs. 34% for ABV and 55% for DOXIL vs. 31% for BV). With the exception of mucositis which was more frequent among DOXIL patients, fewer and less severe adverse events including nausea/vomiting, alopecia, peripheral sensory neuropathy, and neutropenia (<1000 cells/mm^3) were reported in the DOXIL arm relative to ABV. Less than 1% of all patients reported skin toxicity. Relative to BV, a greater degree of neutropenia was found in DOXIL patients, but less peripheral neuropathy and nausea/vomiting were reported. The rates of opportunistic infections were similar among DOXIL, ABV, and BV patients. DOXIL has a superior tumor response rate relative to both ABV and BV, a superior safety profile than ABV and an equivalent safety profile to BV, but without the common dose-limiting toxicities associated with BV. Single agent DOXIL (20 mg/m^2 every 2–3 weeks) is indicated for first-line treatment for KS and is suitable for long-term therapy.

A third trial compared the activity of DOXIL alone or in combination with bleomycin and vincristine in 48 patients AIDS-KS patients refractory to or intolerant of conventional systemic chemotherapy. DOXIL 20 mg/m^2 was administered every 3 weeks (with cycle length modified depending on adverse events and response) for up to 10 cycles. DOXIL was shown to be as efficacious as the three-drug combination and significantly safer (79).

The findings of these randomized trials confirm the result of the earlier open-label trails and show that single-agent DOXIL is effective in the treatment of advanced KS. Indeed based on these results DOXIL has become the standard of care for the first- and second-line treatment of KS.

Ovarian Carcinoma

In early Phase I-II trails, DOXIL showed substantial activity in advanced ovarian cancer with reported response rates ranging from 20% to 26% (80,81). Moreover, the duration of responses was meaningful (on the order of 6 months) in a group of patients who had failed both platinum- and taxane-based therapy (82).

These response results compare favorably with those reported for single-agent paclitaxel in advanced ovarian cancer. Response rates for paclitaxel ranging from 19% to 40% have been reported in early Phase 1–2 trials, with durations of response on the order of 7 months. In a large series of platinum-resistant ovarian cancer patients, an overall response rate of 22% has been reported for paclitaxel, with a median duration of less than 5 months.

Based on this favorable response data, a large comparative trial of DOXIL vs. Topotecan in platinum- and paclitaxel-refractory ovarian cancer patients was launched in mid-1997 and completed in mid-1999. The objective to demonstrate equivalent safety and efficacy between the two drugs was achieved (83). Equivalence in time to tumor progression in patients receiving DOXIL and Topotecan is the primary endpoint of the trial. A supplemental NDA for platinum and taxane-refractory ovarian cancer was submitted and approved in late 1999.

Breast Carcinoma

Doxorubicin is among the most active single agents used in the treatment of advanced breast cancer. However, the duration of treatment with doxorubicin is limited by acute toxicities including nausea and vomiting, and subacute toxicities such as alopecia, mucositis, and bone marrow suppression. Treatment-related neutropenia is frequently encountered with doxorubicin at the dose intensities needed to achieve a meaningful response rate. Severe neutropenia can lead to febrile episodes and occasionally to septic infections that can be life-threatening. Moreover, these agents have the potential to cause irreversible cardiac damage as cumulative doses reach and exceed about 500 mg/m^2.

Treatment strategies for advanced breast cancer vary geographically and among treatment centers in the same country. The goals of systemic chemotherapy in this setting can range from palliation to an intent to cure. Many medical oncologists believe that improved survival is not a realistic objective for systemic chemotherapy, particularly in patients who present with a high disease burden and/or multiple visceral metastatic sites. In some treatment centers patients with advanced breast cancer are not treated with chemotherapy, but rather given supportive care to palliate the signs and symptoms of the disease. This debate has raised awareness of the importance of patients' quality of life and has driven a search for treatment options which provide responses in a significant number of patients and/or delay disease progression for a meaningful period of time with minimal toxicity. In this context, a single-agent regimen would be preferred to combinations of drugs—provided that it is well tolerated and provides a respectable tumor response rate.

Preclinical studies of DOXIL and experience in the treatment of KS patients detailed above suggested that pegylated liposomes deliver a greater proportion of an injected dose of doxorubicin to tumor sites relative to unencapsulated doxorubicin. If this were also the case in breast cancer, one might reasonably expect DOXIL to have similar anti-tumor activity to that of doxorubicin, but at a lower dose intensity, and thus would produce less severe toxicity. This expectation provides a rationale for developing DOXIL as single-agent therapy for advanced breast cancer. The goal would be a tumor response rate comparable to doxorubicin but with a dose and schedule of DOXIL that minimizes the frequency and severity of nausea, vomiting, neutropenia, mucositis, alopecia, and cardiotoxicity.

Indeed DOXIL has been shown in a Phase II trial to provide a response rate and duration of response comparable to doxorubicin but at half the dose intensity (84). At a dose rate of 45 mg/m^2 every 4 weeks, treatment with DOXIL produced objective responses in 31% of patients. This response rate compares favorably to literature values for single agent doxorubicin at doses ranging from 60 to 90 mg/m^2 every 3 weeks. The potential advantages of DOXIL in this setting include improved tolerance (reduced alopecia, nausea, and vomiting, myelosuppression and cardiotoxicity) with comparable efficacy.

In 2003, DOXIL received marketing authorization in the European Union as monotherapy for patients with metastatic breast cancer where there is increased risk of cardiotoxicity (85).

5. CONCLUSIONS AND PERSPECTIVES

This case study of DOXIL provides several important lessons. Despite a good amount of skepticism, liposomes can be engineered to stably encapsulate doxorubicin (and probably other drugs), to recirculate for periods of several days after injection without releasing the drug, to penetrate into tumor tissues and to release encapsulated drug within the tumor (13). This is no small achievement. Moreover, such liposomes can be produced at large scale and reproducibly and enjoy a shelf-life typical of other injectable pharmaceutical products.

The accelerated approval path taken in the clinical development of DOXIL has an ample precedent in the cases of cancer and AIDS drugs. The requirement to demonstrate safety and patient benefit in well-controlled clinical trials applies across the board, regardless of whether the product is a new chemical entity (NCE) or a liposomal drug product. The requirement for Phase IV confirmation of benefit is understandable from both scientific and regulatory perspectives.

DOXIL received approval in the United States in less than 5 years of its development, a remarkable achievement when compared to the 10-year period of time typically needed to register an NCE. This observation points out perhaps the most important lesson of all. Many of the risks associated with the development of NCEs do not apply to liposome formulations of existing drugs with proven biological activity. Liposomes will likely introduce differences in tissue distribution and pharmacokinetics and may affect the safety and efficacy profiles of the drug (positively or negatively). But there is little risk that the intrinsic biological activity of the encapsulated drug will be changed. After all, the drug is not chemically modified, it is simply encapsulated.

The prospects of creating additional approvable products based on liposome technology are brighter than ever. Many of the presumed obstacles to the development of liposome pharmaceuticals such as instability, lack of reproducibility, scale-up difficulties and high cost of goods have been overcome in the case of DOXIL and other liposome-based products as well. Fairly standard pharmaceutical engineering was all that was needed.

The key challenges that face the developers of liposome products are less related to the pharmaceutical engineering aspects of the technology as once believed, but are those facing the developers of other drug delivery technologies, i.e., whether the liposome encapsulated version of the drug is demonstrably superior to the unencapsulated form in terms of benefit to patients.

1. Uptake of radiolabeled liposomes (expressed a percent of the total does of radioactivity injected) in subcutaneously implanted solid tumors 24 h after tail vein injection of various formulations in mice. All were of equivalent size (~100 nm). PG, phosphatidyl-glycerol; PC, phosphatidylcholine; C, cholesterol; DSPC, distearoylphosphatidylcholine; S, sphingo-myelin, GM1, brain derived ganglioside GM1; DPPC, dipalmitoylphosphatidylcholine; PI, hydrogenated phosphatidylinositol; PEG-PE, methyl-polyethylene glycol derivative of distearoylphosphatidylethanola-mine (45).

2. Plasma clearance over 24 hr period of three types of liposomes injected intravenously in cancer patients (statistics given in primary references). D-99 (also known as Evacet and Myocet) liposomes contain dox-orubicin, are 150–200 nm in diameter and are com-posed of egg phosphatidylcholine and cholesterol (12). DaunoXome liposomes contain daunorubicin, are 50 nm in diameter and are composed of distear-oylphosphatidylcholine and cholesterol (6). DOXIL contains doxorubicin, are 85–100 nm in diameter and the composition is listed in Table 3 (13).

3. Illustration of a DOXIL liposome. A single lipid bilayer membrane composed of hydrogenated soy phosphatidyl choline (HSPC), and cholesterol sepa-rates an internal aqueous compartment from the external medium. Doxorubicin is encapsulated in the internal compartment. Polymer groups (linear 2000 Da segments of polyethylene glycol) are grafted to the liposome surface (although not shown, the

polymer also extends from the inner monolayer of the membrane). The mean diameter of the particle, including the PEG layer, is 85 nm.

4. Chemical equilibria driving external doxorubicin accumulation into liposomes containing ammonium sulfate (86).

5. Tumor distribution of doxorubicin following single tail vein injection of equal doses of either DOXIL or Adriamycin (10 mg/kg) in mice bearing subcutaneously implanted C26 colon tumors (mean values, $n = 10$ mice per group).

6. Flow diagram of DOXIL production process.

7. Flow Diagram of DOXIL registration strategy. The first NDA submission was based on a cohort of refractory KS patients who had no treatment options. Accelerated approval was sought and granted. Post-marketing trials in the first-line KS setting were conducted and published, but a supplemental NDA (sNDA) was delayed in preference for a submission for ovarian cancer, the second indication sought for approval. Breast cancer is the likely third indication for DOXIL.

8. Clearance over a 1 week period of total vs. encapsulated doxorubicin after a single 50 mg/m^2 dose of an early formulation of DOXIL (containing 150 nm ammonium sulfate) in cancer patients. Data points represent mean values ± standard deviation for 14 patients in the DOXIL group and four patients in the doxorubicin group (adapted from Ref. 13). The separation method has been described by Druckmann et al. (87).

9. Comparative pharmacokinetics of DOXIL and Adriamycin. Plasma clearance of doxorubicin after a single 50 mg/m^2 dose of an early formulation of DOXIL (containing 150 nm ammonium sulfate) in cancer patients. Data points represent mean values ± standard deviation for 14 patients in the DOXIL group and four patients in the doxorubicin group (adapted from Ref. 13).

REFERENCES

1. Gabizon A. Liposomal anthracyclines. New Drug Ther 1994; 8:431–450.

2. Gabizon A, Chisin R, Amselem S, et al. Pharmacokinetic and imaging studies in patients receiving a formulation of liposome-associated adriamycin. Br J Cancer 1991; 64(6): 1125–1132.

3. Liu D, Liu F, Song YK. Recognition and clearance of liposomes containing phosphatidylserine are mediated by serum opsonin. Biochim Biophys Acta 1995; 1235(1):140–146.

4. Gregoriadis G, Senior J. Control of fate and behaviour of liposomes in vivo. Prog Clin Biol Res 1982; 102(A):263–279.

5. Proffitt RT, Williams LE, Presant CA, et al. Tumor-imaging potential of liposomes loaded with In-111-NTA: biodistribution in mice. J Nucl Med 1983; 24(1):45–51.

6. Gill PS, Espina BM, Muggia F, et al. Phase II clinical and pharmacokinetic evaluation of liposomal daunorubicin. J Clin Oncol 1995; 13(4):996–1003.

7. Papahadjopoulos D, Gabizon A. Liposomes designed to avoid the reticuloendothelial system. Prog Clin Biol Res 1990; 343:85–93.

8. Gabizon A, Shiota R, Papahadjopoulos D. Pharmacokinetics and tissue distribution of doxorubicin encapsulated in stable liposomes with long circulation times. J Natl Cancer Inst 1989; 81(19):1484–1488.

9. Gabizon A, Papahadjopoulos D. Liposome formulations with prolonged circulation time in blood and enhanced uptake by tumors. Proc Natl Acad Sci USA 1988; 85(18):6949–6953.

10. Allen TM, Chonn A. Large unilamellar liposomes with low uptake into the reticuloendothelial system. FEBS Lett 1987; 223(1):42–46.

11. Gabizon AA, Barenholz Y, Bialer M. Prolongation of the circulation time of doxorubicin encapsulated in liposomes containing a polyethylene glycol-derivatized phospholipid: pharmacokinetic studies in rodents and dogs. Pharm Res 1993; 10(5):703–708.

12. Cowens JW, Creaven PJ, Greco WR, et al. Initial clinical (phase I) trial of TLC D-99 (doxorubicin encapsulated in liposomes). Cancer Res 1993; 53(12):2796–2802.

13. Gabizon A, Catane R, Uziely B, et al. Prolonged circulation time and enhanced accumulation in malignant exudates of doxorubicin encapsulated in polyethylene-glycol coated liposomes. Cancer Res 1994; 54(4):987–992.

14. Johnson JR, Williams G, Pazdur R. End points and United States Food and Drug Administration approval of oncology drugs. J Clin Oncol 2003; 21(7):1404–1411.

15. Sleel RT. Chemotherapeutic and biologic agents. In: Skeel RT, ed. Handbook of Cancer Chemotherapy. Boston: A Little Brown, 1991:77–140.

16. Fischl MA, Krown SE, O'Boyle KP, et al. Weekly doxorubicin in the treatment of patients with AIDS-related Kaposi's sarcoma. AIDS Clinical Trials Group. J Acquir Immune Defic Syndr 1993; 6(3):259–264.

17. Gazit Y, Baish JW, Safabakhsh N, et al. Fractal characteristics of tumor vascular architecture during tumor growth and regression. Microcirculation 1997; 4(4):395–402.

18. Yuan F, Leunig M, Huang SK, et al. Microvascular permeability and interstitial penetration of sterically stabilized (stealth) liposomes in a human tumor xenograft. Cancer Res 1994; 54(13):3352–3356.

19. Jain RK. Transport of molecules across tumor vasculature. Cancer Metastasis Rev 1987; 6(4):559–593.

20. Seymour LW. Passive tumor targeting of soluble macromolecules and drug conjugates. Crit Rev Ther Drug Carrier Syst 1992; 9(2):135–187.

21. Hortobagyi GN. Anthracyclines in the treatment of cancer. An overview. Drugs 1997; 54(suppl 4):1–7.

22. Anders RJ, Shanes JG, Zeller FP. Lower incidence of doxorubicin-induced cardiomyopathy by once-a-week low-dose administration. Am Heart J 1986; 111(4):755–759.

23. Shapira J, Gotfried M, Lishner M, et al. Reduced cardiotoxicity of doxorubicin by a 6-hour infusion regimen. A prospective randomized evaluation. Cancer 1990; 65(4):870–873.

24. Valdivieso M, Burgess MA, Ewer MS, et al. Increased therapeutic index of weekly doxorubicin in the therapy of non-small cell lung cancer: a prospective, randomized study. J Clin Oncol 1984; 2(3):207–214.

25. Lum BL, Svec JM, Torti FM. Doxorubicin: alteration of dose scheduling as a means of reducing cardiotoxicity. Drug Intell Clin Pharm 1985; 19(4):259–264.

26. Goren D, Gabizon A, Barenholz Y. The influence of physical characteristics of liposomes containing doxorubicin on their pharmacological behavior. Biochim Biophys Acta 1990; 1029(2):285–294.

27. Olson F, Mayhew E, Maslow D, et al. Characterization, toxicity and therapeutic efficacy of adriamycin encapsulated in liposomes. Eur J Cancer Clin Oncol 1982; 18(2):167–176.

28. Gill PS, Wernz J, Scadden DT, et al. Randomized phase III trial of liposomal daunorubicin versus doxorubicin, bleomycin, and vincristine in AIDS-related Kaposi's sarcoma. J Clin Oncol 1996; 14(8):2353–2364.

29. Kanter PM, Bullard GA, Pilkiewicz FG, et al. Preclinical toxicology study of liposome encapsulated doxorubicin (TLC D-99): comparison with doxorubicin and empty liposomes in mice and dogs. In Vivo 1993; 7(1):85–95.

30. Gabizon AA. Liposomal anthracyclines. Hematol Oncol Clin North Am 1994; 8(2):431–450.

31. Conley BA, Egorin MJ, Whitacre MY, et al. Phase I and pharmacokinetic trial of liposome-encapsulated doxorubicin. Cancer Chemother Pharmacol 1993; 33(2):107–112.

32. Presant CA, Proffitt RT, Turner AF, et al. Successful imaging of human cancer with indium-111-labeled phospholipid vesicles. Cancer 1988; 62(5):905–911.

33. Gregoriadis G, Senior J. The phospholipid component of small unilamellar liposomes controls the rate of clearance of entrapped solutes from the circulation. FEBS Lett 1980; 119(1): 43–46.

34. Juliano RL, Stamp D. The effect of particle size and charge on the clearance rates of liposomes and liposome encapsulated drugs. Biochem Biophys Res Commun 1975; 63(3):651–658.

35. Presant CA, Scolaro M, Kennedy P, et al. Liposomal daunorubicin treatment of HIV-associated Kaposi's sarcoma. Lancet 1993; 341(8855):1242–1243.

36. Lasic DD, Martin FJ, Gabizon A, et al. Sterically stabilized liposomes: a hypothesis on the molecular origin of the extended circulation times. Biochim Biophys Acta 1991; 1070(1): 187–192.

37. Allen TM, Hansen C, Martin F, et al. Liposomes containing synthetic lipid derivatives of poly(ethylene glycol) show prolonged circulation half-lives in vivo. Biochim Biophys Acta 1991; 1066(1):29–36.

38. Northfelt DW, Martin FJ, Working P, et al. Doxorubicin encapsulated in liposomes containing surface-bound polyethylene glycol: pharmacokinetics, tumor localization, and safety in patients with AIDS-related Kaposi's sarcoma. J Clin Pharmacol 1996; 36(1):55–63.

39. Amantea MA, Forrest A, Northfelt DW, et al. Population pharmacokinetics and pharmacodynamics of pegylated-liposomal doxorubicin in patients with AIDS-related Kaposi's sarcoma. Clin Pharmacol Ther 1997; 61(3):301–311.

40. Clerc S, Barenholz Y. Loading of amphipathic weak acids into liposomes in response to transmembrane calcium acetate gradients. Biochim Biophys Acta 1995; 1240(5):257–265.

41. Szoka F, Olson F, Heath T, et al. Preparation of unilamellar liposomes of intermediate size (0.1–0.2 mumol) by a combination of reverse phase evaporation and extrusion through polycarbonate membranes. Biochim Biophys Acta 1980; 601(3): 559–571.

42. Lasic DD, Frederik PM, Stuart MC, et al. Gelation of liposome interior. A novel method for drug encapsulation. FEBS Lett 1992; 312(2–3):255–258.

43. Huang SK, Martin FJ, Jay G, et al. Extravasation and transcytosis of liposomes in Kaposi's sarcoma-like dermal lesions of

transgenic mice bearing the HIV tat gene. Am J Pathol 1993; 143(1):10–14.

44. Huang SK, Lee KD, Hong K, et al. Microscopic localization of sterically stabilized liposomes in colon carcinoma-bearing mice. Cancer Res 1992; 52(19):5135–5143.

45. Papahadjopoulos D, Allen TM, Gabizon A, et al. Sterically stabilized liposomes: improvements in pharmacokinetics and antitumor therapeutic efficacy. Proc Natl Acad Sci USA 1991; 88(24):11460–11464.

46. Huang S, Mayhew E, Gilani S, et al. Pharmacokinetics and therapeutics of sterically stabilized liposomes in mice bearing C-26 colon carcinoma. Cancer Res 1992; 52:6774–6781.

47. Vaage J, Barbera-Guillem E, Abra R, et al. Tissue distribution and therapeutic effect of intravenous free or encapsulated liposomal doxorubicin on human prostate carcinoma xenografts. Cancer 1994; 73(5):1478–1484.

48. Vaage J, Donovan D, Mayhew E, et al. Therapy of human ovarian carcinoma xenografts using doxorubicin encapsulated in sterically stabilized liposomes. Cancer 1993; 72(12): 3671–3675.

49. Williams SS, Alosco TR, Mayhew E, et al. Arrest of human lung tumor xenograft growth in severe combined immunodeficient mice using doxorubicin encapsulated in sterically stabilized liposomes. Cancer Res 1993; 53(17):3964–3967.

50. Gabizon A, Catane R, Uziely B, et al. Prolonged circulation time and enhanced accumulation in malignant exudates of doxorubicin encapsulated in polyethylene-glycol coated liposomes. Cancer Res 1994; 54:987–992.

51. Northfelt D, Martin F, et al. Pharmacokinetics (PK), tumor localization (TL), and safety of Doxil (liposomal doxorubicin) in AIDS patients with Kaposi's sarcoma (AIDS-KS). Proc Am Soc Clin Oncol 1993; 12.

52. Northfelt DW. Stealth liposomal doxorubicin (SLD) delivers more doxorubicin (DOX) to AIDS-Kaposi's sarcoma (AIDS-KS) lesions than to normal skin (Meeting abstract). Proc Am Soc Clin Oncol 1994; 13.

53. Northfelt D, Kaplan L, et al. Pharmacokinetics and tumor localization of DOX-SL (Stealth liposomal doxorubicin) by comparison with adriamycin in patients with AIDS and Kaposi's sarcoma. In: Lasic D, Martin F, eds. Stealth Liposomes. Boca Raton, FL: CRC Press, 1995.

54. Northfelt DW, Martin FJ, et al. Doxorubicin encapsulated in liposomes containing surface-bound polyethylene glycol: pharmacokinetics, tumor localization, and safety in patients with AIDS-related Kaposi's sarcoma. J Clin Pharmacol 1996; 36(1): 55–63.

55. Sturzl M, Zietz C, Eisenburg B, et al. Liposomal doxorubicin in the treatment of AIDS-associated Kaposi's sarcoma: clinical, histological and cell biological evaluation. Res Virol. 1994; 145(3–4):261–269.

56. Bogner JR, Kronawitter U, Rolinski B, et al. Liposomal doxorubicin in the treatment of advanced AIDS-related Kaposi sarcoma. J Acquir Immune Defic Syndr 1994; 7(5):463–468.

57. Gruenaug M, Bogner JR, Loch O, et al. Liposomal doxorubicin in pulmonary Kaposi's sarcoma: improved survival as compared to patients without liposomal doxorubicin. Int Conf AIDS 1996; 11(1):303 (abstract no. Tu.B2221).

58. Goebel FD, Bogner JR, Spathling S, et al. Efficacy and toxicity of liposomal doxorubicin in advanced AIDS-related Kaposi sarcoma (KS). An open study. Int Conf AIDS 1993; 9(1):58 (abstract no. WS-B15-6).

59. Goebel FD, Liebschwager M, Held M, et al. Successful treatment of advanced Kaposi sarcoma (KS) with liposomal doxorubicin–short term observations. Int Conf AIDS 1992; 8(2):B105 (abstract no. PoB 3108).

60. Goebel FD, Gruenaug M, Bogner JR. Survival times of patients with AIDS-related Kaposi's sarcoma with pulmonary involvement using liposomal doxorubicin (Meeting abstract). Proc Annu Meet Am Soc Clin Oncol 1996; 15.

61. Stewart S, Jablonowski H, Goebel FD, et al. Randomized comparative trial of pegylated liposomal doxorubicin versus bleomycin and vincristine in the treatment of AIDS-related Kaposi's sarcoma. International Pegylated Liposomal Doxorubicin Study Group. J Clin Oncol 1998; 16(2):683–691.

62. Harrison M, Tomlinson D, Stewart S. Liposomal-entrapped doxorubicin: an active agent in AIDS-related Kaposi's sarcoma. J Clin Oncol 1995; 13(4):914–920.

63. Northfelt D, Stewart S. DOXIL (pegylated liposomal doxorubicin) as first-line therapy of AIDS-related Kaposi's sarcoma (KS): integrated efficacy and safety results from two comparative trials. Fourth Conf Retro and Opportun Infect 1997; 200 (abstract no. 736).

64. Symon Z, Peyser A, et al. Selective delivery of doxorubicin to patients with breast carcinoma metastases by stealth liposomes. Cancer 1999; 86(1):72–78.

65. Harrington KJ, et al. Effective targeting of solid tumors in patients with locally advanced cancers by radiolabeled pegylated liposomes. Clin Cancer Res 2001; 7(2):243–254.

66. Koukourakis MI, Koukouraki S, et al. High intratumoral accumulation of stealth liposomal doxorubicin in sarcomas—rationale for combination with radiotherapy. Acta Oncol 2000; 39(2):207–211.

67. Schueller J, C MMC, et al. Serum and tissue kinetics of liposomal encapsulated doxorubicin (Caelyx) in advanced breast cancer patient. Proceedings of the 24th Annual San Antonio Breast Cancer Symposium, 2001; Abstract 436.

68. Koukourakis MI, Koukouraki S, et al. High intratumoural accumulation of stealth liposomal doxorubicin (Caelyx) in glioblastomas and in metastatic brain tumours. Br J Cancer 2000; 83(10):1281–1286.

69. Mayer LD, Hope MJ, Cullis PR. Vesicles of variable sizes produced by a rapid extrusion procedure. Biochim Biophys Acta 1986; 858(1):161–168.

70. Vaage J, Barbera-Guillem E, Abra R, et al. Tissue distribution and therapeutic effect of intravenous free or encapsulated liposomal doxorubicin on human prostate carcinoma xenografts. Cancer 1994; 73:1478–1484.

71. Amantea M, Forrest A, Northfelt D, et al. Population pharmacokinetics and pharmacodynamics of pegylated-liposomal doxorubicin in patients with AIDS-related Kaposi's sarcoma. Am Soc Clin Pharmacol Ther 1997; 61:301–311.

72. Gbizon A, Shmeeda H, et al. Pharmacokinetics of pegylated liposomal Doxorubicin: review of animal and human studies. Clin Pharmacokinet 2003; 42(5):419–436.

73. McGuire S. Saquinavir gets accelerated approval. Posit Aware 1996; 7(1):8.

74. Majchrowicz M. DOX-SL: KS treatment in limbo. PI Perspect; 1995(16):13–14.

75. Baker R. Early approval for two lipid-based drugs. Beta 1995:4.

76. Berry G, Billingham M, Alderman E, et al. The use of cardiac biopsy to demonstrate reduced cardiotoxicity in AIDS Kaposi's sarcoma patients treated with pegylated liposomal doxorubicin. Annal Oncol 1998; 9:711–716.

77. Northfelt DW, Dezube B, Thommes J, et al. Randomized comparative trial of Doxil vs. adriamycin, bleomycin, and vincristine (ABV) in the treatment of severe AIDS-related Kaposi's sarcoma (AIDS-KS). 3rd Conf Retro and Opportun Infect 1996:123.

78. Stewart S, Jablonowski H, Goebel FD, et al. Randomized comparative trial of pegylated liposomal doxorubicin versus bleomycin and vincristine in the treatment of AIDS-related Kaposi's sarcoma. International Pegylated Liposomal Doxorubicin Study Group. J Clin Oncol 1998; 16(2):683–691.

79. Mitsuyasu R, yon Roenn J, Krown S, et al. Comparison study of liposomal doxorubicin (DOX) alone or with bleomycin and vincristine (DBV) for treatment of advanced AIDS-associated Kaposi's sarcoma (AIDS-KS): AIDS Clinical Trial Group (ACTG) protocol 286 (Meeting abstract). Proc Annu Meet Am Soc Clin Oncol 1997; 16.

80. Muggia FM, Hainsworth JD, Jeffers S, et al. Phase II study of liposomal doxorubicin in refractory ovarian cancer: antitumor activity and toxicity modification by liposomal encapsulation. J Clin Oncol 1997; 15(3):987–993.

81. Uziely B, Jeffers S, Isacson R, et al. Liposomal doxorubicin: antitumor activity and unique toxicities during two complementary phase I studies. J Clin Oncol 1995; 13(7):1777–1785.

82. Gordon AN, et al. Phase II study of liposomal doxorubicin in platinum- and paclitaxel-refractory epithelial ovarian cancer. J Clin Oncol 2000; 18(17):3093–3100.

83. Gordon A, et al. Interim analysis of a Phase III randomized trial of Doxil/Caelyx (D) versus Topotecan (T) in the treatment of patients with relapsed ovarian cancer. Proc Am Soc Clin Oncol 2000; 19:abst 504, 129a.

84. Ranson MR, Carmichael J, O'Byrne K, et al. Treatment of advanced breast cancer with sterically stabilized liposomal doxorubicin: results of a multicenter phase II trial. J Clin Oncol 1997; 15(10):3185–3191.

85. Wigler N, Inbar M, et al. Reduced cardiac toxicity and comparable efficacy in a phase III trial of pegylated liposomal doxorubicin (Caelyx/Doxil) vs. doxorubicin for first-line treatment of metastatic breast cancer. Proc Am Soc Clin Oncol 2002; 21:abs. 177.

86. Haran G, Cohen R, Bar LK, et al. Transmembrane ammonium sulfate gradients in liposomes produce efficient and stable entrapment of amphipathic weak bases [published erratum appears in Biochim Biophys Acta 1994; Feb 23;1190(1):197]. Biochim Biophys Acta 1993; 1151(2):201–215.

87. Druckmann S, Gabizon A, Barenholz Y. Separation of liposome-associated doxorubicin from non-liposome-associated doxorubicin in human plasma: implications for pharmacokinetic studies. Biochim Biophys Acta 1989; 980(3):381–384.

15

Case Study: AmBisome—A Developmental Case Study of a Liposomal Formulation of the Antifungal Agent Amphotericin B

JILL P. ADLER-MOORE

Department of Biological Sciences, California State Polytechnic University–Pomona, Pomona, California, U.S.A.

RICHARD T. PROFFITT

RichPro Associates, Lincoln, California, U.S.A.

1. DEFINITION OF THE PROBLEM

Systemic fungal infections have long been recognized as a significant cause of death in immunocompromised patients. As early as 1964, Hutter et al. (1) found that aspergillosis was a contributing factor to the cause of death in 37% of cancer patients in a single institution. Shortly thereafter, Bodey (2) reported an increasing incidence of fungal infections in acute leukemia patients at the National Cancer Institute between

481

the years 1959 and 1965. Similarly, Maksymiuk et al. (3) summarized the incidence of fungal infections at the M.D. Anderson Hospital from 1976 to 1980. Overall, there were 233 patients with documented fungal infections out of 27,681 total registrants. Thus, the overall rate of fungal infections was 8.4 per 1000 patients, with the highest frequency of fungal infections (14.7%) occurring in acute leukemia patients.

In addition to cancer patients, surgical patients are also at increased risk for contracting systemic fungal infections because of many factors including antibacterial therapy, corticosteroid therapy, indwelling catheters, total parenteral nutrition, and length of stay in intensive care units (4). Cornwell et al. (5) found that 5.7% of patients in the surgical intensive care unit (SICU) of a large teaching hospital were culture-positive for fungi, of which 43% had systemic fungal infections. Furthermore, these investigators reported that the mortality for SICU patients with positive fungal cultures was more than twice that of patients who were culture-negative (23.3% vs. 10.5%, $p < 0.001$).

With the increasing numbers of bone marrow and solid organ transplantations, the AIDS epidemic and the intensive use of antineoplastic therapies, the number of systemic fungal infections reported in immunocompromised patients has continued to grow (6). Wald et al. (7) reported a doubling of the incidence of invasive aspergillosis over a six-and-a-half year period in a major bone marrow transplantation unit. Allogeneic transplant patients over 40 years of age undergoing corticosteroid therapy for graft-vs.-host disease were at highest risk for infection. In the AIDS setting, *Candida albicans* infections of mucosal membranes (esophageal and oropharyngeal candidiasis) and cryptococcal meningitis occur in a high percentage of patients (8). The advent of highly active antiretroviral combination therapies, however, has greatly reduced the incidence of opportunistic fungal infections in AIDS patients. In contrast, the trend toward more aggressive antileukemic regimens has led to longer periods of neutropenia and increased levels of intestinal mucosal damage. These variables are positively associated with an increased incidence of invasive fungal infections in this population (9).

In recent years, there has been an alarming increase in the rate of fungal infections in hospitalized patients, as well as an increase in the death rate of such patients. For example, Fisher-Hoch and Hutwagner (10) found that the rate of disseminated *Candida* infections increased 11-fold over the period of 1980–1989. Likewise, the rate of invasive Aspergillosis has increased substantially, and is now a major cause of death at leukemia treatment and bone marrow transplantation centers (11). This situation has been exacerbated by the recent identification of less common fungi as the causative agents of these infections including *Fusarium* sp. (12), non-*Candida albicans* spp. (13), *Trichosporon* spp. (14), and *Penicillium marneffei* (15). In general, these fungi are less susceptible than the more common pathogen, *C. albicans*, to the frequently used antifungal azoles, and polyenes (16).

The fungistatic azoles commonly used to treat systemic fungal infections include fluconazole for *Candida* and *Cryptococcus* infections (17) and itraconazole primarily for *Candida, Aspergillus* and *Histoplasma* infections (18,19). Voriconazole is a second generation triazole that was recently approved as primary treatment of acute invasive aspergillosis (20). Voriconazole has been reported to be 4- to 16-fold more active than fluconazole and two- to eightfold more active than itraconazole against *Candida* spp., including *Candida krusei* and *Candida glabrata* (21,22). In addition, voriconazole has shown good activity against *Cryptococcus neoformans* (21), dimorphic fungi such as *Histoplasma capsulatum*, and emerging pathogens such as *Fusarium* spp., dematiaceous molds, and *P. marneffei* (22,23). These three azole drugs are minimally toxic and can be administered either orally or intravenously, but azoles are potent inhibitors of the cytochrome P450 (CYP) 3A4 oxidizers and can give rise to serious drug–drug interactions with other drugs, such as cyclosporin A, which is metabolized by the CYP3A4 system (24,25).

The fungicidal polyene amphotericin B has been extensively used to treat a number of systemic fungal infections, including those caused by yeast (*Candida* and *Cryptococcus*), mycelia (*Aspergillus, Mucor, Rhizopus*), and dimorphic fungi (*H. capsulatum, Coccidioides immitis, Blastomyces dermatitidis, Paracoccidioides brasiliensis*) (26). Thus, amphotericin B has a broad spectrum of fungicidal activity and is administered intravenously as a colloidal dispersion in sodium deoxycholate. In this formulation, amphotericin B at therapeutic doses may also cause a multitude of severe, acute toxicities including fever, chills, rigors, nausea, vomiting, and hypokalemia which require that it be administered as a 4–6 hr intravenous infusion (27). Its acute toxic side effects can lead to modifications of treatment regimen and termination of treatment. Amphotericin B can also cause transient or permanent nephrotoxicity which limits the total amount of drug that may be administered. This toxicity is of primary importance in immunocompromised patients who are receiving other nephrotoxic, immunosuppressive and antimicrobial agents to manage their disease (17).

Despite these limitations, amphotericin B has for many years been considered the drug of choice for treating life-threatening systemic fungal infections (27). For this reason, in the past 20 years there have been many attempts to develop amphotericin B formulations that circumvent these toxicity problems, while maintaining the drug's potent antifungal activity. One of the solutions to this problem has been the development of the liposome formulation known as AmBisome®, which is the topic of this case study.

2. OVERVIEW OF AMPHOTERICIN B/LIPID FORMULATIONS

Since amphotericin B is amphiphilic and insoluble, it readily associates with lipophilic molecules and the resulting lipid/amphotericin B preparations are less toxic when administered in vivo (28). This approach has been used to develop various less toxic formulations of amphotericin B including amphotericin B/intralipid mixtures, Abelcet®, Amphotec® and AmBisome®. The extent of this decrease in toxicity varies with the type of lipid/drug associations formed.

Intralipid is made up of soybean oil, egg yolk phospholipids, and glycerin. This combination of lipids has been mixed in varying ratios with the colloidal dispersion of amphotericin B to form a drug suspension. Clinical investigators have reported some reduction in infusion-related toxicities with the intralipid/amphotericin B combination (29). Other lipids, such as dimyristoyl phosphatidylglycerol and dimyristoyl phosphatidylcholine, can be complexed with amphotericin B in a 10:7:3 molar ratio to form the preparation referred to as Abelcet (30). The large (1–6 μm), ribbon-like complexes characteristic of Abelcet produce less nephrotoxicity than treatment with conventional amphotericin B (31). Other investigators have complexed amphotericin B with cholesteryl sulfate in a molar ratio of 1:1 to form 122 nm discs referred to as Amphotec (32). When injected intravenously, these disc-like structures also reduced the nephrotoxicity associated with amphotericin B treatment (33).

Further reductions in both acute and chronic toxicity, as well as improved amphotericin B pharmacokinetics, were achieved when amphotericin B was incorporated into the small unilamellar liposome formulation known as AmBisome (34). These liposomes are less than 100 nm in diameter, and are less nephrotoxic than amphotericin B (35) or Abelcet (36) producing significantly fewer infusion-related adverse events. Unlike non-liposomal amphotericin B, AmBisome avoids rapid uptake by the liver and spleen characteristic of the other lipid/drug formulations (37) and thus circulates for extended periods of time in the blood stream (8.6–10.4 hr) (38). Due to the small size of the liposomes and their extended half-life in plasma, AmBisome readily penetrates into sites of inflammation and infection (39).

3. DEVELOPMENT OF AMBISOME FORMULATION

3.1. Formulation Criteria

The development of AmBisome proceeded in several stages beginning with a list of criteria that needed to be met to initially formulate the liposomes. These criteria included the following:

- quantitative association of the drug with the liposome;
- liposome size compatible with sterile filtration and suitable for IV administration;

- >30-fold reduction in murine intravenous LD_{50} compared to conventional drug;
- physical and chemical stability to justify a 1–2 year shelf-life;
- long-term blood circulation;
- in vivo antifungal efficacy comparable to, or better, than non-liposomal drug.

The first challenge was to achieve a stable, long-term association of the drug with the liposome. Since the lipid to drug ratio in a formulation can alter the amount of drug incorporated into the liposome, many liposome formulations with varying lipid to drug ratios were prepared and tested. Another goal of the project was to improve the pharmacokinetics of the drug. As a result, the liposomes had to be made less than 100 nm in size to ensure their long-term circulation in the blood and delayed uptake by the reticuloendothelial system (39,40). At this stage of development, probe sonication was used to make the laboratory scale (1–20 mL) preparations while selecting for the appropriate lipid composition.

3.2. Selection of Lipids

The choice of lipids focused on inclusion of cholesterol as well as phospholipids with saturated long-chain fatty acids [dipalmitoyl phosphatidylcholine (DPPC) and distearoyl phosphatidylcholine (DSPC)]. These phospholipids have gel-to-liquid crystalline transition temperatures above 37°C, which ensures liposome integrity after intravenous injection. Also, cholesterol was included in the formulations studied because amphotericin B has an affinity for cholesterol, but has a higher binding affinity for ergosterol, the sterol in fungal cell membranes (41). This association with cholesterol would favor the drug remaining with the liposome until it came into contact with fungi, thus minimizing the drug interaction with non-target tissues. The addition of distearoyl phosphatidylglycerol (DSPG) to the lipid composition provided a negatively charged moiety which, at physiological pH, would bind with the positively charged aminosaccharide group of the amphotericin B molecule, further ensuring drug association with the liposome bilayer. Hydrogenated, rather than non-hydrogenated, phospholipids were also used because of the chemical stability associated with saturated hydrocarbon side chains and the increased physical stability of liposomes prepared from saturated phospholipids. Selection of acidic pH, low ionic strength, and the presence of sucrose in the hydration buffer promoted a stable drug/lipid association and avoided the problems of liposome aggregation.

3.3. Toxicity Screen

When it was determined that a particular lipid composition had entrapped more than 50% of the drug, and the liposome was less than 100 nm in diameter, the murine intravenous LD_{50} in C57BL/6 mice was assessed for this preparation. C57Bl/6 mice were selected for the screening step because of their marked in vivo sensitivity to the toxic effects of amphotericin B (42). This approach ultimately resulted in the selection of an optimized liposome formulation consisting of hydrogenated soy phosphatidylcholine (HSPC), cholesterol, DSPG, and amphotericin B in the molar ratio of 2:1:0.8:0.4 hydrated in a 10 mM succinate, 9% sucrose buffer. The acute intravenous murine LD_{50} of this formulation was greater than 50 mg/kg (43), a 20-fold increase in the safety margin compared to amphotericin B with sodium deoxycholate.

At this juncture, it was necessary to scale-up the production of the amphotericin B liposomes. A micro-emulsification technique was selected since it produced liposomes similar to those made by probe sonication. This procedure could be scaled-up to 100 L batches (44). The micro-emulsification process produced small liposomes with less toxicity (acute intravenous murine $LD_{50} > 125$ mg/kg) and a more consistently homogeneous size distribution compared to probe sonicated liposomes (45–80 nm).

3.4. Raw Materials

Another problem which had to be addressed was making available large amounts of raw material that met the precise specifications needed to make the liposomes reproducibly. Chemical, physical, and biological assays had to be developed by the supplier and by the company to ensure consistent quality for the material from batch to batch. Since the demand for large quantities of very pure phospholipids was unique in the industry, the suppliers and the company had to work together to develop the standards by which these chemicals had to be manufactured and assessed.

To increase the shelf-life of the liposomes, conditions were defined for the lyophilization of the micro-emulsified product. Following refinement of these procedures, the stability of the physical, chemical, and biological properties of the lyophilized product was assured. The rehydrated product had the same properties as the original liposome dispersion. At present, the shelf-life of the product is 3 years.

3.5. Pre-Clinical In Vitro Efficacy and Toxicity Testing

Once the liposome formulation had been optimized, in vitro efficacy studies had to be conducted to determine the range of antifungal activity of this formulation. In general, the in vitro activity of AmBisome and Fungizone (sodium deoxycholate formulation of amphotericin B) was similar against yeast and molds using standard 24 or 48 hr incubations (45). Studies by Anaissie et al. (46) showed that the in vitro antifungal activities of AmBisome and Fungizone were similar for over 100 strains of *Candida*, *Cryptococcus* and *Aspergillus*. However, the AmBisome MIC for other strains of *Candida* (47) and the yeast form of one strain of *P. brasiliensis* (48) were higher than that of Fungizone.

In vitro cytotoxicity studies were used to help establish safe dosing regimens. When human red blood cells were incubated for 2 hr with AmBisome at concentrations up to $100 \mu g/mL$, there was only 5% hemolysis. In comparison, Fungizone at concentrations as low as $1 \mu g/mL$ caused 92% hemolysis (34). This rapid and extensive damage to red blood cells by Fungizone, but not AmBisome, could also be correlated with potassium release from rat red blood cells incubated with these drugs (49). Similar reductions in toxicity have been reported in other cell types, such as primary (Langerhans cells) and established (canine kidney and murine macrophage) cell lines, using electron microscopic (50) and viability assays (51), respectively.

3.6. Pre-Clinical In Vivo Toxicity Testing

The reduced in vivo toxicity of AmBisome compared to non-liposomal amphotericin B was demonstrated in a variety of single and repeated dosing studies in mice, rats, and dogs. Fungizone had an LD_{50} of 2.3 mg/kg whereas reconstituted AmBisome had a single dose LD_{50} in C57BL/6 mice that was above 125 mg/kg (43). Repeated dosing toxicity of AmBisome was tested at 1, 3, 9, and 20 mg/kg/day for 30 days in rats (52), and 1, 4, 8, and 16 mg/kg in dogs for 30 days (53). In the rat study, 12 of 25 female Sprague–Dawley Crl:CD (SD) BR rats from Charles River receiving 20 mg/kg died or were sacrificed after two doses due to acute liver toxicity. The remaining female rats and all of the male rats given 20 mg/kg AmBisome survived to the end of the study (day 30), but had significantly lower weight gain than control rats ($p < 0.01$). Some nephrotoxicity for AmBisome was confirmed by a significant dose-related increase in BUN ($p < 0.01$) for both sexes, but without concomitant rises in serum creatinine. In dogs, blood chemistries of animals receiving 4 mg/ kg AmBisome were compared with published values for dogs receiving 0.6 mg/kg Fungizone (54). Both AST and ALT levels were normal for the AmBisome group, but about 7- and 21-fold higher, respectively, for the Fungizone comparison group. Also, BUN levels were about four times higher for dogs given 0.6 mg/kg Fungizone (229 mg/dL) compared with those given 4 mg/kg AmBisome (58 mg/dL), and creatinine levels were more than 2.5 times higher for Fungizone-treated animals (6.0 mg/dL) compared with those receiving AmBisome (2.3 mg/dL). Thus, in dogs repeated treatments with AmBisome resulted in only mild effects on the kidney while liver enzymes remained within normal ranges.

3.7. Pharmacokinetic Testing

Having established the toxicity profile for AmBisome in several different animal species, biodistribution studies were conducted to optimize the therapeutic dosing of this product. Like other long-circulating, stable liposomes (55), the AmBisome pharmacokinetic studies showed high peak plasma levels, high plasma AUC, and extended plasma elimination (see Table 1). AmBisome also demonstrated a saturable, non-linear distribution from the plasma, most likely due to uptake by the reticuloendothelial system. Thus, when the dose was increased from 1 to 5 mg/kg, the volume of distribution (V_d) and the total clearance (Cl) decreased.

Tissue distribution studies in uninfected rats with 1 month dosing at 1, 3, and 5 mg/kg AmBisome or Fungizone (1 mg/kg) were conducted to compare drug localization over time. At 1 mg/kg most of the AmBisome accumulated in the liver and spleen, showing levels 2–3 times higher than the comparable dose of Fungizone. In contrast, kidney

Table 1 Mean Single Dose Pharmacokinetics of AmBisome in Animals Following Intravenous Administration

Species	n (group)	Dose (mg/kg)	C_{max} (μg/mL)	$AUC_{0-\infty}$ (μg h/mL)	$t_{1/2}$ (hr)	V_d (L/kg)	Cl (mL/hr/kg)
Mouse	3	1	8	36	17	0.68	28
		5	50	1080	24	0.16	4.6
Rat	6	1	7.2	64	9.5	0.21	16
		3	30.3	374	7.9	0.10	8.4
		9	141	1140	8.0	0.10	8.0
		20	235	1810	13.6	0.20	12.6
Rabbit	3–4	1.0	26	60	3.6	0.09	16
		2.5	53	210	5.2	0.09	12
		5	132	840	5.5	0.05	5.3
		10	287	2220	7.7	0.05	4.2
Dog	10	0.25	0.21	2.6	7.0	1.66	110
		1	1.9	11	9.3	0.96	79
		4	18	164	8.4	0.29	26
		8	72	990	11.0	0.14	10
		16	174	2600	11.6	0.07	6.0

and lung drug levels were 6 and 2.5 times lower, respectively. At a dose of 5 mg/kg AmBisome, the relative percent uptake by the liver and spleen decreased, and there was redistribution of the drug into other organs including the kidneys and lungs (43). Shown in Figs. 1 and 2 are kidney and lung tissue concentrations of amphotericin B determined up to 28 days after dosing was discontinued. In both tissues, higher drug levels persisted after treatment with 5 mg/kg AmBisome vs. 1 mg/kgFungizone suggesting that AmBisome could be utilized in prophylactic regimens (56).

Brain accumulation of the drug in uninfected rats was lower than for the other organs, but AmBisome treatment at 5 mg/kg produced at least twofold higher brain levels than were achieved with Fungizone at 1 mg/kg (43). In comparison, when uninfected or Candida infected rabbits were given

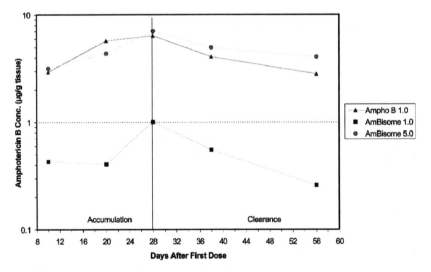

Figure 1 Harlan–Sprague–Dawley rats (female) were given daily intravenous treatments for 28 days with either amphotericin B as Fungizone (1 mg/kg) or AmBisome (1 or 5 mg/kg). At various time points during (drug accumulation days 10, 20, and 28) and after (drug clearance days 38 and 56) treatment, kidney tissues were removed and assayed by HPLC for the concentration of amphotericin B per gram tissue.

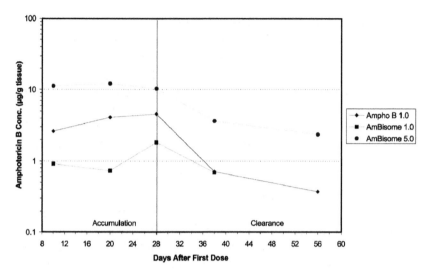

Figure 2 Harlan–Sprague–Dawley rats (female) were given daily intravenous treatments for 28 days with either amphotericin B as Fungizone (1 mg/kg) or AmBisome (1 or 5 mg/kg). At various time points during (drug accumulation days 10, 20, and 28) and after (drug clearance days 38 and 56) treatment, lung tissues were removed and assayed by HPLC for the concentration of amphotericin B per gram tissue.

5 mg/kg of AmBisome, Abelcet, or Amphotec, brain levels of drug achieved with AmBisome were 4–7 times higher than with any of the other formulations (57).

3.8. Pre-Clinical In Vivo Efficacy Testing

With this background of information on the toxicity and pharmacokinetics of AmBisome, pre-clinical efficacy studies were expanded to include testing of AmBisome for the treatment of both extracellular and intracellular fungal infections in immunocompetent and immunosuppressed animals. The liposomal form of the drug was found to be very effective (Table 2). The models of infections utilized included systemic candidiasis (56,58,59,61), pulmonary and systemic aspergillosis (62–64), mucormycosis (65), meningeal cryptococcosis (66), pulmonary blastomycosis (67), coccidioidomycosis (68), histo-plasmosis (69), pulmonary paracoccidoidomycosis (48), tri-chosporonosis (70), and leishmaniasis (71,72). In all models, AmBisome could be administered at much higher daily (3–30 mg/kg) and total doses of amphotericin B compared with Fungizone because it was much less toxic. As a result, fungi were completely eradicated from infected tissues in up to 80% of the animals with infections including blastomycosis, paracoccidoidomycosis, cryptococcal meningitis, and candidia-sis. Of particular relevance is the capability of AmBisome to eliminate (57) or markedly reduce (66,68) fungal infection in the brain. In some models, equal doses of Fungizone and AmBisome (1 mg/kg or less) produced comparable efficacy, e.g., treatment of infections due to *C. albicans* (58), *C. neofor-mans* (58), *C. immitis* (73), *Leishmania donovani* (72), and *H. capsulatum* (69).

Since the pharmacokinetic data indicated that following intravenous administration, AmBisome would remain in the tissues for several weeks, prophylactic pre-clinical efficacy stu-dies were also conducted. AmBisome's persistence in the tissue was associated with continued bioavailability and persistent antifungal effect. Thus, after a single intravenous prophylactic dose of AmBisome at 5, 10, or 20 mg/kg, immunocompetent or immunosuppressed mice were protected from a subsequent lethal challenge with *C. albicans* (56) or *H. capsulatum* (56). Also, prophylactic treatment with 6.05 mg/kg AmBisome administered by the aerosol route over 3 days prevented infection with *Aspergillus fumigatus* in immunosuppressed mice (64).

3.9. Mode of Action

The pre-clinical efficacy studies showed that AmBisome could be used to treat infections in many organs including the brain, spleen, liver, kidneys, and lungs. To elucidate AmBisome's antimicrobial mode of action, additional in vivo and in vitro studies were necessary. To follow the in vivo distribution of AmBisome, liposomes were prepared with the fluorescent dye

Table 2 Overview of In Vivo Pharmacology Studies Comparing Amphotericin B with AmBisome

Test species	Micro-organism	Dose regimen	Amphotericin B Dose (mg/kg)	Amphotericin B Major findings	AmBisome Dose (mg/kg)	AmBisome Major findings	Refs.
Mouse C57BL/6J, female ($n = 5$)	*C. albicans*, strain CSPU 39	Multi: 5 cycles	0.7	80% survival at 21 days. 1.0×10^3 CFU/mg kidney tissue ($p < 0.05$)	2.5, 5.0, 10.0	100% survival all doses. Dose related kidney clearance of yeast: 68 CFU/mg at 2.5 ($p < 0.05$); 16 CFU/mg at 5.0 ($p < 0.05$); 9 CFU/mg at 10.0 ($p < 0.05$)	58
Mouse BALB/C; female, Immuno-competent ($n = 10$)	*C. albicans*, ATCC 44858	Multi:5 cycles	0.4	100% survival at 14 days. Significant reduction in CFU/mg of kidney tissue ($p < 0.001$).Prevented relapse	0.4, 7.0	90–100% survival at 14 days. Significant reduction in CFU/mg of kidney tissue ($p < 0.001$). Prevented relapse	59
Mouse, BALB/C; female, leukopenic ($n = 10$)	*C. albicans*, ATCC 44858	Multi: 5 cycles	0.3	100% survival, significant reductions in CFU in liver, spleen and lungs ($p < 0.001$). Growth inhibited only in kidneys. Infection relapsed	0.3 7.0	50% survival 100% survival. Reduction in CFU/mg of kidney ($p < 0.001$). Prevented relapse	59
Mouse, C57BL/6, female, immuno-competent ($n = 5$)	*C. albicans*, strain CSPU 39	Single Prophy-lactic day 7 pre-challenge	1.0	Log CFU = 5.2 at day 7 post-challenge	1, 5, 10 or 20	All doses reduced fungal burden ($p < 0.05$); doses of 5, 10, and 20mg/kg resulted in less weight loss. At 1mg/kg, log CFU = 4.6 at day 7 post-challenge	56

Animal model	Organism, strain	Dosing schedule	Dose	Outcome	Dose (mg/kg)	Results	Ref.
Mouse, C57BL/6, female immunosuppressed ($n = 5$)	*C. albicans,* strain CSPU 39	Single Prophylactic day 7 pre-challenge.	1.0	Log CFU = 4.2 at day 7 post-challenge	1, 5, or 20	All doses reduced fungal burden ($p < 0.05$); doses of 5 or 20 mg/kg resulted in less weight loss. At 1.0 and 5.0 mg/kg, CFU = 3.2 and CFU = 3.5 at day 7 post-challenge	61
Mouse, CF1, male, neutropenic ($n = 20$)	*Candida lusitaniae,* strain 1706	Multi: 3 cycles	1	Significantly prolonged survival ($p < 0.05$); significant reduction in CFU/g kidney tissue ($p < 0.05$)	10, 30	Significantly prolonged survival ($p < 0.05$); significant reduction in CFU/g kidney tissue ($p < 0.05$)	
	C. lusitaniae, strain 524	Multi: 3 cycles	1	Significantly prolonged survival ($p < 0.05$); significant reduction in CFU/g kidney tissue ($p < 0.05$)	10, 15	Significantly prolonged survival ($p < 0.05$); significant reduction in CFU/g kidney tissue ($p < 0.05$)	
	C. lusitaniae, strain 2819	Multi: 3 cycles	1	Survival not significantly prolonged; no significant reduction in CFU/g kidney tissue	10, 30	Significantly reduced CFU/g kidney tissue ($p < 0.05$)	
	C. lusitaniae, strain 5W31 (resistant)	Multi: 3 cycles	1	Survival not significantly prolonged; no significant reduction in CFU/g kidney tissue	10, 30	At 10 mg/kg survival not prolonged; CFU/g kidney tissue not reduced. At 30 mg/kg survival not prolonged; significant reduction in CFU/g kidney tissue ($p < 0.05$)	

(Continued)

Table 2 Overview of In Vivo Pharmacology Studies Comparing Amphotericin B with AmBisome (*Continued*)

Test species	Micro-organism	Dose regimen	Amphotericin B		AmBisome		Refs.
			Dose (mg/kg)	Major findings	Dose (mg/kg)	Major findings	
Rabbit, New Zealand White; granulo-cytopenic ($n = 5$–18)	*Aspergillus fumigatus*, isolate 4215 (intratracheal)	Multi: 10 cycles	1	30% survival at day 10 ($p = 0.1$). Approximately 15-fold reduction in mean number of CFU ($p < 0.001$)	1.0	80–100% survival at day 10 ($p < 0.01$). Approximately eight-fold reduction in mean number of CFU at 1.0 mg/kg ($p < 0.01$)	62
					5.0, 10.0	Approximately 15-fold reduction in mean number of CFU at 5.0 and 10.0 mg/kg ($p < 0.001$)	
Rat, strain R, female, granulo-cytopenic ($n = 15$)	*A. fumigatus*, clinical isolate (left lung intubation)	Multi: 10 cycles	1	Increased survival to 13% ($p = 0.006$ vs. controls). No effect on dissemination to the right lung. Reduced dissemination of infection to liver and spleen in 39% of animals vs. controls (NS)	1	No increase in survival vs. untreated controls. Reduced dissemination of infection to right lung by 33% ($p < 0.01$), and to the liver and spleen in 59% of animals ($p < 0.01$)	63
					10	Increased survival to 27% ($p = 0.027$ vs. controls). Reduced dissemination	

Animal model	Organism	Regimen	Dose	Results	Dose	Results	Ref.
Mouse, ICR, female, immuno-suppressed ($n = 10$)	*A. fumigatus*, ATCC 13073 (intranasal)	Multi 3 cycles, aerosol prior to challenge	6.73 (total dose)	100% survival at day 9 post-challenge. 2 log reduction in lung CFU/gm, 0% of animals cleared of lung infection	6.05 (total dose)	of infection to right lung by 33% ($p < 0.01$). Completely prevented dissemination of infection to liver and spleen in 100% of animals. 100% survival at day 9 post-challenge. 3 log reduction in lung CFU/gm, 80% of animals cleared of lung infection	64
Mouse, BALB/c male, diabetic ($n = 19$–30)	*Rhizopus oryzae*, 99–880 (IV)	Multi 4 cycles	0.75 b.i.d.	No significant improvement in survival	2.5 b.i.d.; 7.5 b.i.d.	No significant mprovement in survival Significant improvement in median survival ($p = 0.01$) and total survival ($p = 0.001$)	65
Mouse, ICR ($n = 9$–10)	*C. neoformans*, clinical isolate 89–98 (intracranial)	Multi: 7 cycles	0.3 (IV), 1.0 (IV), 3.0 (IP), 7.0 (IP)	Survival significantly prolonged ($p < 0.05$). Survival significantly prolonged ($p < 0.05$). Significant reduction in brain CFU ($p < 0.05$)	0.3 (IV), 3.0, 7.0 (IV)	Survival was not prolonged. Survival significantly prolonged ($p < 0.05$). Significant reduction in brain CFU ($p < 0.05$)	66
					1.0, 20, 30 (all doses IV)	Significant reduction in brain CFU ($p < 0.05$). 44% of 20 mg/kg group and 78% of 30 mg/kg group were culture-negative on day 30.	

(Continued)

Table 2 Overview of In Vivo Pharmacology Studies Comparing Amphotericin B with AmBisome (*Continued*)

Test species	Micro-organism	Dose regimen	Amphotericin B		AmBisome		Refs.
			Dose (mg/kg)	Major findings	Dose (mg/kg)	Major findings	
Mouse, CD-1, male (n = 10)	*Blastomyces dermatitidis*, ATCC 26199 (intranasal)	Multi: 6 cycles	0.6	100% survival ($p < 0.001$). Significant reduction in CFU/lung ($p < 0.001$). No animal cleared of infection	1.0, 3.0, 7.5, 15.0	90–100% survival ($p < 0.001$); significant dose-dependent reduction in CFU/lung ($p < 0.001$); 70–80% cleared of infection at top two dosages	67
Rabbit, New Zealand white, (n = 8–10)	*Coccidioides immitis*, strain: Salviera (intracisternal)	Multi: 9 Cycles	1.0	100% survival; no animals cleared of infection ($p < 0.05$ vs. control). About 1.0 log reduction in CFU in brain	7.5, 15.0, 22.5	100% survival in all groups ($p < 0.05$ vs. control). About 1.9–2.7 log reduction in CFU in brain	68
Mouse, athymic *nu/nu* (n = 10–20)	*Histoplasma capsulatum*, isolate 93–255	Multi: 6 Cycles	0.3, 0.6, 1.0 (IV), 3.0 (IP)	Survival prolonged significantly ($p < 0.005$); Survival prolonged significantly ($p < 0.001$)	0.3, 0.6 (IV), 1.0 (IV), 3.0 (IV)	Survival prolonged significantly ($p < 0.005$). Survival prolonged significantly ($p < 0.001$); more effective than 1.0 mg/kg amphotericin B ($p < 0.02$). Survival prolonged significantly ($p < 0.001$).	

Animal model	Infection	Regimen	Dose	Result	Dose	Result	Ref.
Mouse, male BALB/c (n = 14–15)	*Paracoccidioides brasiliensis*, isolate Gar (intranasal)	Multi: 6 cycles	0.6	47% survival ($p < 0.05$) at day 40	0.6; 5.0, 15.0, 30.0	7% survival ($p < 0.05$) at day 40; 67–86% survival ($p < 0.05$–0.0001); significant reduction in CFU/lung ($p < 0.01$–0.001)	48
Mouse, CF1, male, immunosuppressed (n = 20)	*Trichosporon beigelii*, isolate 009 (resistant)	Multi: 10 cycles	1 (IP)	No significant reduction of CFU/g of kidney tissue	1, 5; 10	No significant reduction of CFU/g of kidney tissue; Significant reduction ($p < 0.05$) of CFU/g of kidney tissue	70
	T. beigelii, isolate 008 (Partially resistant)	Multi: 10 cycles	1 (IP)	Significant reduction ($p < 0.05$) of CFU/g of kidney tissue	1, 5, 10	Significant dose-dependent reduction ($p < 0.05$) of CFU/g of kidney tissue for all dosages	
Mouse, female BALB/c, (n = 5)	*L. donovani*, strain MHOM/FR/91/LEM2259V	Multi: 6 cycles	0.8	1–2 log decrease in liver and spleen CFU; 2–3 log reduction in lung CFU	0.8; 5, 50	4–6 log decrease in liver and spleen CFU; complete clearance of lung CFU for 14 weeks; Complete clearance of liver, spleen, and lung CFU for 14 weeks	71; 72
Mouse, female BALB/c (n = 5)	*L. donovani*, MHOM/ET/67/L82	Single	0.04; 0.2; 1.0	5.3% inhibition of amastigotes in liver ($p = 0.26$). 22.0% inhibition of amastigotes in liver ($p = 0.016$). 52.7% inhibition of amastigotes in liver ($p = 0.0003$).	0.04; 0.2; 1.0; 5.0	15.8% inhibition of amastigotes in liver ($p = 0.11$); 41.2% inhibition of amastigotes in liver ($p = 0.001$); 84.5% inhibition of amastigotes in liver ($p < 0.0001$); 99.8% inhibition of amastigotes in liver ($p < 0.0001$)	

sulforhodamine entrapped inside the liposomes. *C. albicans* infected mice were treated with the fluorescently labeled form of AmBisome; controls included liposomes with fluorescent dye but lacking amphotericin B and unlabeled AmBisome. Infected kidneys were collected 17 hr after treatment, frozen, sectioned and examined for localization of red fluorescence, or fixed and stained with Gomori methenamine silver to detect fungi. The sections from the mice administered fluorescent liposomes showed bright fluorescence associated with sites of fungal infection, but the unmodified AmBisome showed only faint, diffuse autofluorescence throughout the tissue. Direct evidence of the interactions between the liposomes and the fungi was obtained from in vitro studies using the same type of fluorescent-labeled AmBisome and non-drug containing fluorescent-labeled liposomes. After incubation with the AmBisome, the fungal cells showed dye distribution throughout the fungal cytoplasm; these cells were all dead. In comparison, the fluorescent dye remained on the surface of viable fungal cells even after 24 hr of incubation when liposomes lacking drug were used (60). These results suggested that AmBisome could bind to the surface of the fungal cells, breakdown, and release their contents into the fungal cytoplasm. Visualization of liposomes, with and without drug, binding to the surface of the fungal cell wall was demonstrated using freeze-fracture analysis of *C. glabrata* incubated with the liposomes (Fig. 3).

The presence of liposomal lipids within the fungal cytoplasm after treatment of *C. glabrata* with AmBisome was detected by incorporating a small amount of gold-labeled phosphatidylethanolamine into the AmBisome or liposomes without drug. The results (Fig. 4) showed that lipid from the liposomes without drug could not penetrate into the fungal cytoplasm whereas the lipid from AmBisome could be seen throughout the cytoplasm. Delivery of amphotericin B into the fungal cytoplasm of cells incubated with AmBisome was visualized by reacting fungal thin sections with anti-amphotericin B antibody (generously provided by Dr. John Cleary, University of Mississippi) followed by treatment with immunogold-labeled anti-antibody (Fig. 5).

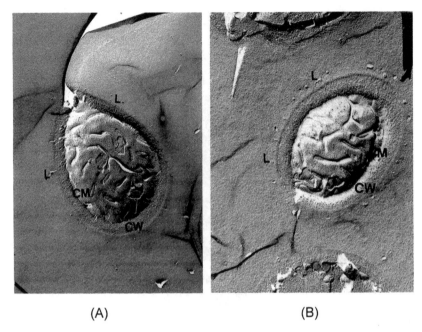

(A)   (B)

Figure 3 *Candida glabrata* was mixed with non-drug containing liposomes with the same lipid composition as AmBisome (A) or with AmBisome (B) and processed for freeze-fracture electron microscopy. With both types of liposome preparations, liposomes (L) were seen adhering to the outer surface of the cell wall (CW) of the yeast, and no intact liposomes were observed within the cell wall or at the cell membrane (CM). Size = 300 nm/0.5 cm. (Photographs courtesy of Kevin Franke, California State Polytechnic University, Pomona, CA.)

In summary, these data suggest that AmBisome can localize at infection sites, and interact directly with the fungus. It appears that following binding of the AmBisome to the fungal cell wall, drug-containing liposomes breakdown, and release their constituents which traverse the cell wall. Upon contact with the fungal cell membrane below the fungal cell wall, the released amphotericin B is able to bind to the ergosterol in the membrane for which it has a $10\times$ higher binding affinity than for the cholesterol in the liposome (41). This proposed mode of action may help explain

(A)

(B)

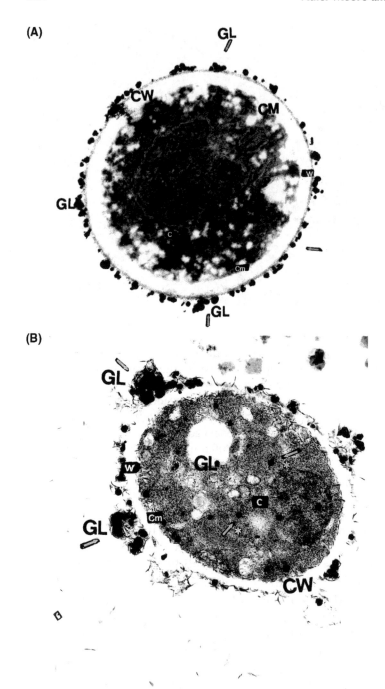

Figure 4   (*Caption on facing page*)

how circulating AmBisome which retains active drug within its lipid bilayer can provide potent antifungal efficacy at the infection site when it comes in contact with the fungus.

3.10. Clinical Testing

Approval of AmBisome for patient use required extensive clinical testing to determine if the compelling pre-clinical safety and efficacy data for the drug were applicable to the human population. This component of the drug development process took many years since the total number of confirmed fungal infections per year is limited by the difficulties involved in diagnosing fungal infections (74,75) and by the complexities of treating patients with fungal infections. The following is a summary of the many clinical trials that were conducted, and which eventually led to the approval of the drug in 48 countries including the United States. These studies and reports include initial compassionate use of the drug, salvage therapy, toxicological evaluation in the presence of other nephrotoxic drugs, prophylactic use, efficacy in different patient populations (including neutropenic, AIDS and

---

Figure 4 (*facing page*) *Candida glabrata* was incubated with gold-labeled lipid in non-drug containing liposomes with the same lipid composition as AmBisome (A) or with gold-labeled lipid in AmBisome (B). After incubation for 24 hr, the samples were fixed. The gold-labeled lipid was enhanced with silver, and the samples were processed for thin section electron microscopy. With the non-drug containing liposome preparation (A), gold-labeled lipid (GL) was seen adhering to the outer surface of the cell wall (w) of the yeast, with no lipid present within the cell wall, at the cell membrane (Cm), or in the cytoplasm (C). With the AmBisome (B), gold-labeled lipid (GL) was seen at the cell wall surface (w) at the cell membrane (Cm) and in the cytoplasm (C). Size = 300 nm/0.5 cm. (Photographs courtesy of Kevin Franke, California State Polytechnic University, Pomona, CA.)

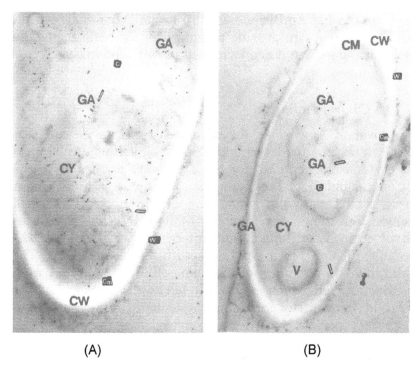

(A)   (B)

Figure 5 *Aspergillus fumigatus* was incubated with non-liposomal amphotericin B as Fungizone (A) or AmBisome (B). After incubation for 10 hr (Fungizone) or 14 hr (AmBisome), the samples were fixed, processed for thin section electron microscopy, sectioned and incubated with anti-amphotericin B antibody (generously provided by Dr. John Cleary of University of Mississippi), followed by incubation with gold-labeled anti-antibody. After incubation with Fungizone, some gold-labeled anti-antibody (GA) was seen adhering to the outer surface of the cell wall (CW) of the fungus, but most of the anti-antibody was present throughout the disrupted cytoplasm (CY) of the fungus. After 14 hr incubation with AmBisome, gold-labeled anti-antibody (GA) was seen at the cell wall surface (CW) and in the disrupted cytoplasm (CY) of the fungus. Size = 300 nm/0.5 cm. (Photographs courtesy of Kevin Franke, California State Polytechnic University, Pomona, CA.)

pediatric patients), safety and pharmacokinetic evaluation, and empirical use.

The first clinical use of AmBisome occurred in 1987 when a heart transplant patient developed pulmonary aspergillosis that could not be treated with conventional amphotericin B due to nephrotoxicity (76). After 34 days of treatment with AmBisome at 1 mg/kg/day, the infection was eradicated, and no evidence of recurrence was reported during a 16 month follow-up period. Also, during the treatment period, kidney function improved and there were no acute side effects such as fever and chills. At 16 months after the completion of therapy the patient was alive with no evidence of recurrent infection. This initial success provided substantial impetus for broader safety and efficacy testing of AmBisome.

3.10.1. Early Clinical Trials of AmBisome in Europe

The earliest safety data on AmBisome were reported by Meunier et al. (77). This multicenter study included 126 patients receiving 133 episodes of AmBisome treatment. The majority of these patients had failed previous conventional amphotericin B therapy due to toxicity or renal insufficiency. The mean duration of AmBisome administration was 21 days at an average dose of 2.1 mg/kg (range = 0.45–5.0 mg/kg). Hypokalemia was the most common side effect and was observed in 24 cases. Nausea, vomiting, fever, chills and rigors were observed in a total of five instances. Serum chemistries were monitored closely throughout the study for indications of organ toxicity. Although many patients entered this study with elevated creatinine levels, in only 11 cases did creatinine levels become elevated during AmBisome treatment. In 17 episodes, creatinine was initially high but returned to normal. Glutamyl oxaloacetate transaminase became elevated in 19 instances, and elevation in alkaline phosphatase was observed in 22 cases. However, there were no reports of discontinuation of AmBisome therapy due to adverse side effects. Thus, AmBisome was well tolerated even in severely ill patients.

The efficacy evaluation of the same patients described above was published in 1991 (78). AmBisome was analyzed as a salvage therapy in immunocompromised patients with suspected or proven fungal infections who had failed previous antifungal therapy, or had renal insufficiency or toxicity. Out of a total of 126 patients, there were 64 cases with proven invasive infections. Of these, 37 (58%) were cured, 12 (19%) improved, and 15 (23%) failed to respond. From these results, it was concluded that AmBisome was an effective salvage agent in the majority of patients with invasive (78) fungal infections. Although this was not the traditional randomized comparative trial usually required to gain marketing approval, it was sufficient in some countries to gain limited regulatory approval of AmBisome for salvage therapy in patients with severe fungal infections who could not be treated effectively with any other available agent.

As a result of its unique safety profile including very low kidney toxicity, AmBisome was further evaluated in a series of 187 transplant recipients who were also receiving cyclosporin (79). Cyclosporin is an immunosuppressive drug that is frequently administered to transplant patients to prevent graft rejection. However, cyclosporin has significant nephrotoxicity which often precludes the use of conventional amphotericin B in those patients who have contracted fungal infections. AmBisome was administered to these patients at dose levels between 1 and 4 mg/kg/day for a median of 11 days (range 1–112 days). The side effects attributed to AmBisome therapy were observed in only 7% of the cases, and resulted in its discontinuation in six cases. The serum creatinine increase, which was 20% overall, was not statistically significant indicating that AmBisome had minimal effect on kidney function. Other side effects possibly related to AmBisome included elevated serum urea and alkaline phosphatase which rapidly normalized after AmBisome therapy was discontinued. Thus, in the context of this study, when patients received a variety of potentially toxic drugs, the AmBisome side-effect profile was mild and manageable in the vast majority of cases.

AmBisome has also been investigated in neutropenic patients with suspected or confirmed fungal infections. In one report (80), 116 patients who had failed, or could not tolerate, conventional amphotericin B were treated with AmBisome. The median duration of AmBisome therapy was 12 days (range: 2–96 days) and the median total dose administered was 1684 mg (range: 180–10,440 mg). Adverse events were rare, even in patients receiving 5 mg/kg/day. No clinically significant deterioration in serum urea or creatinine resulted from AmBisome therapy, although electrolyte replacement was required in some patients. Abnormal hepatic function possibly attributable to AmBisome was noted in 17% of the treatment episodes, but was severe enough to warrant discontinuation of therapy in only two cases. Overall, 61% of the patients treated with AmBisome recovered. In 21 patients with proven aspergillosis, 13 (62%) obtained complete or partial resolution of the clinical and radiological signs of infection. In this study, AmBisome was shown to be an effective agent for the treatment of *Aspergillus* infections, and was far less toxic than conventional amphotericin B.

3.10.2. Prophylactic Studies

Since fungal infections in immunosuppressed patients are often fatal, clinicians have long sought an effective method to prevent such infections in high-risk patients such as bone marrow and solid organ transplant recipients. Although the broad activity spectrum of amphotericin B makes it an ideal prophylactic agent, its severe toxicities have precluded its use at dose levels that would assure success. Tollemar et al. (81), having recognized the value of AmBisome's low toxicity profile, were the first to study AmBisome in a randomized prospective prophylactic trial in bone marrow transplant recipients (82). Their results showed that AmBisome at 1 mg/kg per day significantly reduced fungal colonization compared to placebo, but the incidence of suspected (5 AmBisome vs. 7 placebo patients) and proven fungal infections (1 AmBisome vs. 3 placebo patients) were not significantly different. Nevertheless, AmBisome was well tolerated suggesting that prophylactic clinical trials with greater numbers of patients and at higher doses as indicated in the pre-clinical studies might show statistically significant efficacy for AmBisome.

Several years later, Tollemar et al. (83) published a paper on the prophylactic potential of AmBisome in liver transplant recipients. In this randomized prospective

double-blind trial, AmBisome at 1 mg/kg /day completely prevented invasive fungal infections, while 6 of 37 patients (16%)
in the placebo control group developed such infections
($p < 0.01$). However, there was no significant difference in
30-day survival between the two groups. Two AmBisome
and three placebo patients developed suspected fungal infections (NS) based on the presence of *Candida* antigen. AmBisome was well tolerated, although backache in one patient,
thrombocytopenia in another patient, and suspected nephrotoxicity and transient thrombocytopenia in a third patient
were observed. Interestingly, an economic analysis of this
study showed that the cost of AmBisome for prophylaxis
was less than the cost of treatment for the verified fungal
infections in the placebo control group.

Another prophylaxis study in neutropenic patients compared AmBisome given three times weekly at 2 mg/kg with
placebo (84). In this study, no patients receiving AmBisome
developed a proven fungal infection compared to three in
the placebo arm (NS) and suspected fungal infections defined
as fever for greater than 96 hr while on broad-spectrum antibiotics occurred in 42 and 46% of the AmBisome and placebo
patients, respectively (NS). In comparison, significantly
fewer patients in the AmBisome arm became colonized with
fungus compared to the placebo arm during the study (15 vs.
35, respectively, $p < 0.05$). As in other prophylactic studies,
AmBisome was well tolerated, but despite positive trends, this
regimen did not lead to a significant reduction in fungal
infection or reduce the requirement for systemic antifungal
therapy.

In a randomized prospective study, AmBisome (3 mg/kg,
three times per week) was compared to a combination of
fluconazole (200 mg every 12 hr) and itraconazole (200 mg
every 12 hr) as prophylaxis for fungal infections in patients
who were undergoing induction chemotherapy for acute myelogenous leukemia or myelodysplastic syndrome (85). Both
treatment arms of the study showed similar antifungal prophylaxis, although AmBisome treatment was associated
with significantly high rates of increase in serum bilirubin
($P = 0.012$) and serum creatinine ($P = 0.021$).

3.10.3. Therapeutic Studies

AmBisome treatment of opportunistic fungal infections in AIDS patients also provided encouraging results. One multicenter study (86) described the use of AmBisome in AIDS patients with cryptococcosis. Patients were treated at 3 mg/kg/day for at least 42 days. Of the 23 enrollments evaluable for clinical efficacy, 18 (78%) were either cured or improved. *Cryptococcus neoformans* was eradicated in 14 (67%) of 21 patients that were mycologically evaluable. AmBisome was well tolerated in these patients. Leenders et al. (87) conducted a randomized prospective study of cryptococcosis to compare the efficacy of AmBisome at 4 mg/kg vs. standard amphotericin B at 0.7 mg/kg daily for 3 weeks, each followed by oral Fluconazole (400 mg per day) for 7 weeks. Fifteen patients received AmBisome and 13 received amphotericin B. Although time to clinical response was the same in both treatment groups, cerebral spinal fluid (CSF) sterilization was achieved with AmBisome within 7 days in 6 of 15 patients compared to 1 in the amphotericin B group. By the end of intravenous therapy, 11 of 15 (73%) AmBisome patients had achieved CSF conversion compared to 3 of 8 (37.5%) evaluable patients in the amphotericin B group. Thus, AmBisome therapy resulted in significantly earlier CSF culture conversion than conventional amphotericin B therapy and was significantly less nephrotoxic.

AmBisome has also proven to be an effective treatment for leishmaniasis. Several clinical studies were done to test the efficacy of high dose AmBisome for treatment of drug-resistant visceral leishmaniasis (88,89). In one pilot study, a single dose of AmBisome at 5 mg/kg or five daily doses of AmBisome at 1 mg/kg were found to be curative in 91 and 93%, respectively, of visceral leishmaniasis patients at the 6 month follow-up (88). A more extensive multicenter study was performed in a total of 203 patients (89). In this study, a single 7.5 mg/kg dose of AmBisome was curative in 96% of the cases after 1 month. At the 6 month follow-up examination, 90% of the patients remained disease-free. Thus, AmBisome was shown to be a suitable treatment for drug-resistant visceral leishmaniasis.

AmBisome was evaluated for the treatment of fever of unknown origin in a prospective randomized clinical trial conducted in Europe (35). In this study, 338 neutropenic pediatric and adult patients with fever of unknown origin were randomized to receive one of the following regimens: amphotericin B (1 mg/kg), AmBisome (1 mg/kg), or AmBisome (3 mg/kg). Overall, 64% of patients in the amphotericin B arm experienced adverse events, which was significantly higher than in either the 1 or 3 mg/kg AmBisome arms (36 and 43%, respectively). Likewise, kidney toxicity, defined as at least a doubling of serum creatinine, was significantly higher in the amphotericin B group. With the exception of lower serum potassium in the amphotericin B group, there were no other significant differences in blood chemistry values. From an efficacy point of view, patients were defined as responders if they had resolution of fever ($<38°C$) and recovery of neutrophils ($\geq 0.5 \times 10^9/L$) for three consecutive days. Discontinuation of therapy, addition of another antifungal therapy, or proven breakthrough fungal infection were failure criteria. There was a 49% response rate in the amphotericin B arm compared to 58% ($p = 0.09$) and 63% ($p = 0.03$) response rates for AmBisome at 1 and 3 mg/kg, respectively. Thus, this study provided evidence that AmBisome was significantly safer while being equivalent to conventional amphotericin B at 1 mg/kg and significantly better at 3 mg/kg with regard to resolution of fever of unknown origin.

Based on AmBisome's encouraging safety profile, additional evaluations of this product for pediatric applications were done (90,91). In a study of 15 immunosuppressed infants, with suspected fungal infections or severe *Candida* colonization, who were about to have bone marrow transplantation for primary immunodeficiency, AmBisome could be administered at daily doses up to 6 mg/kg without serious side effects (90). These patients had minimal side effects despite the fact that many were receiving concomitant nephrotoxic agents such as cyclosporin. However, some patients ($n=3$) required increased potassium supplementation and one patient showed a significant rise in creatinine level. In another study, Scarcella et al. (91) reported on their experience with AmBisome in 44 pre-term or full term neonates with fungal infections. AmBisome was given at 1–5 mg/ kg/day for up to 49 days with essentially no side effects. This treatment cured 32 (72.7%) of those treated. It confirmed the drug's safety and efficacy in this population. In addition, AmBisome (2.5–7.0 mg/kg/day) has been evaluated prospectively in 24 very low birth weight infants with systemic candidiasis (92). Twenty (83%) infants were considered clinically cured at the end of treatment. No major adverse effects were recorded, although one patient developed increased bilirubin and hepatic transaminases levels during therapy. Thus, it was concluded that AmBisome is an effective, safe, and convenient antifungal therapy for systemic fungal infections in very low birth weight infants.

AmBisome (5 mg/kg) has also been compared to conventional amphotericin B (1 mg/kg) for the treatment of proven or suspected invasive fungal infections in a randomized trial (93). A total of 66 patients were eligible for analysis, 32 on AmBisome and 34 on amphotericin B. After 14 days of therapy, the complete plus partial response rate in the AmBisome treatment arm was significantly greater than in the amphotericin B arm ($p = 0.03$). This trend was also evident at the completion of therapy (median duration: 14.5 days for AmBisome and 16.5 days for amphotericin B), but not statistically significant ($p = 0.09$). This trial showed that AmBisome was more effective than the maximum tolerated dose of conventional amphotericin B in a prospective randomized comparative trial against confirmed fungal infections.

3.10.4. Clinical Studies in the United States

The studies summarized thus far were conducted outside of the United States. The clinical investigation of AmBisome in the United States began with a prospective evaluation of its safety, tolerance, and pharmacokinetics at doses of 1–7.5 mg/kg/day in febrile neutropenic patients (37). At each dose level, 8–12 patients were enrolled. Infusion related side effects (fever, chills, or rigors) were only associated with 5%

of the 331 infusions that were assessed, although two patients required premedication for infusion-related side effects. Looking at longer-term safety, no significant changes in serum creatinine, potassium, magnesium, or transaminase levels were reported. However, these investigators did observe increases in serum bilirubin and alkaline phosphatase that did not require discontinuation of therapy. Pharmacokinetically, both the peak plasma concentration (C_{max}) and the area under the curve (AUC) values for AmBisome followed a nonlinear dose relationship. Thus, as dosages increased, the increases in these parameters were disproportionately greater than the increase in dose, suggesting reticuloendothelial saturation and redistribution. Also, clearance values tended to decrease with increasing dose. Complete pharmacokinetic parameters are presented in Table 3. Finally, while this study was not powered to evaluate efficacy, it was noted that no patient showed evidence of a breakthrough fungal infection.

In a second study conducted in the United States (94), AmBisome (3 mg/kg) was compared to amphotericin B (0.6 mg/kg) in 687 febrile neutropenic patients. Results from this study formed the basis for AmBisome's approval in the United States for use as empiric therapy for presumed fungal infections. Composite efficacy success was defined as: (i) survival for at least 7 days post-study, (ii) resolution of fever during neutropenia, (iii) resolution of any confirmed fungal infection at study entry, (iv) no new emerging fungal infections within 7 days after the last dose, and (v) no premature

Table 3 Mean Single Dose Pharmacokinetic Parameters in Patients Following Intravenous AmBisome

Dose (mg/Kg)	n (group)	C_{max} (μg/mL)	$AUC_{0-\infty}$ (μg.h/mL)	$t_{1/2}$ (hr)	Cl (mL/hr/Kg)	V_d (L/Kg)
1.0	8	7.3	32	10.7	39	0.58
2.5	8	17.2	71	8.1	51	0.69
5.0	12	57.6	294	6.4	21	0.22
7.5	8	83.7	534	8.5	25	0.26

discontinuation due to toxicity or lack of efficacy. The composite success rates were equivalent for both arms of the trial (50%, AmBisome vs. 49%, amphotericin B). Furthermore, comparable responses were observed for survival, resolution of fever, and total incidence of proven plus suspected fungal infections. However, there were fewer proven emergent fungal infections, as well as a lower overall mortality in the AmBisome treatment arm. Finally, there were significantly fewer infusion related events and lower nephrotoxicity in the AmBisome group. Thus, it was concluded that AmBisome was equivalent to amphotericin B for empiric antifungal therapy in neutropenic patients, but superior in reducing proven emergent fungal infections and in reducing acute and chronic treatment related toxicities.

In another prospective, randomized, clinical trial for empiric therapy of presumed fungal infections, AmBisome was compared with voriconazole (95). In this study, 422 febrile neutropenic patients were treated with AmBisome (3 mg/kg/day) and 415 were treated with voriconazole (6 mg/kg every 12 hr for 2 doses and then 200 mg every 12 hr). The composite efficacy success was defined as described above for the AmBisome vs. amphotericin B empiric clinical trial (95). Although numerically fewer breakthrough fungal infections were seen in the voriconazole group (8 patients vs. 21 patients in the AmBisome group), voriconazole did not fulfill the criterion for non-inferiority to AmBisome based on the composite endpoint. Thus, the FDA advisory committee recommended unanimously that voriconazole not be approved for empiric therapy in febrile neutropenic patients (96).

The clinical safety of AmBisome (3 and 5 mg/kg) was compared directly with another lipid-based amphotericin B product, Abelcet (5 mg/kg) in a prospective, blinded trial (36). A total of 244 neutropenic patients with fever unresponsive to broad-spectrum antibacterial therapy were each randomly assigned to one of the three treatment groups. Infusion-related reactions such as chills, rigors, fever, and hypoxia on the first day of treatment were all significantly higher ($p < 0.01$) in the Abelcet group. Although a significantly greater number of Abelcet patients ($p < 0.003$) received pre-medications after day 1 for side effects, chills and rigors continued to be significantly ($p < 0.001$) more frequent in this group. Kidney toxicity, defined as a doubling of serum creatinine level, occurred in 42.3% of the Abelcet patients compared to 14.5% of the total AmBisome patients in both arms of the trial ($p < 0.001$). Finally, Abelcet was discontinued prematurely due to toxicities in 32% of patients compared to 13% of the AmBisome patients. Thus, in conclusion, Abelcet treatment (5 mg/kg) was associated with poorer drug tolerance, more nephrotoxicity, and required the use of more pre-medications than did AmBisome treatment at either 3 or 5 mg/kg.

The existing body of clinical data shows conclusively that AmBisome is safe and effective. The continued growth in worldwide sales of AmBisome and on-going pharmacokinetic and efficacy trials indicates that this drug has a unique and expanding role in the treatment of serious life-threatening fungal infections.

4. SUMMARY

In formulating the drug AmBisome, much new technology needed to be developed to ensure the production of large scale, reproducible, sterile, chemically stable, and biologically active material. There was a paucity of information available in the literature to help do this. The project required innovation, synthesis of information, varying expertise, and a lot of cooperation. It was necessary to have a clear focus on the end-product to know what kinds of physical, chemical, and biological assays had to be developed. Since novel lipids were being used and formulated into a unique carrier, special assays were required. Standards of ruggedness had to be established and maintained for the raw materials, as well as the final product. Since the active agent, amphotericin B, had been an approved drug for over 30 years, it was also necessary to conduct comparative evaluation of AmBisome with the conventional drug in both pre-clinical and clinical testing. As more studies were conducted with AmBisome, it became clear that this formulation had unique pharmacokinetics and an impressive safety profile. This made it possible to consider novel ways of using this product which would not be feasible with the conventional drug. In the future, these novel applications will be studied further. AmBisome is the first liposomal anti-infective agent to be approved for clinical use. Experience with AmBisome and its continued clinical success has opened the way for the introduction of other anti-infective drugs to be formulated as liposomes.

REFERENCES

1. Hutter RVP, Lieberman PH, Collins HS. Aspergillosis in a cancer hospital. Cancer 1964; 17:747.

2. Bodey GP. Fungal infections complicating acute leukemia. J Chronic Dis 1966; 19:667.

3. Maksymiuk AW, Thongprasert S, Hopfer R, Luna M, Fainstein V, Bodey GP. Systemic candidiasis in cancer patients. Am J Med 1984; 77:20.

4. Lortholary O, Dupont B. Antifungal prophylaxis during neutropenia and immunodeficiency. Clin Microbiol Rev 1997; 10:477.

5. Cornwell EE III, Belzberg H, Berne TV, Dougherty WR, Morales IR, Asensio J, Demetriades D. The pattern of fungal infections in critically ill surgical patients. Amer Surgeon 1995; 61:847.

6. Fridkin SK, Jarvis WR. Epidemology of nosocomial fungal infections. Clin Microbiol Rev 1996; 9:499.

7. Wald A, Leisenring W, van Burik J, Bowden RA. Epidimology of *Aspergillus* infections in a large cohort of patients undergoing bone marrow transplantation. J Inf Dis 1997; 175:1459.

8. Hood S, Denning DW. Treatment of fungal infections in AIDS. J Antimicrob Chemother 1996; 37(suppl B):71.

9. Walsh TJ, Hiemenz JW, Anaissie E. Recent progress and current problems in treatment of invasive fungal infections in neutropenic patients. Infect Dis Clin North Amer 1996; 10:365.

10. Fisher-Hoch SP, Hutwagner L. Opportunistic candidiasis: an epidemic of the 1980s. Clin Infect Dis 1995; 21:897.

11. Denning D. Invasive aspergillosis. Clin Infect Dis 1998; 26:781.

12. Anaissie E, Kantarjian H, Ro J, Hopfer R, Rolston K, Fainstein V, Bodey GP. The emerging role of *Fusarium* infections in patients with cancer. Medicine 1988; 67:77.

13. Abi-Said D, Anaissie E, Uzun O, Raad I, Pinzcowski H, Vartivarian S. The epidemiology of hematogenous candidiasis caused by different *Candida* species. Clin Infect Dis 1997; 24:1122.

14. Walsh TJ, Newman KR, Moody M, Wharton RC, Wade JC. Trichosporonosis in patients with neoplastic disease. Medicine 1986; 65:268.

15. Supparatpinyo K, Nelson KE, Merz WG, Breslin BJ, Cooper CR Jr, Kamwan C, Sirisanthana T. Response to antifungal therapy by human immunodeficiency virus-infected patients with disseminated *Penicillium marneffei* infections and in vitro susceptibilities of isolates from clinical specimens. Antimicrob Agents Chemother 1993; 37:2407.

16. Espinel-Ingroff A, Dawson K, Pfaller M, Anaissie E, Breslin B, Dixon D, Fothergill A, Paetznick V, Peter J, Rinaldi M, Walsh T. Comparative and collaborative evaluation of standardization of antifungal susceptibility testing for filamentous fungi. Antimicrob Agents Chemother 1995; 39:314.

17. Walsh TJ, Hiemenz JW, Anaissie E. Recent progress and current problems in treatment of invasive fungal infections in neutropenic patients. Infect Dis Clin North Amer 1996; 10:365.

18. Warnock DW. Fungal infections in neutropenia: current problems and chemotherapeutic control. J Antimicrob Chemother 1998; 41(suppl D):95.

19. Wheat J, Hafner R, Korzun AH, Limjoco MT, Spencer P, Larsen RA, Hecht FM, Powderly W. Itraconazole treatment of disseminated histoplasmosis in patients with the acquired immunodeficiency syndrome. Am J Med 1995; 98:336.

20. Denning DW, Ribaud P, Milpied N, Caillot D, Herbrecht R, Thiel E, Haas A, Ruhnke M, Lode H. Efficacy and safety of voriconazole in the treatment of acute invasive aspergillosis. Clin Infect Dis 2002; 34:563.

21. Nguyen MH, Yu CY. Voriconazole against fluconazole-susceptible and resistant *Candida* isolates: in-vitro efficacy compared with that of itraconazole and ketoconazole. J Antimicrob Chemother 1998; 42:253.

22. Ruhnke M, Schmidt-Westhausen A, Trautmann M. *In vitro* activities of voriconazole (UK-109,486) against fluconazole-susceptible and resistant *Candida albicans* isolates from oral cavities of patients with human immunodeficiency virus infection. Antimicrob Agents Chemother 1997; 41:575.

23. McGinnis MR, Pasarell L, Sutton DA, Fothergill AW, Cooper CR Jr, Rinaldi MG. In vitro activity of voriconazole against selected fungi. Med Mycol 1998; 36:239.

24. Dresser GK, Spence JD, Bailey DG. Pharmacokinetic-pharmacodynamic consequences and clinical relevance of cytochrome P450 3A4 inhibition. Clin Pharmacokinet 2000; 38:41.

25. Kramer MR, Marshall SE, Denning DW, Keogh AM, Tucker RM, Galgiani JN, Lewiston NJ, Stevens DA, Theodore J. Cyclosporine and itraconazole interaction in heart and lung transplant recipients. Ann Intern Med 1990; 113:327.

26. Khoo SH, Bond J, Denning DW. Administering amphotericin B: a practical approach. J Antimicrob Chemother 1994; 33:203.

27. Lyman CA, Walsh TJ. Systemically administered antifungal agents. A review of their clinical pharmacology and therapeutic applications. Drugs 1992; 44:9.

28. Brajtburg J, Bolard J. Carrier effects on biological activity of Amphotericin B. Clin Microbiol Rev 1996; 9:512.

29. Moreau P, Milpied N, Fayette N, Ramee JF, Harousseau JL. Reduced renal toxicity and improved clinical tolerance of amphotericin B mixed with intralipid compared with conventional amphotericin B in neutropenic patients. J Antimicrob Chemother 1992; 30:535.

30. Janoff AS, Boni LT, Popescu MC, Minchey SR, Cullis PR, Madden TD, Taraschi T, Gruner SM, Shyamsunder MW, Tate E, Mendelsohn R, Bonner D. Unusual lipid structures selectively reduce the toxicity of amphotericin B. Proc Natl Acad Sci USA 1988; 85:6122.

31. Walsh TJ, Hiemenz JW, Seibel NL, Perfect JR, Horwith G, Lee L, Silber JL, DiNubile MJ, Reboli A, Bow E, Lister J, Anaissie EJ. Amphotericin B lipid complex for invasive fungal infections: analysis of safety and efficacy in 556 cases. Clin Infect Dis 1998; 26:1383.

32. Guo LS, Fielding RM, Lasic DD, Hamilton RL, Mufson D. Novel antifungal drug delivery: stable amphotericin B-cholesteryl sulfate discs. Int J Pharmacol 1991; 75:45.

33. Bowden RA, Cays M, Gooley T, Mamelok RD, van Burik JA. Phase I study of amphotericin B colloidal dispersion for the treatment of invasive fungal infections after marrow transplant. J Infect Dis 1996; 173:1208.

34. Adler-Moore JP, Proffitt RT. Development, characterization, efficacy and mode of action of AmBisome, a unilamellar liposomal formulation of amphotericin B. J Liposome Res 1993; 3:429.

35. Prentice HG, Hann IM, Herbrecht R, Aoun M, Kvaloy S, Catovsky D, Pinkerton CR, Schey SA, Jacobs F, Oakhill A, Stevens RF, Darbyshire PJ, Gibson BE. A randomized comparison of liposomal versus conventional amphotericin B for the treatment of pyrexia of unknown origin in neutropenic patients. Br J Haematol 1997; 98:711.

36. Wingard JR, White MH, Anaissie EJ, Rafalli JT, Goodman JL, Arrieta AC. A randomized, double-blind comparative trial evaluating the safety of liposomal amphotericin B versus amphotericin B lipid complex in the empirical treatment of febrile neutropenia. Clin Infect Dis 2000; 31:1155.

37. Walsh TJ, Yeldandi V, McEvoy M, Gonzalez C, Chanock S, Freifeld A, Seibel NI, Whitcomb PO, Jarosinski P, Boswell G, Bekersky I, Alak A, Buell D, Barret J, Wilson W. Safety, tolerance, and pharmacokinetics of a small unilamellar liposomal formulation of amphotericin B (AmBisome) in neutropenic patients. Antimicrob Agents Chemother 1998; 42:2391.

38. Boswell GW, Buell D, Bekersky I. AmBisome (liposomal amphotericin B): a comparative review. J Clin Pharmacol 1998; 38:583.

39. Senior JH. Fate and behavior of liposomes *in vivo*: a review of controlling factors. Crit Rev Therapeut Drug Carrier Syst 1987; 3:123.

40. Storm G, Woodle MC. Long circulating liposomes therapeutics: from concept to clinical reality. In: Woodle MC, Storm G, eds. Long Circulating Liposomes: Old Drugs, New Therapeutics. New York: Springer-Verlag, 1998:3.

41. Readio JD, Bittman R. Equilibrium binding of amphotericin B and its methyl ester and borate complex to sterols. Biochim Biophys Acta 1982; 685:219.

42. Brajburg J, Elberg S, Kobyashi GS, Medoff G. Toxicity and induction of resistance to *Listeria monocytogenes* infection by amphotericin B in inbred strains of mice. Infect Immunity 1986; 54:303.

43. Proffitt RT, Satorius A, Chiang S-M, Sullivan L, Adler-Moore JP. Pharmacology and toxicology of a liposomal formulation of amphotericin B (AmBisome) in rodents. J Antimicrob Chemother 1991; 28(suppl B):49.

44. Gamble RC. Method for preparing small vesicles using microemulsification. US Patent #4.753,788, 1988.

45. Adler-Moore JP, Proffitt RT. AmBisome®: long circulating formulation of amphotericin B. Woodle MC, Storm G, eds. Long Circulating Liposomes: Old Drugs, New Therapeutics. New York: Springer-Verlag, 1998:185.

46. Anaissie E, Paetznik V, Proffitt R, Adler-Moore J, Bodey GP. Comparison of the in vitro antifungal activity of free and liposome-encapsulated amphotericin B. Eur J Clin Microb Infect Dis 1991; 10:665.

47. Johnson EM, Ojwang JO, Szekely A, Wallace TL, Warnock DW. Comparison of *in vitro* antifungal activities of free and liposome-encapsulated nystatin with those of four amphotericin B formulations. Antimicrob Agents Chemother 1998; 42:1412.

48. Clemons KV, Stevens DA. Comparison of a liposomal amphotericin B formulation (AmBisome) and deoxycholate amphotericin B (Fungizone) for the treatment of murine paracoccidioidomycosis. J Med Vet Mycol 1993; 31:387.

49. Jensen GM, Skenes CR, Bunch TH, Weissman CA, Amirgha-hari N, Satorius A, Moynihan K, Eley CGS. Determination of the relative toxicity of amphotericin B formulations: a red blood cell potassium release assay. Drug Deliv 1999; 6:81.

50. Sperry PJ, Cua DJ, Wetzel SA, Adler-Moore JP. Antimicrobial activity of AmBisome and non-liposomal amphotericin B following uptake of *Candida glabrata* by murine epidermal Langerhans cells. Med Mycol 1998; 36:135.

51. McAndrews BJ, Lee MJA, Adler-Moore JP. Comparative toxicities of Fungizone and AmBisome for cultured kidney cells and macrophages. In: Proceedings of the 95th American Society of Microbiology, 1993. Abst. 11.

52. Boswell GW, Bekersky I, Buell D, Hiles R, Walsh TJ. Toxicological profile and pharmacokinetics of a unilamellar liposomal vesicle formulation of amphotericin B in rats. Antimicrob Agents Chemother 1998; 42:263.

53. Bekersky I, Boswell GW, Hiles R, Fielding RM, Buell D, Walsh TJ. Safety and toxicokinetics of intravenous liposomal amphotericin B (AmBisome) in beagle dogs. Pharm Res 1999; 16:1694.

54. Fielding RM, Singer AW, Wang LH, Babbar S, Guo LS. Relationship of pharmacokinetics and drug distribution in tissue to increased safety of amphotericin B colloidal dispersion in dogs. Antimicrob Agents Chemother 1992; 36:299.

55. Woodle MC, Storm G, eds. Long Circulating Liposomes: Old Drugs, New Therapeutics. New York: Springer-Verlag, 1998.

56. Garcia A, Adler-Moore JP, Proffitt RT. Single dose AmBisome (liposomal amphotericin B) as prophylaxis for murine systemic candidiasis and histoplasmosis. Antimicrob Agents Chemother 2000; 44:2327.

57. Groll AH, Giri N, Petraitis V, Petraitiene R, Candelario M, Bacher JS, Piscitelli SC, Walsh TJ. Comparative efficacy and distribution of lipid formulations of amphotericin B in experimental *Candida albicans* infection of the central nervous system. J Infect Dis 2000; 182:274.

58. Adler-Moore JP, Chiang S-M, Satorius A, Guerra D, McAndrews B, McManus EJ, Proffitt RT. Treatment of murine

candidiasis and cryptococcosis with a unilamellar liposomal amphotericin B formulation (AmBisome). J Antimicrob Chemother 1991; 28(suppl B):63.

59. van Etten EWM, van den Heuvel-de Groot C, Bakker-Woudenberg IA. Efficacies of amphotericin B-desoxycholate (Fungizone), liposomal amphotericin B (AmBisome) and fluconazole in the treatment of systemic candidosis in immunocompetent and leucopenic mice. J Antimicrob Chemother 1993; 32:723.

60. Adler-Moore JP. AmBisome targeting to fungal infections. Bone Marrow Transplant 1994; 14(suppl 5):S3.

61. Karyotakis NC, Anaissie EJ. Efficacy of escalating doses of liposomal amphotericin B (AmBisome) against hematogenous *Candida lusitaniae* and *Candida krusei* infection in mice. Antimicrob Agents Chemother 1994; 38:2660.

62. Francis P, Lee JW, Hoffman A, Peter J, Francesconi A, Bacher J, Shelhamer J, Pizzo PA, Walsh TJ. Efficacy of unilamellar liposomal ampho. B in treatment of pulmonary aspergillosis in persistently granulocytopenic rabbits: the potential role of bronchoalveolar D-mannitol and serum galactomannan as markers of infection. J Infect Dis 1994; 169:356.

63. Leenders ACAP, de Marie S, ten Kate MT, Bakker-Woudenberg IA, Verbrugh HA. Liposomal amphotericin B (AmBisome) reduces dissemination of infection as compared with amphotericin B deoxycholate (Fungizone) in a rat model of pulmonary aspergillosis. J Antimicrob Chemother 1996; 38:215.

64. Allen SD, Sorensen KN, Nejdl MJ, Durrant C, Proffitt RT. Prophylactic efficacy of aerolized liposomal (AmBisome) and non-liposomal (Fungizone) amphotericin B in murine pulmonary aspergillosis. J Antimicrob Chemother 1994; 34:1001.

65. Ibraham AS, Avanessian V, Spellberg B, Edwards JE Jr. Liposomal Amphotericin B, and not amphotericin B deoxycholate, improves survival of diabetic mice infected with *Rhizopus oryzae*. Antimicrob Agents Chemother 2003; 47:3343.

66. Albert MM, Stahl-Carroll L, Luther MF, Graybill JR. Comparison of liposomal amphotericin B to amphotericin B for treatment of murine cryptococcal meningitis. J Mycol Med 1995; 5:1.

67. Clemons KV, Stevens DA. Therapeutic efficacy of a liposomal formulation of amphotericin B (AmBisome) against murine blastomycosis. J Antimicrob Chemother 1993; 32:465.

68. Clemons KV, Sobel RA, Williams PL, Pappagianis D, Stevens DA. Efficacy of intravenous liposomal amphotericin B (AmBisome) against coccidioidal meningitis in rabbits. Antimicrob Agents Chemother 2002; 46:2420.

69. Graybill JR, Bocanegra R. Liposomal amphotericin B therapy of murine histoplasmosis. Antimicrob Agents Chemother 1995; 39:1885.

70. Anaissie EJ, Hachem R, Karyotakis NC, Gokaslan A, Dignani MC, Stephens LC, Tin-U CK. Comparative efficacies of amphotericin B, triazoles, and combination of both as experimental therapy for murine trichosporonosis. Antimicrob Agents Chemther 1994; 38:2541.

71. Gangneux JP, Sulahian A, Garin YJ, Farinotti R, Derouin F. Therapy of visceral leishmaniasis due to *Leishmania infantum*: experimental assessment of efficacy of AmBisome. Antimicrob Agents Chemother 1996; 40:1214.

72. Croft SL, Davidson RN, Thornton EA. Liposomal amphotericin B in the treatment of visceral leishmaniasis. J Antimicrob Chemother 1991; 28(suppl B):111.

73. Albert MM, Adams K, Luther MJ, Sun SH, Graybill JR. Efficacy of AmBisome in murine coccidioidomycosis. J Med Vet Mycol 1994; 32:467.

74. Van Burik JA, Myerson D, Schreckhise RW, Bowden RA. Panfungal PCR assay for detection of fungal infection in human blood specimens. J Clin Microbiol 1998; 36:1169.

75. Kaufman L. Laboratory methods for the diagnosis and confirmation of systemic mycoses. Clin Infect Dis 1992; 14(suppl 1):S23.

76. Katz NM, Pierce PF, Anzeck RA, Visner MS, Canter HG, Foegh ML, Pearle DL, Tracy C, Rahman A. Liposomal amphotericin B for treatment of pulmonary aspergillosis in a heart transplant patient. J Heart Transplant 1990; 9:14.

77. Meunier F, Prentice HG, Ringdén O. Liposomal amphotericin B (AmBisome): safety data from a phase II/III clinical trial. J Antimicrob Chemother 1991; 28(suppl B):83.

78. Ringden O, Meunier F, Tollemar J, Ricci P, Tura S, Kuse E, Viviani MA, Gorin NC, Klastersky J, Fenaux P, Prentice HG, Ksionski G. Efficacy of amphotericin B encapsulated in liposomes (AmBisome) in the treatment of invasive fungal infections in immunocompromised patients. J Antimicrob Chemother 1991; 28(suppl B):73.

79. Ringdén O, Andstrom E, Remberger M, Svahn B-M, Tollemar J. Safety of liposomal amphotericin B (AmBisome) in 187 transplant recipients treated with cyclosporin. Bone Marrow Transplant 1994; 14(suppl 5):10.

80. Mills W, Chopra R, Linch DC, Goldstone AH. Liposomal Amphotericin B in the treatment of fungal infections in neutropenic patients: a single-centre experience of 133 episodes in 116 patients. Brit J Haematol 1994; 86:754.

81. Tollemar J, Ringdén O, Tydén G. Liposomal amphotericin B (AmBisome)® treatment in solid organ and bone marrow transplant recipients. Efficacy and safety evaluation. Clin Transplant 1990; 4:167.

82. Tollemar J, Ringdén O, Andersson S, Sundberg B, Ljungman P, Tyden G. Randomized double-blind study of liposomal amphotericin B (AmBisome) prophylaxis of invasive fungal infections in bone marrow transplant recipients. Bone Marrow Transplant 1993; 12:577.

83. Tollemar J, Hockerstedt K, Ericzon BG, Jalanko H, Ringdén O. Liposomal amphotericin B prevents invasive fungal infections in liver transplant recipients. Transplantation 1995; 59:1.

84. Kelsey SM, Goldman JM, McCann S, Newland AC, Scarffe JH, Oppenheim BA, Mufti GJ. Liposomal amphotericin (AmBisome) in the prophylaxis of fungal infections in neutropenic patients: a randomised, double-blind, placebo-controlled study. Bone Marrow Transplant 1999; 23:163.

85. Mattiuzzi GN, Estey E, Raad I, Giles F, Cortes J, Shen Y, Kontoyiannis D, Koller C, Munsell M, Beran M, Kantarjian H. Liposomal amphotericin B versus the combination of fluconazole and itraconazole as prophylaxis for invasive fungal infec-

tions during induction chemotherapy for patients with acute myelogenous leukemia and myelodysplastic syndrome. Cancer 2003; 97:450.

86. Coker RJ, Viviani M, Gazzard BG, Du Pont B, Pohle HD, Murphy SM, Atouguia J, Champalimaud JL, Harris JRW. Treatment of cryptococcosis with liposomal Amphotericin B (AmBisome) in 23 patients with AIDS. AIDS 1993; 7:829.

87. Leenders ACAP, Reiss P, Portegies P, Clezy K, Hop WCJ, Hoy J, Borleffs JCC, Allworth T, Kauffmann RH, Jones P, Kroon FP, Verbrugh HA, de Marie S. Liposomal amphotericin B (AmBisome) compared with amphotericin B both followed by oral fluconazole in the treatment of AIDS-associated cryptococcal meningitis. AIDS 1997; 11:1463.

88. Sundar S, Agrawal G, Rai M, Makharia MK, Murray HW. Treatment of Indian visceral leishmaniasis with single or daily infusions of low dose liposomal amphotericin B: randomised trial. BMJ 2001; 323:419.

89. Sundar S, Jha TK, Thakur CP, Mishra M, Singh VP, Buffels R. Single-dose liposomal amphotericin B in the treatment of visceral leishmaniasis in India: a multicenter study. Clin Infect Dis 2003; 37:800.

90. Pasic S, Flannagan L, Cant A. Liposomal amphotericin (AmBisome) is safe in bone marrow transplantation for primary immunodeficiency. Bone Marrow Transplant 1997; 19:1229.

91. Scarcella, Pasquariello MB, Giugliano B, Vendemmia M, de Lucia A. Liposomal amphotericin B treatment for neonatal fungal infections. Pediatr Infect Dis J 1998; 17:146.

92. Juster-Reicher A, Leibovitz E, Linder N, Amitay M, Flidel-Rimon O, Even-Tov S, Mogilner B, Barzilai A. Liposomal amphotericin B (AmBisome) in the treatment of neonatal candidiasis in very low birth weight infants. Infection 2000; 28:223.

93. Leenders AC, Daenen S, Jansen RL, Hop WC, Lowenberg B, Wijermans PW, Cornelissen J, Herbrecht R, van der Lelie H, Hoogsteden HC, Verbrugh HA, de Marie S. Liposomal amphotericin B compared with amphotericin B deoxycholate in the treatment of documented and suspected neutropenia-

associated invasive fungal infections. Brit J Haematol 1998; 103:205.

94. Walsh TJ, Finberg RW, Arndt C, Hiemenz J, Schwartz C, Bodensteiner D, Pappas P, Seibel N, Greenberg RN, Dummer S, Schuster M, Holcenberg JS. Liposomal amphotericin B for empirical therapy in patients with persistent fever and neutropenia. N Engl J Med 1999; 340:764.

95. Walsh TJ, Pappas P, Winston DJ, Lazarus HM, Petersen F, Raffalli J, Yanovich S, Stiff P, Greenberg R, Donowitz G, Schuster M, Reboli A, Wingard J, Arndt C, Reinhardt J, Hadley S, Finberg R, Laverdiere M, Perfect J, Garber G, Fioritoni G, Anaissie E, Lee J, National Institute of Allergy, Infectious Diseases Mycoses Study Group. Voriconazole compared with liposomal amphotericin B for empirical antifungal therapy in patients with neutropenia and persistent fever. N Engl J Med 2002; 346:225.

96. Powers JH, Dixon CA, Goldberger MJ. Voriconazole versus liposomal amphotericin B in patients with neutropenia and persistent fever. N Engl J Med 2002; 346:289.

16

Case Study: Optimization of a Liposomal Formulation with Respect to Tissue Damage

GAYLE A. BRAZEAU

Departments of Pharmacy Practice and
Pharmaceutical Sciences, School of Pharmacy
and Pharmaceutical Sciences, University at
Buffalo, State University of New York,
Buffalo, New York, U.S.A.

1. INTRODUCTION

It is recognized that the development of injectable dispersed systems requires the pharmaceutical scientist to have an understanding of the physicochemical properties, biopharmaceutical properties, particle and rheological characterization, and release properties of these formulations. These issues are well discussed and outlined by other contributors to this volume. However, a consideration in the development of an

527

injectable dispersed system, that most often may be over-
looked, is the interaction of this formulation or components
with the tissues at the site of injection and the extent to which
a formulation may cause pain upon injection. Even the best
formulated products, with respect to the ideal physicochemical, biopharmaceutical and release properties, could be
found to have limited application if shown in animal or clinical studies to cause tissue damage and/or pain following
injection.

As such, it is critical in parenteral products, from
solutions to suspensions and other dispersed systems, to consider the extent of tissue damage and/or pain upon injection
(that may be caused by the components or final formulations)
early in the product development process. This requires a systematic approach to optimize the final formulation with
respect to the desired therapeutic requirements, formulation
requirements, and physiological constraints as shown in Fig. 1.

Tissue damage can be defined as a formulation-induced
reversible or irreversible change in the anatomy, biochemistry, or physiology at the injection site. Injection sites and tissues where damage can occur include vascular epithelial
cells, red blood cells, or muscle tissue depending upon the
selected route of administration. An interesting observation
is that formulations are reported to cause subcutaneous irritation upon injection. It should be remembered that the subcutaneous space is located beneath the skin and above
skeletal muscle tissue so that there is no definable tissue that
can be studied in this case. As such, if a formulation is
thought to cause subcutaneous damage, it is recommended
to look at the adjacent tissues to quantify the extent of tissue
damage.

The extent of tissue damage is quantitated relatively
easily by comparison with control formulations. Furthermore,
the mechanism of this damage, like the extent of damage, can
be investigated using experimental methods ranging from
sub-cellular organelles to in vivo animal experiments to clinical studies (1). Pain is defined as an unpleasant sensation
associated with the injection of the formulation. Pain can
occur immediately upon injection or may be delayed based

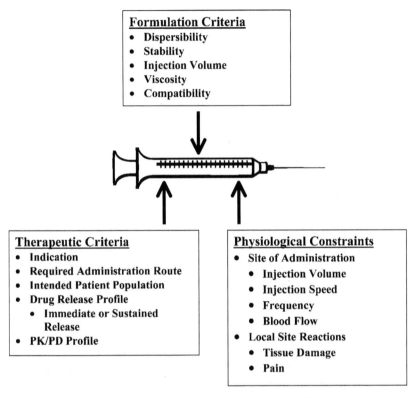

Figure 1   Optimization parameters of parenteral formulations.

upon the mechanism underlying the development of pain. In contrast, quantification of pain upon injection and the mechanism responsible for pain is more difficult to determine experimentally with available methodologies (2). The determination of pain associated with injection will require close collaboration between pharmaceutical scientists and neuroscientists in order to utilize appropriate models to understand how formulations activate nociceptors, the cells responsible for pain perception peripherally and centrally. The specific topic of pain upon injection is not covered in the present discussion.

The present case study will highlight a specific dispersed system, namely a liposomal formulation of loxapine, with respect to understanding and minimizing tissue damage upon injection during the early development of a potential formulation (3). The focus of this case study will be to investigate how to assess systematically for the extent of tissue damage in order to minimize this in a dispersed system. Tissue damage will be defined as the myotoxicity or damage caused to muscle tissue following an intramuscular injection.

2. LIPOSOMAL FORMULATIONS AND INTRAMUSCULAR INJECTIONS

Liposomes have been suggested to be a useful system for drug administration since they can provide either immediate drug release or sustained drug delivery, and can reduce tissue muscle damage upon injection. As early as 1984, Arrowsmith et al. (4) demonstrated that a liposomal formulation could provide sustained release of cortisone hexadecanote. Furthermore, it has been demonstrated by Kadir et al. (5,6) that liposomal formulations can be used to reduce the toxicity of certain drugs following intramuscular administration. While it had been suggested that liposomes are a useful dispersed system for parenteral administration with respect to decreasing tissue damage, it has been unclear as to how specific liposomal characteristics (e.g., liposome charge, fluidity, and size) impact on the extent of tissue damage. Furthermore, it is always critical to ensure that the selected formulation that provides minimal tissue damage is also acceptable with respect to the stability of the liposome, stability of the drug in the liposome, and the intended release property of the drug from the liposomal formulation.

3. INTRAMUSCULAR LOXAPINE FORMULATION

Loxapine is a dibenzoxazepine-based anti-psychotic drug that is available in both an oral and a parenteral formulation (Fig. 2). In the parenteral formulation, the drug has been formulated using the hydrochloride salt at a concentration of 50 mg/mL, 5% polysorbate 80 NF, and 70% propylene glycol NF (7). The compound is slightly soluble in water and moderately

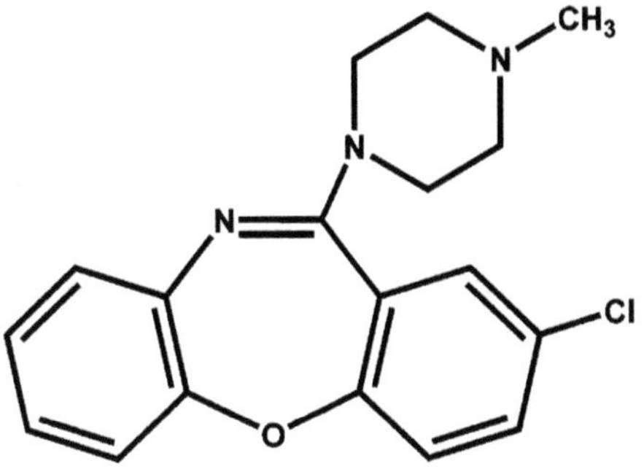

Figure 2 Loxapine structure.

soluble in propylene glycol. This compound is an ideal candidate for a sustained intramuscular formulation due to its short half-life (7), relatively low oral bioavailability (8), and effective plasma levels in the ng/mL range (7). This delivery approach could also avoid the potentially toxic effect of this compound on liver cells (9).

However, the one adverse effect reported by Meltzer et al. (10) following intramuscular injection of the current parenteral formulation was an increase in serum creatine kinase levels. This is an indication that the formulation may be causing muscle damage at the injection site. This finding was not unexpected based upon our earlier findings showing that propylene glycol at a concentration range above 40% was extremely myotoxic using the rodent isolated muscle model and in vivo studies in rabbits (11–13). Furthermore, the surfactant polysorbate 80 at low concentrations has been shown to cause muscle damage, most likely as a function of its detergent action on membranes, namely on membrane phospholipids.

4. STUDY OBJECTIVES

Based upon this background literature data, the overall goal of these studies was to develop a liposomal formulation for loxapine that would show minimal toxicity to muscle tissue following intramuscular injection in vitro and in vivo, while at the same time providing a sustained released profile of the drug. In vitro myotoxicity would be assessed using the previously established isolated rodent muscle model (11–13), in which the cumulative release of the cytosolic enzyme creatine kinase is measured over a 2-hr period into a bath. The in vivo myotoxicity studies of the selected formulations were conducted in a cannulated rodent model.

5. IN VITRO LIPOSOMAL MYOTOXICITY STUDIES

The optimization process during formulation development must include studies early on in the process that test the tissue-damaging potential of the individual components in the formulation. If this process is employed, the formulator is able to select those excipients or formulation factors that will minimize tissue damage in the final formulation. Prior to developing any liposomal drug formulations, it is essential to determine the influence of the selected liposomal composition and characteristics, in the absence of drug, on muscle damage (14). Three specific factors were investigated in these studies: liposome charge (positive or negative), size (large, ranging from 1.5 to 2.0 μm and small, ranging from 0.2 to 0.5 μm) and fluidity. The ratio of the lipid components of the various lipid formulations is discussed below.

Liposomes were prepared using the standard thin film hydration method (3,14). The two sizes of liposomes (1.5–2.0 and 0.2–0.5 μm) were obtained using extrusion through polycarbonate membranes (3,14) followed by size determination using laser light scattering. Negatively charged liposomes were prepared using phosphatidylcholine and phosphatidylglycerol (PC–PG) (7:3 M). Positive liposomes were prepared using phosphatidylcholine and stearylamine (PC–SA) (9:1). The fluidity of the membranes was changed by adding cholesterol to the liposomes. The ratio of phosphatidylcholine: stearylamine:cholesterol (PC–SA–CH) was 7:1:2 M for the positively charged liposomes, while for the negatively charged liposomes the ratio of phosphatidylcholine:phosphatidylglycerol:cholesterol (PC–PG–CH) was 4:3:3 M. The total concentration of lipids in these formulations was 25 mg/mL. Since we injected 15 μl of each formulation, a total of 0.375 μg of lipid was injected into each muscle in these in vitro studies.

Since it will often be necessary to conduct tissue damage or pharmacokinetic studies in separate groups of animals with a given formulation over a period of 1–2 weeks, the investigator must ensure that the prepared liposomes do not change their size with time. The size of all the tested formulations was stable over 6–11 days, with the coefficient of variation of size as a function of time ranging from 3 to 10% (3).

The myotoxicities of these selected formulations are shown in Fig. 3. For comparison, the two negative controls, saline and untreated muscles, and the two positive control formulations, Phenytoin (Dilantin®, a commercially available formulation at a concentration of 50 mg/mL with 40% propylene glycol, 10% ethanol at pH 12) and a muscle sliced in half, are provided as reference points. All of the eight liposomal formulations were determined to be equal to the normal saline negative control in creatine kinase released, but significantly lower than the two positive controls. It could be concluded

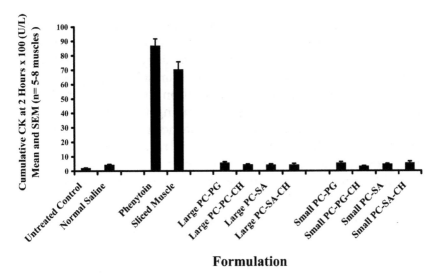

Figure 3 In vitro myotoxicity of empty liposomal formulations.

that these selected liposomal formulations did not result in muscle damage following injection compared to normal saline. Furthermore, the extent of muscle damage in these selected formulations appeared to be independent of liposome size, charge, or fluidity. These findings provided further evidence to demonstrate the compatibility of these dispersed systems containing phosphatidylcholine, phosphatidylglycerol stearylamine, and cholesterol with muscle tissue.

Another important consideration is the pH of the final formulation. In separate studies using buffer solutions, it was demonstrated that muscle myotoxicity was related to pH with more acidic preparations causing more myotoxicity relative to formulations with pH values near physiological or muscle pH (6.0). Furthermore, myotoxicity was related to proton concentration and buffer capacity. It was found that myotoxicity was greater in those formulations at low pH with higher buffer capacity compared to those with lower buffer capacity (15).

6. LOXAPINE LIPOSOMAL FORMULATIONS AND IN VITRO MYOTOXICITY STUDIES

Once the dispersed system without the drug has been shown not to cause tissue damage, the investigator must formulate the desired therapeutic agent in this system such that it achieves the requisite concentration (or solubility) and stability of the active ingredient. In addition, the final formulation itself must be stable for storage during the study period. It becomes critical to ensure that the active drug is stable in the liposomal formulation both in vitro and in vivo. It also becomes critical to have a sensitive and selective assay that can differentiate between the active agent, metabolites and components of the dispersed system.

Loxapine liposomes were formulated using phosphatidylcholine:phosphatidylglycerol (7:3 M) at a drug to lipid ratio of 1:2 M and a lipid concentration of 100 mg/mL. The aver -age liposome diameter was 1.1 μm and the encapsulation efficiency was $62 \pm 11\%$ (SD). The initial concentration of loxapine in the final formulation for all studies was between 12.2 and 12.4 mg/mL. This was the highest concentration that could be achieved in this formulation; however, it was considered that this would be suitable for future studies based upon the loxapine serum levels and the available HPLC assay methodology. In vitro release studies demonstrated that $71 \pm 4.7\%$ (SD) and 61% of the loxapine were released into isotonic phosphate buffer 7.4 and a pH 6.0 muscle homogenate, respectively (3).

It now becomes important to determine the myotoxicity of the final formulation to determine if the liposome formulation can reduce tissue damage. In the next study, the myotoxicity of loxapine at a concentration of 12.2 mg/mL in phosphatidylcholine:phospatidylglycerol liposome was compared with: the commercially available loxapine product (50 mg/mL) with propylene glycol and polysorbate 80; the commercial formulation diluted to the same concentration as the liposomal formulation; and the drug free loxapine solvent system (70% propylene glycol and 5% polysorbate 80). The last two treatments are important because this will allow a direct comparison with the liposomal formulation and will also indicate the extent to which these two excipients in the commercial formulation contribute to the myotoxicity of the final formulation, respectively. In any myotoxicity study, it is always recommended to include the solvent controls and to ensure that the concentrations of all ingredients are the same as myotoxicity is most often concentration dependent. It would also be important to include as a control the blank liposomal formulation, without drug, as a comparison. In earlier studies and as shown in Fig. 3, the liposomal formulation to be used in these studies, 7.3 M phosphatidylcholine:phosphatidylglycerol and a size of $1.1\,\mu m$ (classified as a large liposome) was no more myotoxic than normal saline. As such, this liposomal formulation appears to be non-myotoxic.

The results of the myotoxicity studies for these formulations are shown in Fig. 4. Similar to Fig. 3, the positive and negative controls have been provided as reference points. Interestingly, the myotoxicity of the undiluted commercially available loxapine solution was markedly higher than the

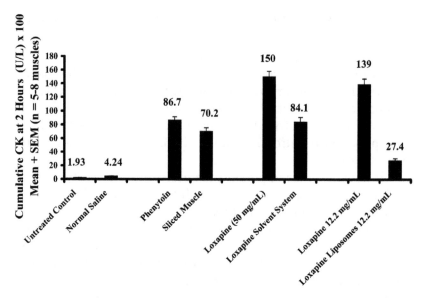

Figure 4   In vitro myotoxicity of liposomal loxapine (12.2 mg/mL), the commercially available loxapine formulation at the same concentration (12.2 mg/mL), the commercially available loxapine formulation at 50 mg/mL, and the solvent system for the commercially available loxapine formulation. Mean values are shown above each bar graph.

two positive controls. This is not surprising as the phenytoin formulation only contains 40% propylene glycol compared to 70% propylene glycol in the loxapine solution. Furthermore, this formulation does not contain any surfactant. The solvent system for the commercially available formulation appeared to have contributed significantly to the myotoxicity of the final formulation [84.1 ± 6.83 (SEM) compared to 150 ± 8.28 (SEM)]. It is often difficult to discriminate between the myotoxicity caused by the drug and that caused by the solvent system because of the difficulty in determining the myotoxicity of the drug alone due to limited aqueous solubility. In the present studies, it was also impossible to determine whether loxapine precipitation occurring at the site of injection caused the myotoxicity.

Furthermore, when the original formulation was diluted to the same concentration as the formulated liposomal formulation (12.2 ng/mL), the extent of myotoxicity was not remarkably reduced compared to the undiluted solution [139 ± 8.4 (SEM) vs. 150 ± 8.28 (SEM)]. When the loxapine solution is diluted, the myotoxicity could have been caused by precipitation of the drug at the injection site as both the drug and solvent system concentration in the formulation is reduced. As such, the solvent system at this lower concentration does not have adequate solubilizing power to keep the drug in solution. These findings suggest that simply diluting the commercially available loxapine formulation does not reduce the extent of myotoxicity.

In contrast, loxapine encapsulated in a liposomal formulation significantly ($p < 0.05$) reduced muscle damage by 80% compared to the commercially available formulation (including drug and the solvent system) diluted to the same concentration. However, the tissue damage was approximately seven times higher compared to normal saline. It appeared that this dispersed system formulation was able to significantly reduce tissue damage associated with loxapine administration. This suggests that this formulation might be useful for intramuscular administration from the perspective of reducing tissue damage. However, it still needs to be determined if the same results are observed following intramuscular administration in an animal model.

7. IN VIVO MYOTOXICITY OF A LOXAPINE LIPOSOMAL FORMULATION

The previous in vitro studies indicate the potential of this liposomal formulation to reduce the degree of muscle damage following intramuscular injection. However, this model only looks at the acute short-time effects and specific interactions of the formulation with the muscle tissue. It is therefore critical for the investigator to determine in animal experiments whether the presence of an intact blood flow system and the drug absorption process may alter the findings. It is possible that the toxicity could be enhanced if the dispersed system causes changes in the vascular permeability, thus allowing the potential for inflammatory mediators to reach the injection site. Alternatively, it is possible that blood flow causing absorption from the site of injection could reduce the extent of tissue damage as the formulation components could be diluted at the injection site.

The rabbit and rodent have been the primary in vivo models used to assess tissue damage following injection. The markers of tissue damage that have been used primarily are the release of cytosolic enzymes such as creatine kinase or lactate dehydrogenase and histological evaluation. The specific experimental considerations, advantages, and disadvantages of using these two animal models have been previously discussed (16).

The cannulated rodent model may be the preferred model in the testing of dispersed systems for their in vivo myotoxicity because of the ease in injection, handling, and blood sampling.

The volume of formulation needed for the injection procedure is smaller than for other animals and myotoxicity can be easily assessed over a relatively short period of time compared to larger animals (usually no longer than 12–36 hr). In these studies, the carotid artery was cannulated for blood sample determination and the test formulations (0.3 mL) were injected into the thigh muscle (musculus rectus femoris). Myotoxicity was assessed by measuring plasma creatine kinase levels over a 12 hr period and calculating the area under the curve using the trapezoidal rule. In preliminary work, we have found that plasma creatine kinase levels peak at approximately 2 hr postinjection for all formulations and return to baseline serum levels by 12–24 hr.

The formulations investigated in this study were similar to the in vitro studies: phenytoin; normal saline; loxapine 50 mg/mL; loxapine 10.2 mg/mL; and loxapine liposomes 10.2 mg/mL. The hypothesis was to test whether a liposomally encapsulated loxapine would be less myotoxic than the commercial formulation or the commercial formulation diluted to the same concentration as the liposomal formulation. It was predicted that the liposomal formulation would be less myotoxic because of the potential to limit the exposure of the muscle tissue to the drug by slowing the release of the drug over time at the site of injection and/or due to the absence of the solvent system. Furthermore, in the diluted solution, there could be the potential for the loxapine to pre-cipitate at the site of injection, which could contribute to the myotoxicity. In this study, the concentration of the diluted loxapine solution was matched to the concentration of the lox-apine in the formulated liposomal treatment (10.2 mg/mL) in order to compare the two formulations at the same drug con-centration. The results of these studies are shown in Fig. 5. As in the previous studies, the undiluted commercially available loxapine formulation was more myotoxic, in this case six times, than the positive control phenytoin. The higher myo-toxicity associated with loxapine formulations compared to phenytoin could be a function of differences in the solvent vehicles. The phenytoin (Dilantin) formulation used in these studies contained 40% propylene glycol and 10% ethanol at a pH of 12. The phenytoin formulation has been shown to have the potential to precipitate at the injection site (17). It

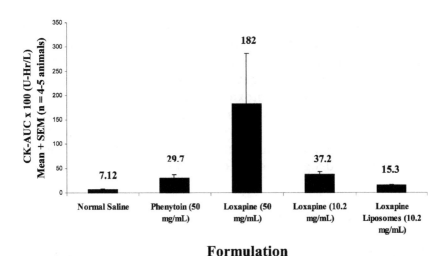

Figure 5 In vivo myotoxicity of liposomal loxapine (10.2 mg/mL), the commercially available loxapine formulation diluted to the same concentration, the commercially available loxapine formulation at 50 mg/mL, phenytoin (50 mg/mL), and normal saline. Mean values are shown above each bar graph.

is unknown as to whether the loxapine formulation precipitated at the site of injection.

The diluted commercially available loxapine formulation was similar in the extent of the myotoxicity to that of phenytoin. In contrast, there was only a two-fold difference between the in vivo myotoxicity of the liposomal formulation of loxapine compared to normal saline. These findings suggest that the liposomal encapsulation of loxapine causes less myotoxicity compared to the commercially available formulation associated with injection. It is unclear as to the specific mechanism responsible for this reduced myotoxicity. It could be associated with the lack of a solvent system in this formulation. Organic co-solvents have been shown to cause myotoxicity (11–13). Alternatively, it could be due to the slow release of the loxapine from the liposomes, thus minimizing the drug concentration that muscle fibers are exposed to at the injection site and any potential for precipitation.

In summary in vitro and in vivo studies indicated that this liposomal formulation could reduce tissue damage at the injection site. In subsequent pharmacokinetic intravenous and intramuscular studies with this dispersed system, it was suggested that this liposomal formulation could provide sustained drug delivery compared to the loxapine solution following an intramuscular injection (3). Possible mechanisms to explain the reduced myotoxicity include the absence of the co-solvent propylene glycol and the surfactant polysorbate 80 from the formulation and/or the slow release of loxapine from the liposomal formulation, thus minimizing the concentration of the drug at the injection site.

8. CONCLUDING REMARKS

In this case study, we have demonstrated the importance of the formulator determining the extent of potential tissue damage caused by the drug and/or other formulation components early during the development process. The type of tissue damage is certainly a function of whether the drug will be administered intravenously, subcutaneously, intramuscularly, or via some other parenteral route. The selected route of administration, in turn, will determine what type of in vitro studies should be conducted during the formulation development phase to determine the extent of tissue damage caused by the drug and/or other formulation components (1). The isolated rodent muscle model has been shown to be a useful and rapid system to optimize a given formulation with respect to minimizing tissue damage for numerous drugs and routes of administration. These in vitro studies do not preclude studies in animals to further test for tissue damage and the relationship to the drug absorption and pharmacokinetics. However, it should allow the formulator a rational means to determine what would be the most appropriate formulation to utilize in subsequent animal and clinical studies.

REFERENCES

1. Gupta PK, Brazeau GA, eds. Injectable Drug Development. Techniques to Reduce Pain and Irritation. Denver, CO: Interpharm Press, 1999.

2. Brazeau GA, Cooper BC, Svetic KA, Smith CL, Gupta P. Current perspectives on pain upon injection of drugs. J Pharm Sci 1998; 87:667.

3. Al-Suwayeh SA. Development of an intramuscular liposomal formulation for the antipsychotic drug loxapine. Thesis, University of Florida, 1997.

4. Arrowsmith M, Hadgraft J, Kelleway IW. The in vivo release of cortisone esters from liposomes and the intramuscular clearance of liposomes. Int J Pharm 1984; 20:347.

5. Kadir F, Elling WMC, Abrahams D, Zuidema J, Crommelin JA. Tissue reaction after intramuscular injection of liposomes in mice. Int J Clin Pharmacol Ther Toxicol 1992; 30:374.

6. Kadir F, Oussoren C, Crommelin DJA. Liposomal formulations to reduce irritation of intramuscularly and subcutaneously administered drugs. In: Gupta PK, Brazeau GA, eds. Injectable Drug Development. Techniques to Reduce Pain and Irritation. Denver, CO: Interpharm Press, 1999:337.

7. American Hospital Formulary Service Drug Information, American Society of Health-System Pharmacists, Inc., Bethesda, MD, 1999:2013.

8. Simpson M, Cooper TB, Lee JH, Young MA. Clinical and plasma level characteristics of intramuscular and oral loxapine. Psychopharmacology 1978; 56:225.

9. Munyon WH, Salo R, Briones DF. Cytotoxic effects of neuroleptic drugs. Psychopharmacology 1987; 91:182.

10. Meltzer Y, Cola PA, Parsa M. Marked elevations of serum creatine kinase activity associated with antipsychotic drug treatment. Neuropsychopharmacology 1996; 15:395.

11. Brazeau GA, Fung H-L. An in vitro model to evaluate muscle damage following intramuscular injections. Pharm Res 1989; 6:167.

12. Brazeau GA, Fung H-L. Use of an in-vitro model for the assessment of muscle damage from intramuscular injections: in vitro-in vivo correlation and predictability with mixed solvent systems. Pharm Res 1989; 6:766.

13. Brazeau GA, Fung H-L. The effect of organic cosolvent-induced muscle damage on the bioavailability on intramuscular [14]C –Diazepam. J Pharm Sci 1990; 79:113.

14. Al-Suwayeh SA, Tebbett IR, Wielbo D, Brazeau GA. In vitro–in vivo myotoxicity of intramuscular liposomal formulation. Pharm Res 1996; 13:1384.

15. Napaporn, Thomas M, Svetic K, Shahrokh Z, Brazeau GA. Assessment of the myotoxicity of pharmaceutical buffers using an in vitro muscle model: effect of pH, capacity, tonicity and buffer type. Pharm Dev Technol 2000; 5:123–130.

16. Brazeau GA. A primer on in vitro and in vivo cytosolic enzyme release methods. In: Gupta PK, Brazeau GA, eds. Injectable Drug Development Techniques to Reduce Pain and Irritation. Denver, CO: Interpharm Press, 1999:155.

17. Wilensky J, Lowden JA. Inadequate serum levels after intramuscular administration of diphenylhydantoins. Neurology 1973; 23:318.

17

Case Study: In Vitro/In Vivo Release from Injectable Microspheres

BRIAN C. CLARK and
PAUL A. DICKINSON
Pharmaceutical and Analytical R&D,
AstraZeneca, Macclesfield, U.K.

IAN T. PYRAH
Safety Assessment, AstraZeneca,
Macclesfield, U.K.

1. INTRODUCTION

This case study describes in vitro and in vivo characterization performed to support manufacturing scale-up, from laboratory to pilot scale, of an experimental microsphere formulation.

The experimental formulation consists of an active pseudo-decapeptide encapsulated in a poly(lactide-co-glycolide) matrix at a target loading of approximately 8% w/w. Microspheres were manufactured in a continuous manner by a proprietary process (1) involving the formation of an oil-in-water emulsion, extraction of the organic phase, drying and collection of microspheres in the size range 25–125 μm. Mannitol is

543

added to the extraction phase to facilitate subsequent handling. Three formulation variants, based on polymers differing in lactide:glycolide ratio and molecular weight distribution, were investigated in pre-clinical studies and in vivo and in vitro release characteristics were determined. A preliminary in vitro–in vivo correlation was developed for using a rat model, supporting the use of the in vitro release test to select batches for pre-clinical use.

The formulation was designed for parenteral administration by the subcutaneous or intramuscular routes, and hence was required to be sterile. The feasibility of terminal sterilization by γ-irradiation using a 25 kGy cycle was investigated, with reference to impurity levels, active agent concentration and long-term stability.

Microspheres were filled into vials for long-term storage. Immediately prior to use, each vial was shaken to break up agglomerated material, an aliquot of suspending medium added, and the vial shaken again to suspend the microspheres. A unit dose was then drawn into a hypodermic syringe and administered by the chosen route. The short-term stability of the formulation in the sus-pending medium was investigated, with particular emphasis on extraction of active substance into the suspending medium.

2. IN VITRO STUDIES

2.1. Characterization Studies

The morphology and chemical homogeneity of active and placebo microspheres were investigated using scanning electron microscopy (SEM), mercury intrusion porosimetry, infrared spectroscopy (IR), and static secondary ion mass spectrometry (SSIMS). The changes that take place during in vitro dissolution testing were studied using these techniques and in addition nuclear magnetic resonance spectroscopy (NMR) and gel permeation chromatography (GPC) were used.

Freshly manufactured microspheres were all similar in that they comprised spherical or near-spherical particles in the size range 25–125 μm (Fig. 1).

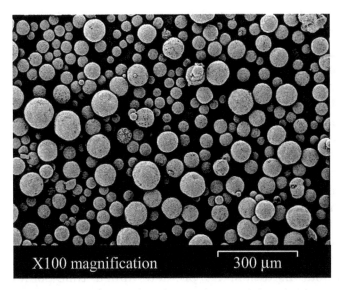

Figure 1 SEM image showing freshly manufactured active microspheres (×100 magnification).

SEM images at high magnification (Fig. 2) showed the presence of a discontinuous surface coating of rod-like crystals, determined by IR and SSIMS to be mannitol (containing residual drug in the case of the active microspheres). Furthermore, the active microspheres, but not the placebo, exhibited voids that were confined to the surface layer of the particles. These were distributed randomly, in patches or in bands,

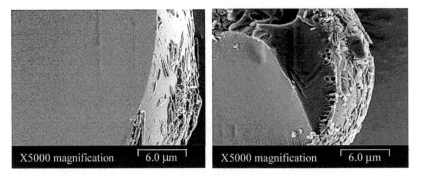

Figure 2 SEM images of placebo (left) and active (right) microsphere sections (×5000 magnification).

and were shown by mercury intrusion porosimetry to be non-porous.

Cross-sections of active microspheres, embedded in epoxy resin, were examined, in a north–south and east–west direction at 10 μm intervals to determine the distribution of drug throughout the polymer matrix (Fig. 3). The data, expressed as peak area for each absorption band examined, show a homogeneous distribution of drug associated with polymer throughout the sample.

Further SEM analysis was performed on cross-sections of the agglomerated mass evident following 7 days exposure to dissolution medium in a preliminary in vitro release test. In this test, 10 mg of microspheres were placed in a vial, 10 mL of pre-heated isotonic phosphate-buffered saline (PBS, pH 7.4) was added, and the vial sealed and incubated at 37°C without agitation. At pre-determined time-points (typically 1,4,7,11,14,18,21 and 28 days), 5 mL samples of supernatant liquid were withdrawn and the concentration of dissolved drug measured by UV spectroscopy, and 5 mL of pre-heated replacement buffer were added to the dissolution vial to maintain constant volume. Sink conditions may be

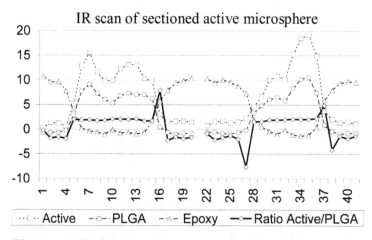

Figure 3 IR imaging of a sectioned active microsphere, showing chemical homogeneity for both north–south (left hand group, 1–20 μm) and east–west (right hand group, 22–41 μm) traverses, as evidenced by constant active/PLGA ratio.

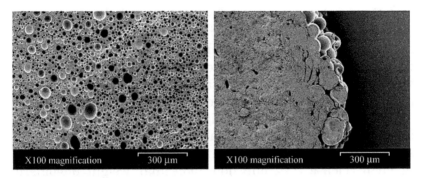

Figure 4 SEM images (×100 magnification) of placebo (left) and active (right) microspheres after 7 days of the preliminary in vitro release test (isotonic PBS, pH 7.4, 37°C).

assumed as the solubility of the active substance is greater than 100 mg/mL in the dissolution medium.

Low magnification images (Fig. 4) reveal extensive agglomeration of microspheres; the image for the active sample shows the presence of partially agglomerated microspheres at the edge of the mass. Also apparent is the development of porosity, extensive in the case of the placebo sample.

High magnification images (Fig. 5) show that extensive porosity is also apparent in the active sample, pores generally being small in comparison with those in the placebo sample.

SSIMS and IR spectroscopy showed that the drug was still distributed throughout the bulk of these samples and that the coating of mannitol was no longer present.

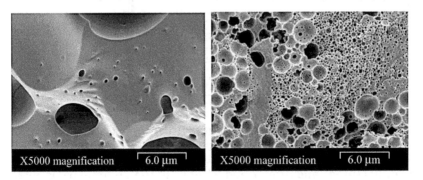

Figure 5 SEM images as for Fig. 4, but ×5000 magnification.

Further investigations after 21 days of the preliminary in vitro release test established that the remaining undissolved material was semi-liquid and therefore not amenable to SEM analysis. A combination of NMR, IR, SSIMS, and GPC was used to investigate chemical changes in the polymer that may have occurred during in vitro testing. GPC clearly shows a reduction in its molecular weight after 7 days of in vitro testing and this reduction was greater for the placebo than for the samples containing the drug. After 21 days of the preliminary in vitro release test, only low molecular weight oligomeric species were found by GPC (Fig. 6 and Fig. 7).

NMR revealed that the lactide content had increased, in comparison to glycolide, but no monomeric species was detected. Furthermore SSIMS did not detect degradation products or additional contamination on the "day 7" samples.

IR spectroscopy showed that some of the peaks in the IR spectra of the drug occurred at slightly different positions when the drug was incorporated into the microspheres (specifically a shift in the peptide backbone carbonyl signal from 1645 to 1670

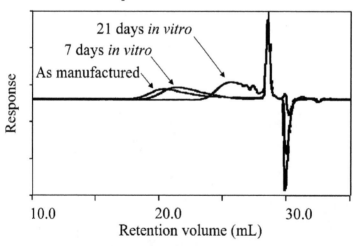

Figure 6 GPC volume-elution curves for active microspheres "as manufactured" and after 7 and 21 days of the preliminary in vitro release test (isotonic PBS, pH 7.4, 37°C).

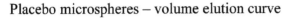

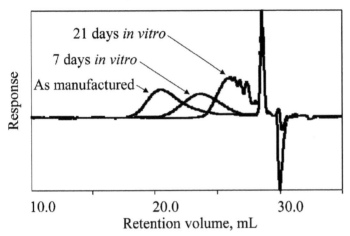

Placebo microspheres – volume elution curve

Figure 7 GPC volume-elution curves for placebo microspheres "as manufactured" and after 7 and 21 days of the preliminary in vitro release test (isotonic PBS, pH 7.4, 37°C).

cm^{-1}). This indicates an interaction (possibly hydrogen bonding) between the drug and the PLGA copolymer.

The glass transition temperature, T_g, of the active microspheres was found to vary from 43.3 to 45.6°C for freshly manufactured microspheres. After exposure to dissolution medium for 24 h, the T_g was re-measured and values were found to vary from 25.3 to 27.7°C, indicating extensive hydration.

In summary, it appears that active microspheres are chemically homogeneous save for a discontinuous surface coating of mannitol enriched with drug. The loss of this coating is rapid, and the surface drug is likely to contribute significantly to the initial burst in the in vitro release profile. It is considered likely that the surface voids apparent in the SEM images were formed in the later stages of the microsphere-hardening process due to the loss of solvent from the globule surface, and are absent from the placebo due to reduced viscosity. The alternative hypothesis that voids are due to loss of active drug into the aqueous phase during hardening is not supported, as IR spectroscopy gave no indication of surface drug depletion.

Annotated *in vitro* release profile for experimental microspheres

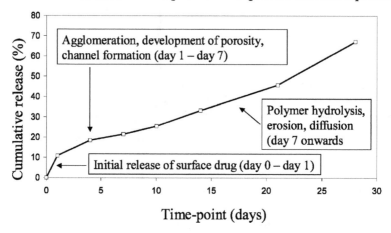

Figure 8 Annotated in vitro release profile for the experimental microspheres in the preliminary in vitro release test (isotonic PBS, pH 7.4, 37°C), indicating the predominant factors for each phase of drug release.

Drug release in the preliminary in vitro release test is characterized by agglomeration/amalgamation of microspheres into a continuous mass with a network of voids. Continuous release of drug occurs by a process involving hydrolytic degradation, reduction in molecular weight, erosion, channel formation and diffusion, as indicated in Fig. 8.

2.2. Stability Studies

Stability studies on active microspheres (pre- and post-irradiated) were carried out and the results are summarized in Tables 1 and 2.

2.2.1. Summary of Stability Results

A preliminary stability study encompassing light exposure [>1.2 million lux-hours, alongside a light-protected (dark) control to assess the effect of temperature variations within the light cabinet], postulated long-term storage conditions (4°C, 25°C/60% RH), accelerated conditions (30°C/60% RH)

Table 1   Stability Results for Non-Irradiated Active Microspheres (6 min Time-Point)

Parameter	Initial	4°C	Light	Dark control	25°C/60 % RH	30°C/60 % RH	40°C	50°C
Appearance	A homogeneous, white, free-flowing solid						An off-white aggregate	
Active agent	100.0	98.7	102.6	106.6	100.0	101.3	98.7	94.7
Total impurities, %	4.81	4.36	4.45	4.74	4.29	4.53	5.68	10.9
Water content, %w/w	0.7	0.6	1.0	1.0	0.7	0.9	0.5	0.4
Residual solvents, %w/w	1.9	2.0	1.8	1.7	1.5	0.3	1.5	1.6
Weight-average molecular weight (Mw), kDa	N/T	10.0	N/T	N/T	11.2	10.7	7.5	4.9
Burst, %	N/T	0.9	N/T	N/T	0.5	0.1	19.1	65.3

N/T: not tested.

Table 2   Stability Results for γ-Irradiated Active Microspheres (6 min Time-Point)

Parameter	Initial	4°C	Light	Dark control	25°C/60 % RH	30°C/60 % RH	40°C	50°C
Appearance	A homogeneous, white, free-flowing solid						An off-white aggregate	
Active agent	96.1	97.4	90.8	98.7	96.1	94.7	92.1	84.2
Total impurities, %	6.80	6.53	6.84	6.25	6.72	9.43	8.05	13.8
Water content, %w/w	0.8	0.7	1.0	0.8	0.8	1.0	0.7	0.4
Residual solvents, %w/w	1.8	1.8	1.7	1.6	1.3	0.2	1.0	1.1
Weight-average molecular weight (Mw), kDa	N/T	10.5	N/T	N/T	9.4	9.1	6.7	4.4
Burst, %	N/T	1.9	N/T	N/T	0.5	3.2	27.5	78.4

N/T: not tested.

and stress conditions (40°C, 50°C), was conducted over a period of 6 months. Samples were stored in unopened vials, within a secondary aluminum container except for the light exposure samples. All vials were stored inverted with the exception of the control samples at 4 °C and 25°C/60% RH, in order to assess the potential for interaction with the closure system. A summary of the results is presented in Table 1 (non-irradiated active microspheres) and Table 2 (γ-irradiated microspheres). All active agent results are expressed as a percentage of the initial assay for the non-irradiated microsphere so that degradation due to γ-irradiation and storage can be assessed.

The results for active microspheres demonstrate that terminal sterilization by γ-irradiation at a standard dose of 25 kGy gives rise to immediate degradation leading to an approximate 4% loss of active agent and a corresponding approximate 2% increase in total impurities; the poor mass-balance was not investigated but is assumed to be due to chromophore degradation and/or the formation of drug-polymer adducts not eluted by the chromatographic assay procedure. Non-irradiated microspheres appear to be stable at conditions up to and including 30°C/60% RH, and to be unaffected by light exposure, whereas γ-irradiated microspheres exhibit some degradation in light and at 30°C/60% RH.

At higher temperatures, degradation is significant and temperature related, and is more pronounced for γ-irradiated microspheres. Degradation is indicated by aggregation, a reduction in active agent assay, a reduction in polymer molecular weight, an increase in total impurities, and an increase, very marked for the 40 °C and 50°C storage conditions, in the in vitro burst. The subsequent release profile is much less affected by storage conditions, as shown in Figs. 9 and 10. The increase in the in vitro burst correlates with the observed decrease in weight average molecular weight, and is considered to arise partly due to an increase in erosion (loss of polymer and associated drug from the surface of the dissoluting material) and partly due to an increase in drug mobility arising from the reduction in polymer viscosity.

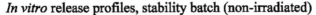

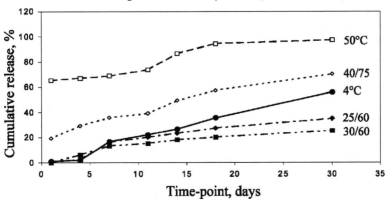

Figure 9 In vitro release profiles (isotonic PBS, pH 7.4, 37°C) for the non-irradiated stability batch after 6 months storage at 4°C, 25°C/60% RH ('25/60'), 30°C/60% RH ('30/60'), 40°C/75% RH ('40/75'), and 50°C.

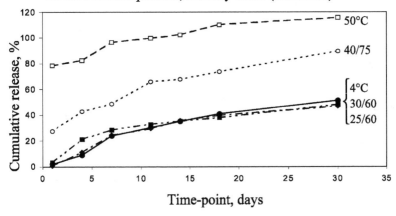

Figure 10 In vitro release profiles (isotonic PBS, pH 7.4, 37°C) for the γ-irradiated stability batch after 6 months storage at 4°C, 25°C/60% RH ('25/60'), 30°C/60% RH ('30/60'), 40°C/75% RH ('40/75') and 50°C.

The results of the exploratory stability study are considered to support a storage life, for γ-irradiated microspheres, of 6 months when stored at 2–8°C in the dark.

2.3. Comparison of In Vitro Release Profiles

Microspheres manufactured at laboratory (F2) and pilot (F1) scale were compared using the preliminary dissolution test (see Sec. 2.1) (Fig. 11). No formal comparison of release profiles could be made, as the available data were insufficient to support the development of a mathematical model, and the use of different test time-points prohibited the determination of similarity/difference factors. However, the correspondence of the middle section (7–21 days) of the release profiles for F1 and F2 was considered sufficient to support the use of pilot scale material for in vivo studies.

2.4. Preliminary In Vitro–In Vivo Correlation

The preliminary in vitro release test described in Sec. 2.1 was considered deficient in that microspheres are uncontained,

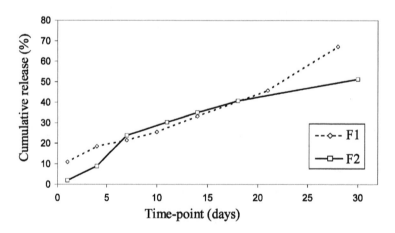

Figure 11 Comparison of in vitro release profiles in the preliminary in vitro release test (isotonic PBS, pH 7.4, 37°C) for laboratory (F2) and pilot (F1) scale microsphere batches, showing correspondence between 7 and 21 days.

leading to sampling difficulties and poor reproducibility at early time-points; release after 28 days was low in comparison with in vivo data, perhaps as a result of sample agglomeration in the non-agitated dissolution vessel, and the test did not discriminate between formulation variants expected to exhibit different release profiles.

An improved in vitro release test was developed, using a sequential factorial experimental design (FED) approach to maximize discrimination between two batches known to differ with regard to in vivo release profile.

The factors investigated in the FED studies are shown in Table 3.

Temperature and pH of the dissolution medium were identified as being the most important factors, followed by the chemical composition and osmolarity of the buffer system, sample size and mixing by agitation. Other factors were of lesser importance.

The following test parameters were selected for further investigation: 100 mg of microspheres were placed in a 4 cm length of dialysis tubing and the ends clipped. The tubing was placed in a 60 mL amber glass vial, 25 mL of pre-heated

Table 3 Variable Factors Investigated in FED Optimization Studies for the In Vitro Release Method

Variable	Range investigated	Significance
Buffer system	Acetate, phosphate, citrate/phosphate/borate	Intermediate
Buffer pH	5–9.6	High
Buffer osmolarity	200–500 mOsmol/L	Intermediate
Osmolarity adjuster	NaCl, Na_2SO_4	Intermediate
Temperature	30–45°C	High
Catalyst	Piperidine 0–1.0%	Low
Surfactant	Polyethylene glycol, 0–0.5%	Low
Sample containment	None, dialysis membrane (8000 Da cut-off)	Low
Agitation	Y/N	Intermediate (after day 12)
Sample size	10–100 mg	Intermediate

isotonic citrate-phosphate-borate buffer (pH 9.5) containing
0.5% surfactant added (this was done to aid wetting of the
sample, even though the FED study did not identify surfactant as an important variable), the vial sealed, and incubated
at 37°C with continuous agitation. At pre-determined time-
points (typically 1,4,7,11,14,18,21 and 28 days), 5 mL samples
of supernatant liquid were withdrawn and the concentration
of the dissolved drug measured by UV spectroscopy, and
5 mL of pre-heated replacement buffer was added to the dissolution vial to maintain constant volume. Sink conditions may
be assumed, even within the dialysis tubing, as the solubility
of the active substance is greater than 100 mg/mL of dissolution medium.

   This dialysis sac diffusion method was applied to three
active microsphere batches formulated to give low (approximately 1%), intermediate (approximately 10%), and high
(approximately 20%) burst in vivo using a rat model; in vivo
and in vitro release profiles are compared in Figs. 12–14.

   The purpose of the comparison was primarily to establish
whether the in vitro release method was capable of discrimi-

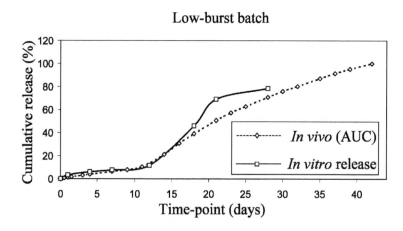

Figure 12 Comparison of in vivo and in vitro release for a low-
burst batch (dialysis sac diffusion method, citrate-phosphate-borate
buffer, pH 9.5, 37°C).

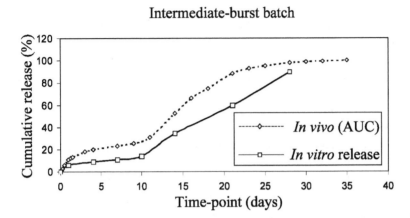

Figure 13 Comparison of in vivo and in vitro release for an intermediate batch (dialysis sac diffusion method, citrate-phosphate-borate buffer, pH 9.5, 37°C).

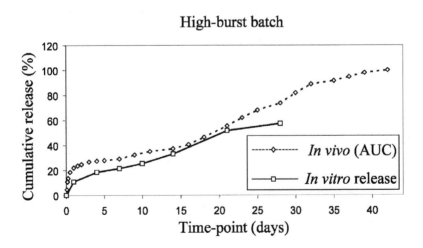

Figure 14 Comparison of in vivo and in vitro release for a high-burst batch (dialysis sac diffusion method, citrate-phosphate-borate buffer, pH 9.5, 37°C).

nating between batches differing in initial burst, and therefore in vitro testing was not continued to 100% release. The in vitro method places the batches in the correct rank order with regard to in vivo release. Pooling data for the low-, intermediate-, and high-burst batches and performing a regression analysis resulted in a second-order polynomial fit with an R^2 value of 0.88, as shown in Fig. 15.

The in vitro–in vivo correlation is considered preliminary as the measurement of in vitro release was not continued until 100% release; the in vivo and in vitro sampling regimes differ and it is possible that the in vivo measurements do not adequately model the initial burst release. For the correlation to be useful in assuring the performance of a marketed product, these issues would need to be addressed and studies repeated in human subjects. However, the preliminary in vitro/in vivo correlation was considered adequate to support the use of the in vitro release method for the selection of batches for toxicological and pharmacokinetic studies in animal models.

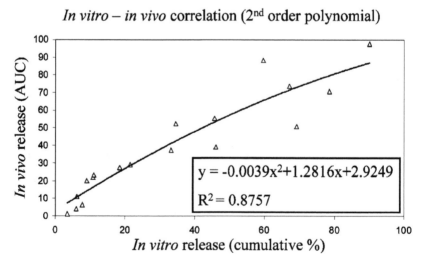

Figure 15 Preliminary IVIVC for high-, intermediate-, and low-burst batches (dialysis sac diffusion method, citrate-phosphate-borate buffer, pH 9.5, 37°C).

2.5. Suspending Medium

The experimental formulation is stored dry and is mixed with an aliquot of suspending medium immediately prior to administration. The suspending medium is an isotonic solution of sodium carboxymethylcellulose (viscosity improver), polysorbate 80 (surfactant), and sodium chloride (tonicity adjuster) designed to hold microspheres in a homogeneous suspension to allow withdrawal and administration of an accurate dose.

2.5.1. Investigation of Dose Homogeneity for Repeat Dosing

Microspheres for administration in pre-clinical studies were supplied in multi-dose vials and the viability of sequential withdrawal of individual doses at 5 min intervals was investigated. The results are presented in Table 4.

The results exhibit some variability but were considered to support the use of this dosing regimen in pre-clinical studies.

2.5.2. Release of Drug into Suspending Medium

It was intended that repeat dosing would take place over a period of about 30 min; a study to assess the extent of release of drug into the suspending medium at room temperature, prior to injection, was performed and the results are presented in Table 5.

Table 4 Homogeneity of Suspension

Dose number	Nominal dose (%)
1	71.2
2	102.4
3	96.9
4	95.0
5	93.4
6	81.3
7	103.0
Average	91.9
RSD (%)	12.7

Table 5   Drug Release from Microspheres into Suspending
Medium

	Drug release (% nominal)		
	---	---	---
Months	1.0 g/2.5 mL dilution	1.0 g/3.4 mL dilution	1.0 g/4.5 mL dilution
5	0.04	0.31	0.35
10	0.04	0.34	0.39
30	0.05	0.43	0.45
60	0.06	0.50	0.53

The extent of drug release into the suspending medium under the conditions of the study was considered to be acceptable for all dilutions, and this administration procedure was applied in pre-clinical dosing.

2.5.3.   Stability of Suspending Medium

While the suspending medium was shown to be stable under normal conditions (Table 6), some degradation was evident at high temperatures (40°C and above) and on exposure to light. Degradation was characterized by a reduction in pH, an increase in osmolality, and a reduction in viscosity. After 6 months at 50°C, slight turbidity was observed. These changes are thought to be due to breakdown of sodium carboxymethylcellulose and polysorbate 80 leading to the formation of acidic contaminants.

The data are considered to support an interim storage life of 12 months at 20–25°C in the designated container.

3.   IN VIVO STUDY

This study was primarily performed to ensure that the larger (pilot scale) batch and new suspending media produced a formulation with a similar release profile in vivo to that produced by an earlier smaller (laboratory scale) batch and using a different suspending medium. The effect of dose volume (microsphere concentration) was also investigated.

Table 6 Stability Results for Suspending Medium (6 m Time-Point)

Parameter	Initial	4°C	Light	Dark control	25°C/60% RH	30°C/60% RH	40°C	50°C
Appearance	Clear colorless solution free from extraneous matter							Slightly opaque solution
pH	5.4	5.3	5.1	5.3	5.2	5.0	4.8	4.5
Volume average (mL)	5.0	5.0	5.0	5.0	5.0	5.0	5.0	5.0
Osmolality (mOsm/kg)	302	302	303	302	302	303	304	312
Viscosity (cP)	2.0	2.0	1.6	2.1	2.0	2.0	1.7	1.4

3.1. Experimental Procedures

3.1.1. Formulations

Larger batch manufacture (F1) in suspending medium 1 (lower
   viscosity):

Injection volume = 0.62 mL   32 mg mL^{-1} drug concentration
                                (40% w/v microspheres)
Injection volume = 0.83 mL   24 mg mL^{-1} drug concentration
                                (31% w/v microspheres)
Injection volume = 1.25 mL   16 mg mL^{-1} drug concentration
                                (21% w/v microspheres)

Smaller scale manufacture (F2) in suspending medium 2 (higher
   viscosity):

Injection volume = 1.5 mL   13.3 mg mL^{-1} drug concentration
                                (19% w/v microspheres)

3.1.2. Study Design

Four dosing groups of 12 male Sprague–Dawley rats (48 in total,
390–453 g) received a single subcutaneous injection in the scruff
of the neck containing 20 mg of drug formulated as described in
Sec. 3.1.1. Each dosing group was sub-divided into four sampling groups for removal of blood samples at various times up
to 42 days post-dose. On termination, a representative number
of injections sites were dissected and any formulation remaining
assayed for drug. Four sampling groups were employed to
ensure that sample volumes did not exceed accepted limits as
discussed in Sec. 3.2.2 of Chapter 4. This approach means that
only mean data can be reported. Rats were chosen as the least
neurophysiological-sensitive animal suitable to meet the aims
of the study. The use of this rat data in an attempt to start to
establish IVIVCs suffers from the fact that mean data must be
used and the reference formulation must be dosed to a separate group of animals (see Chapter 5). Blood was collected into
tubes containing protease inhibitors that had previously been
stored in iced water; the plasma was then immediately separated by centrifugation at 4°C and the plasma transferred to
tubes and frozen immediately.

   Drug plasma concentrations were determined by high performance liquid chromatography and electrospray ionization

tandem mass spectroscopy after solid phase extraction of the drug. The amount of drug remaining encapsulated at the injection site was determined by HPLC-UV after extraction. All data manipulations were carried out using standard methods. The concentrations were adjusted for drug dose and rat weight as appropriate. When the drug concentration was below the limit of quantification it was considered to be $0 \, ng \, mL^{-1}$ in animals which received a full dose and for animals that did not receive the full dose, that data point was removed from the calculation of mean concentration for that time-point. The area under the plasma concentration (AUC) time curve was determined using the linear trapezoidal rule on mean data due to sparse sampling points for individual animals. The burst, defined as the release over 1.5 days as a proportion of the release over 30 days, was approximated by comparing AUC for these time intervals. The fraction absorbed vs. time was also approximated by comparing the AUC for the time interval to the AUC from 0 to 42 days for that formulation or to the AUC from 0 to 42 days measured for the F1 injection volume $= 0.83 \, mL$ formulation. It should be noted that approximating the absorption rate using AUC, for these data, might underestimate the rate of absorption at the earlier time-points (days 1–3). As this was an early development study, sophisticated IVIVC approaches (inclusion of reference formulation) were not pursued. Later studies did incorporate these approaches.

3.2. Results and Discussion

Mean drug plasma concentration times curves after administration of drug encapsulated in microspheres from the larger and smaller batches are shown in Fig. 16. Pharmacokinetic parameters for the data are given in Table 7.

Larger batch formulations produced a greater initial drug plasma concentration than that from the smaller scale formulation. The plasma concentration then declined to a nadir at days 4–5 which was earlier and lower than for the smaller scale batch. The plasma concentration rose to a "plateau" value at days 7–9 and then tracked the smaller scale formulation. Comparison across the three F1 formulations suggests that, as injection volume increased (from 0.62 to 1.25 mL) the

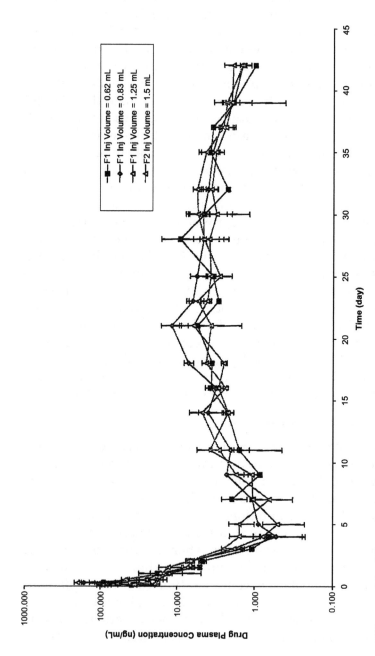

Figure 16 Mean drug plasma concentration time curves after administration of 20 mg of drug encapsulated in microspheres manufactured on two different scales ($n = 3$, mean $\pm$ SEM for each time-point, 12 animals in total).

Table 7 Mean PK Parameters after SC Administration of 20 mg of Drug Encapsulated in Microspheres

Pharmacokinetic parameter	Formulation F1 Inj. vol. = 0.627 mL	Formulation F2 Inj. vol. = 0.83 mL	Inj. vol. = 1.25 mL	Inj. vol. = 1.5 mL
$AUC_{(0-1.5d)}$ (ng mL^{-1} day)[a]	64.8	85.8	99.9	61.23
$AUC_{(0-4d)}$ (ng mL^{-1} day)[a]	74.2	98.4	113.7	78.9
$AUC_{(0-28d)}$ (ng mL^{-1} day)[a]	191.0	273.5	223.2	196.4
$AUC_{(0-30d)}$ (ng mL^{-1} day)[a]	213.6	288.5	234.0	212.1
$AUC_{(0-42d)}$ (ng mL^{-1} day)[a]	265.5	345.5	285.2	279.4
C_{max} (ng mL^{-1})[a]	133.5	242.8	311.4	73.5
t_{max} (day)	0.25	0.25	0.25	0.5
$C_{ss,av}$ 7–28 days (ng mL^{-1})[a]	5.56	7.60	4.83	5.11
Drug remaining at injection site (mg)	1.54[b]	0.38[b]	n.d.[c]	0.34[b]
Burst (%)	30.3	29.7	42.7	28.9

[a]Values normalized for 272 g rat weight.
[b]Mean $n = 2$.
[c]Not determined.

release rate (absorption rate) of the drug increased (Fig. 17). That is, F1 injection volume 1.25 mL gave the greatest burst and then the lowest plateau concentration ($C_{ss,av}$ 7–28 days) suggesting that the rapid early release meant that there was less drug for release later on (Table 7). The total amount released was similar to that for the F1 injection volume 0.62 mL as $AUCs_{(0-42d)}$ were comparable for both formulations. The F1 injection volume 0.83 mL gave the same percent burst as the F1 injection volume 0.62 mL solids formulation; however, the $C_{ss,av}$ 7–28 days was higher as was the $AUC_{(0-42d)}$ suggesting prolonged faster release (Table 7). This is supported by the apparently lower quantity of drug left at the injection site for the F1 injection volume 0.83 mL formulation relative to the F1 injection volume 0.62 mL formulation (Table 7), although the small sample size should be noted. These observations are probably explained by the injection volume which led to different extents of dispersion within the subcutaneous tissue (see Sec. 3.3). That is, the larger injection volumes produced greater dispersion and therefore greater formulation surface area, which might be expected to produce greater drug release. The relative surface area differences due to dispersion extent may have been accentuated by the agglomeration of microspheres, which was also seen in the in vitro tests (Fig. 4). The nature of the injection vehicle may be important as, although the F2 injection volume 1.5 mL formulation was dosed in a volume most similar to the F1 injection volume 1.25 mL formulation, the subcutaneous dispersion and pharmacokinetic parameters were most similar to the F1 injection volume 0.83 mL formulation. Possibly the greater viscosity of the F2 media reduced dispersion in the subcutaneous tissue.

3.3. Injection Site

Injection of the formulation produced the expected tissue reaction (see Sec. 3 of Chapter 4). For the F1 injection volume 0.62 mL formulation, the subcutaneous mass was well defined (Fig. 18), the formulation being encapsulated within a foreign body reaction. As the injection volume increased for the larger batch microspheres, more, smaller masses were found dispersed within the subcutaneous tissue (Figs. 19 and 20).

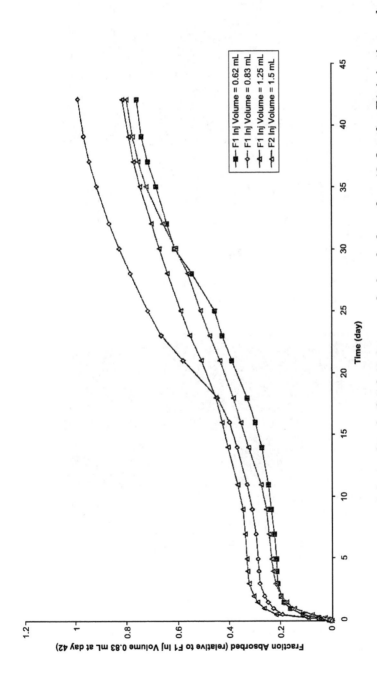

Figure 17 Fraction of drug absorbed relative to total absorbed at day 42 for the F1 injection volume 0.83 mL formulation, as approximated by AUC, after administration of 20 mg of drug encapsulated in microspheres manufactured on two different scales.

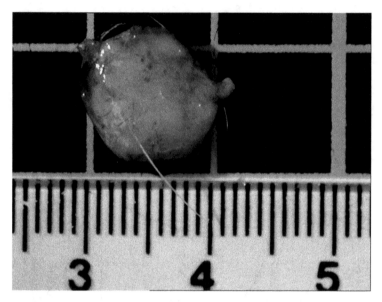

Figure 18 Subcutaneous masses removed from a rat on day 42 which had received the F1 injection volume 0.62 mL formulation.

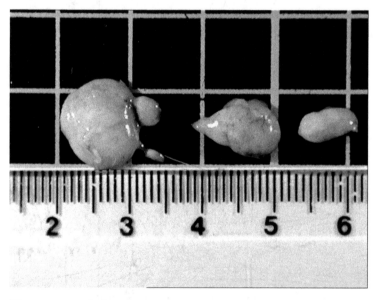

Figure 19 Subcutaneous masses removed from a rat on day 42 which had received the F1 injection volume 0.83 mL formulation.

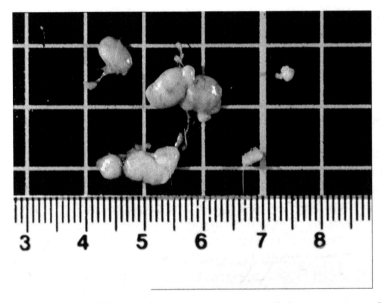

Figure 20 Subcutaneous masses removed from a rat on day 42 which had received the F1 injection volume 1.25 mL formulation.

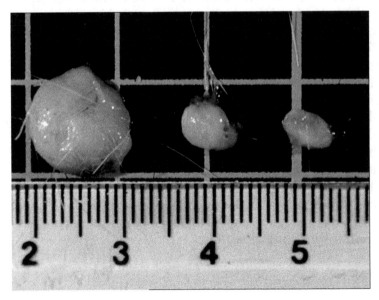

Figure 21 Subcutaneous masses removed from a rat on day 42 which had received the F2 injection volume 1.5 mL formulation.

The F2 formulation produced masses similar to the F1 injection volume 0.83 mL formulation (Fig. 21).

4. CONCLUSIONS

Based on the similarity of the in vitro and in vivo release profiles for microspheres manufactured at pilot scale to those manufactured at laboratory scale, alongside manufacturing performance, the manufacture of active microspheres was considered to have been successfully scaled up from laboratory- to pilot-scale. Terminal sterilization by γ-irradiation at a dose of 25 kGy is feasible, provided the product is stored under appropriate conditions. Particular care must be taken to avoid even short-term temperature excursions, such as may be experienced during transportation of material, as a significant increase in burst release may be expected to arise. An in vitro release method capable of discriminating between low-, intermediate-, and high-burst batches has been developed, and the preliminary in vitro/in vivo correlation for the rat model supports the use of the in vitro release method for batch selection. The use of multi-dose vials in which microspheres are suspended in an appropriate medium, and repeat doses withdrawn for injection, is acceptable for pre-clinical studies of these formulations.

It has been shown elsewhere (Sec. 3 of Chapter 4) that pre-clinical species are likely to give similar release pharmacokinetics as those seen in the clinic for "preformed" controlled release systems and should be sufficiently good for formulation development work. However, this case study has shown that, even within a species, differences can occur depending on the dosing technique/dose volume. Therefore, care should be taken when extrapolating to the clinical situation and development of the optimal or absolute pre-clinical model will require some empirical development based on feedback from initial clinical studies.

REFERENCE

1. Tice TR, Gilley RM. Microencapsulation process and products therefrom. United States Patent Number 5,407,609 (1995).

18

Case Study: Biodegradable Microspheres for the Sustained Release of Proteins

Guidelines for Formulating and Processing Dry Powder Pharmaceutical Products

MARK A. TRACY

Formulation Development, Alkermes, Inc.,
Cambridge, Massachusetts, U.S.A.

1. INTRODUCTION

In developing dry powder pharmaceutical products, there are important general principles that provide valuable guidance in formulating and processing the product. This article will review two important guidelines and provide examples using the development of biodegradable microsphere products as a case study.

The first guideline is to maximize product stability by minimizing molecular mobility. Minimizing molecular mobility can be achieved by selecting formulation components and processing unit operations and conditions that reduce the probability that formulation components will interact and change from the desired form. Examples include processing proteins at reduced temperature to prevent denaturation and reducing solvent levels in the product to prevent drug reactions or polymer relaxation.

The second guideline is to understand the role of particle structure and morphology in product function and stability. Particle properties can vary tremendously depending on variables such as size, porosity, and morphology. All too often the performance of a pharmaceutical powder formulation depends on the powder microstructure which can be controlled by formulation and processing. Examples include the effect of drug particle size on initial release from microspheres and the effects of protein lyophilizate particle size and morphology on protein stability.

Biodegradable microspheres for the controlled release of drugs, including proteins, provide an excellent illustration of the importance of understanding and utilizing these guidelines. Microspheres provide a means of delivering drugs for periods of days, weeks, or months with a single injection. They consist primarily of a biodegradable polymer, for example poly(lactide-co-glycolide) (PLG), and the encapsulated drug as well as appropriate stabilizers or release modifiers.

2. GUIDELINE #1: MINIMIZE MOLECULAR MOBILITY TO MAXIMIZE STABILITY

An excellent example of this principle is provided by the formulation and processing of biodegradable microspheres containing proteins (1). Proteins are relatively fragile macromolecules whose tertiary structure is important for activity. The free energy difference between the native and unfolded state at room temperature is only of the order of the strength of one hydrogen bond (a few kilocalories/mol) (2). Therefore, it does not take much energy input at room temperature to disrupt a protein's native state. Thus, to produce pharmaceutical products for these drugs, formulation and processing approaches are required which minimize the molecular mobility during processing, after administration, and during storage.

A process has been developed and commercialized to make a long-acting form of the protein human growth hormone (hGH) in which the protein is encapsulated in biodegradable polymeric microspheres (Fig. 1) (1,3,4). The process first involves preparing solid particles of the protein, for example, by lyophilization. The particles are encapsulated by suspending them in a polymer solution containing an organic solvent, spray-freezing the suspension into liquid nitrogen to form nascent microspheres, extracting the polymer solvent in a polymer non-solvent at sub-zero temperatures, and finally drying the product under vacuum to minimize residual solvents.

The solid state form of the protein can be created by spray freeze-drying (SFD). The resulting intermediate is a dry powder itself in which protein mobility is reduced by the removal of water. This powder is exposed to organic

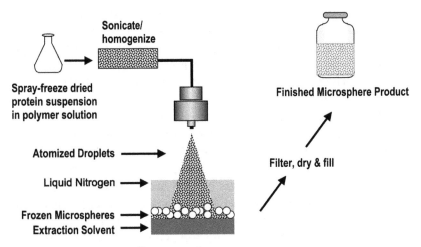

Figure 1 ProLease® encapsulation process steps.

solvents during the encapsulation process. Protein native structure is not thermodynamically favored in organic solvents. Protein stability is maintained in organic solvents because the solid state provides a kinetic trap that maintains the stability (5,6). Key process factors that secure the kinetic trap are the low processing temperature and an anhydrous environment.

The encapsulation process keeps protein mobility minimized by encapsulating the protein in the solid state at low temperatures in the absence of water. Furthermore, the low processing temperature, well below the glass transition temperature of the polymer, minimizes polymer relaxation as well after the microspheres are formed.

Maintaining drug stability and activity after administration is another challenge since the drug has to remain intact for the duration of release (days to months) at physiological conditions (37°C, pH 7). Maintaining protein stability over prolonged time periods has been achieved again using the principle of minimizing mobility. One approach for some proteins, including hGH and others, is to complex them with metal ions such as zinc (3,4,6,7). Zinc-protein complexes are more stable to denaturation (7). Figure 2 shows size exclusion chromatograms for hGH released from microspheres. The integrity of the protein released from microspheres containing zinc-hGH was similar to unencapsulated hGH and better than hGH released from microspheres without zinc (3,4). In essence the zinc binds to the protein forming a solid precipitate which provides the protein in a more rigid, longer lasting form. Another approach is to use salting out additives like ammonium sulfate (6).

Pharmaceutical products are typically required to have a shelf life of 2 years at the target storage temperature. For most products this is either 2–8°C or room temperature. In order to maximize storage stability, it is important to minimize molecular mobility. This can be accomplished by storing the product sufficiently below its glass transition temperature (T_g) to prevent interactions that can adversely affect the product stability over short time periods such as days or weeks to longer periods of months or years. The rule of thumb is to

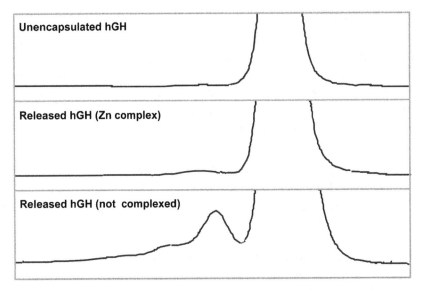

Figure 2 Effect of zinc complexation on hGH stability during release from microspheres in vitro. The top chromatogram represents unencapsulated hGH, the middle hGH released from microspheres with the stabilizer zinc added to form a zinc–hGH complex, and the bottom hGH released without complexing with zinc. The largest peak represents hGH monomer. Peaks to the left of it represent hGH dimers and oligomers. The buffer contained 50 mM HEPES and 10 mM KCl, pH 7.2. Release was carried out at 37°C. (Adapted from Ref. 3.)

store the product at least 30–50°C below its T_g (8,9). For microsphere products, the polymer, poly(lactide-co-glycolide), is the major component and has a T_g about 40°C. In manufacturing, residual solvents in sufficient amounts can act as plasticizers and reduce the T_g promoting mobility. A reduction in T_g can impact the product stability during storage. It is thus important to minimize residual organic solvents in microsphere products to minimize mobility and maximize the product shelf life. This can be accomplished by developing a suitable process for drying microspheres. Protein and peptide microsphere products have been developed with shelf lives of at least 2 years.

2.1. Summary

For protein microsphere products, stability is maximized by formulating the protein in the solid state, encapsulating under a low temperature, anhydrous environment, and minimizing residual solvents. All of these steps have the goal of minimizing the mobility of the protein or polymer during preparation, administration, and storage.

3. GUIDELINE #2: UNDERSTAND THE ROLE OF PARTICLE STRUCTURE AND MORPHOLOGY IN PRODUCT FUNCTION AND STABILITY

The size and microstructure of a dry powder formulation can make the difference between an effective and ineffective product. Particle size or porosity for example can affect dissolution properties of the powder or, as for inhalation powders, their ability to deposit in the appropriate part of the lung for absorption (10). Similarly in biodegradable microspheres, the ability to control size and morphology is critical in obtaining desirable release characteristics and optimal stability.

As noted above, microspheres produced by the process described contain protein particles encapsulated in a polymer matrix. The mechanism of protein release involves first the hydration of the microspheres, followed initially by dissolution and diffusion of protein at or with access via pores to the surface (burst), and finally by release through additional pores and channels created by polymer degradation (11–13). The relative size of the protein particles to the microsphere is an important parameter in controlling the initial release or burst. The particle size of the protein powder can be controlled by varying SFD process parameters (14). Protein particles with sizes from several microns down to less than 1 µm are produced by sonicating particles made by SFD with different mass flow ratios (ratio of atomization air/liquid mass flow rates). Figure 3 shows that encapsulating submicron protein particles results in a substantially lower initial release than particles much larger than 1 micron (14).

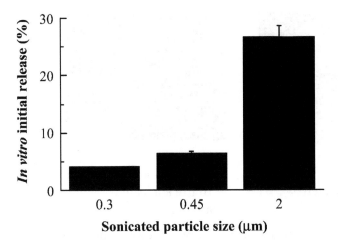

Figure 3   Effects of protein particle size on initial release in vitro. The particle size represents the median protein particle size measured after sonication in a PLG-methylene chloride solution using a Coulter counter. The initial release represents the percentage of protein released from microspheres within the first 24 hr after incubation in a physiological buffer at 37°C.

It is interesting to note how these small protein particles are formed. Spray-freeze drying alone produced particles with a characteristic size of about 10–50 μm, too large to be adequately encapsulated in microspheres whose size is about 50–100 μm. Under sonication or homogenization, these particles break down further into the 0.1–5 μm range suitable for encapsulation. Interestingly, the particles that break down smaller are smaller to start and are characterized by a finer, more friable microstructure (Fig. 4). We hypothesize that the more friable microstructure is created during the spray-freezing step as smaller particles freeze faster creating smaller ice crystals. Upon drying, the smaller ice crystals are sublimated leaving behind the finer microstructure. In fact these powders have been shown to have a higher specific surface area indicative of the finer structure (14).

One possible disadvantage of the finer structure is the potential to affect the stability of the protein. Proteins can denature at the hydrophobic ice interface. We have observed

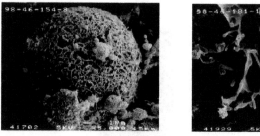 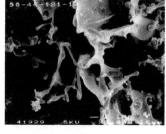

~ 15 μm $D_{V,50}$ unsonicated	~ 45 μm $D_{V,50}$ unsonicated
~ 0.2 μm sonicated	~ 10 μm sonicated
Ratio ~ 75	Ratio ~ 4.5

Figure 4 The morphology of SFD protein powders prepared using different SFD processing parameters. The microstructure of the smaller-sized powder on the left is finer than that on the right. As a result, under sonication, it breaks down into a proportionally smaller particle as indicated by the unsonicated/sonicated particle size ratios given. (The scanning electron micrographs are from Ref. 14.)

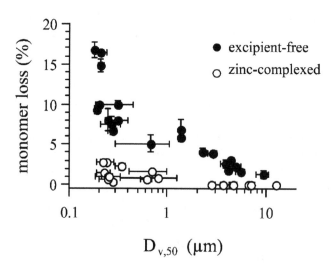

Figure 5 Improving integrity of SFD protein powders. The effect of zinc on the percentage of monomer loss is shown vs. median protein particle size. (Adapted from Ref. 14.)

a correlation between monomer loss, particle size, and specific surface area for BSA in the absence of any stabilizers. However, using the principle of minimizing mobility, proteins like BSA can be stabilized to prevent denaturation. For example, complexing BSA with zinc resulted in a significant enhancement in stability for submicron particles (14). Figure 5 shows that by adding zinc to BSA the monomer loss was greatly reduced in preparing protein particles for encapsulation by SFD. The effect was especially marked for the submicron protein particles. Formulating with sugars also stabilized the protein in the small particles (15).

3.1. Summary

The protein particle size and morphology impacts the release from microspheres and protein stability. Process and formulation variables must both be balanced to optimize particle size, morphology, and stability to produce a microsphere product with optimal release characteristics.

4. CONCLUSIONS

Two key themes have emerged from the development of biodegradable microsphere products for proteins. These themes apply in general to dry powder pharmaceutical products.

1. Maximize powder and drug stability by minimizing molecular mobility through formulation and processing.
2. Understand the effects of particle size and morphology on product function. Identify key process and formulation variables that affect product characteristics and control them.

REFERENCES

1. Tracy MA. Devolpment and scale-up of a microsphere protein New York system. Biotechnol Progr 1998; 14:108–115.

2. Creighton TE. Protein Structures and Molecular Properties. New York: Freeman and Co., 1984.

3. Johnson OL, Cleland JL, Lee HJ, Charnis M, Duenas E, Jaworowicz W, Shepard D, Shahzamani A, Jones AJS, Putney SD. A month-long effect from a single injection of microencapsulated Human Growth Hormone. Nat Med 1996; 2: 795–799.

4. Johnson OL, Jaworowicz W, Cleland JL, Bailey L, Charnis M, Duenas E, Wu C, Shepard D, Magil S, Last T, Jones AJS, Putney SD. The stabilization and encapsulation of human growth hormone into biodegradable microspheres. Pharm Res 1997; 14:730–735.

5. Griebenow K, Klibanov AM. On Protein Denaturation in Aqueous-Organic Mixtures but not in Pure Organic Solvents. J Am Chem Soc 1996; 118:11695–11700.

6. Putney SD, Burke PA. Improving protein therapeutics with sustained release formulations. Nat Biotechnol 1998; 16:1–6.

7. Cunningham BC, Mulkerrin MG, Wells JA. Dimeration of human growth hormone by Zinc. Science 1991; 253:545–548.

8. Hancock BC, Zografi G. Characteristics and significance of the amorphous state in pharmaceutical systems. J Pharm Sci 1997; 86:1–12.

9. Hancock BC, Shamblin SL, Zografi G. Molecular mobility of amorphous pharmaceutical solids below their glass transition temperatures. Pharm Res 1995; 12:799–806.

10. Edwards DA, Hanes J, Caponetti G, Hrkach J, Ben-Jebria A, Eskew ML, Mintzes J, Deaver D, Lotan N, Langer R. Large porous particles for pulmonary drug delivery. Science 1997; 276:1868–1871.

11. Siegel RA, Langer R. Controlled release of polypeptides and other macromolecules. Pharm Res 1984; 1:2–10.

12. Bawa R, Siegel RA, Marasca B, Karel M, Langer R. An explanation for the controlled release of maromolecules from polymors. J Contr Rel 1985; 1:259–267.

13. Saltzman WM, Langer R. Transport rates of proteins in porous materials with known microgeometry. Biophys J 1989; 55:163–171.

14. Costantino HR, Firouzabadian L, Hogeland K, Wu C, Beganski C, Carrasquillo KG, Cordova M, Griebenow K, Zale SE, Tracy MA. Protein spray-freeze drying. Effect of atomization conditions on particle size and stability. Pharm Res 2000; 17:1374–1383.

15. Costantino HR, Firouzabadian L, Wu C, Carrasquillo KG, Griebenow K, Zale SE, Tracy MA. Protein spray-freeze drying. 2. Effect of formulation variables on particle size and stability. J pharm Sci 2002; 91:388–395.

19

Injectable Dispersed Systems: Quality and Regulatory Considerations

JAMES P. SIMPSON

Regulatory and Government Affairs, Zimmer, Inc., Warsaw, Indiana, U.S.A.

MICHAEL J. AKERS

Pharmaceutical Research and Development, Baxter Pharmaceutical Solutions LLC, Bloomington, Indiana, U.S.A.

1. INTRODUCTION AND SCOPE

The aim of this chapter is to provide the reader with an appreciation of the principles and requirements for registering and marketing injectable drug products as dispersed systems. These dosage forms offer unique characteristics having certain distinct advantages over more conventional solid and liquid sterile products. Such unique characteristics also present special challenges in the manufacturing and control of these dosage forms.

2. REGULATORY REQUIREMENTS

2.1. Drug/Device Development and Application for Approval

Whether a manufacturer intends to market products worldwide or only within the United States, numerous regulations must be satisfied before market entry is permitted. These regulations have been established to protect the public safety. Generally, these regulations focus on assuring that the intended drug or device has the safety, identity, strength, quality, and purity (drugs) it purports to have and that devices are safe and effective for their intended use. Requirements for the applications of new drugs and devices can be voluminous and complicated. Regulatory requirements vary depending on the clinical indication and mode of use. To improve the timeliness of the product development and regulatory approval processes, a basic understanding of regulatory requirements is recommended.

2.2. International Considerations

Major new products are being considered for global markets as well as US domestic markets. To sell products abroad, there is usually a product approval or registration process for market entry. Understanding international quality and regulatory standards and requirements can improve speed-to-market and reduce frustration among the professionals responsible for new product approval. These requirements usually come from the country's regulatory body responsible for assuring the safety of the product. A few of these key regulatory bodies are noted in Table 1.

Unfortunately, there is little standardization of product approval requirements by the major regulatory bodies of the world. There are, however, harmonization efforts underway with most major regulatory bodies participating. The International Conference on Harmonization (ICH) has made significant progress in setting global specifications. The intent of these global specifications is that eventually all regulatory bodies, worldwide, will recognize and apply these specifications consistently. The ICH has published guidelines

Table 1 Regulatory Agencies by Country

Country	Regulatory agency
United States	Food and Drug Administration
United Kingdom	Medicines and Healthcare products Regulatory Agency (MHRA)
Japan	Koseishio
Canada	Health Protections Branch
France	Agence du médicament
Germany	Bundesinstitut für Arzneimittel und Medizinprodukte (BfArM) and the Paul Ehrlich Institut
Italy	Istituto Superiore di Sanità
Spain	Ministerio de Sanidad y Consumo

in the areas of quality, safety, and efficacy (see Appendix 1). These guidelines provide an excellent reference for any individual or organization involved in pharmaceutical product development. The hoped for advantage of regulatory harmonization will be the development of mutual agreements between and among regulatory bodies. These agreements will allow one country to accept the drug and device application already approved by other countries.

2.3. Current Good Manufacturing Practices

In the United States, the Food and Drug Administration (FDA) has been given significant regulatory power under the Food, Drug and Cosmetic Act. This Act is enforceable, under Federal power, and contains key terms such as adulteration and misbranding. Good Manufacturing Practices were developed to establish FDA's expectations on how drug and device manufacturers should comply with the Food, Drug and Cosmetic Act. These regulations can be found in the Code of Federal Regulations as follows:

Title 21 Code of Federal Regulations Part 211 (Drugs)
Title 21 Code of Federal Regulations Part 820 (Devices)

These regulations provide the minimum acceptable requirements for manufacturing drug and device products.

Failure to comply with these regulations constitutes adulteration under the Food Drug and Cosmetic Act. Once a product is
deemed adulterated the product or products involved may be
recalled or seized. Manufacturers can be enjoined and long
court battles can result. While the movement towards regulatory harmonization advances, the FDA continues to operate
under its own set of rules. Exemplified by the sheer number
of detailed regulations coupled with the frequency and intensity of regulatory actions, the FDA is considered the world's
premier regulatory body. The FDA has published many useful
guidance documents* on the development and manufacturing
of drug and device products. FDA guidance documents represent this Agency's current thinking on a particular subject.
According to the FDA they do not create or confer any rights
for or on any person and do not operate to bind the FDA or
the public. The FDA will accept alternative approaches, if
such approaches satisfy the requirements of the applicable
statute, regulations, or both. These guidelines are particularly useful since they elaborate FDA expectations and direct
enforcement activity. Additionally, they have been found useful in the development community when scientists are
actively involved in the development or modification of drug
and device products.

Guidance documents cover key topics such as

- advertising,
- biopharmaceutics,
- chemistry guidances,
- clinical guidances,
- compliance guidances,
- generic drug guidances,
- information technology guidances, and
- labeling guidances.

Other guidance documents come in the form of FDA
investigator inspectional guides. These guides provide the

---

*FDA guidance documents can take the form of guidelines to industry,
letters, inspection guides, etc.

industry with insight into the areas of training and special interest FDA investigators and scientists may have for a particular subject. All of these documents, guidelines, and inspectional references are available in the public domain via the internet (http://www.fda.gov).

2.4. Pre-Approval Inspections

In the United States, most drug and certain diagnostic device products require pre-approval inspections (PAIs) before the FDA will grant product approval. Exceptions to this include minor supplements to existing new drug applications (NDAs) or abbreviated new drug applications (ANDAs) where the manufacturer has had recent inspections and/or a good regulatory history. Major delays in product approval can be encountered when these PAIs do not meet FDA expectations. Additionally, weaknesses found during a PAI can direct FDA investigators into other areas within the firm's quality systems or development activities not originally within the scope of the PAI. For example, a poorly designed new product stability protocol can lead FDA investigators into the firm's entire marketed product stability program. In another example, problems discovered on review of high performance liquid chromatograms can lead into an investigation of a firm's entire analytical chemistry quality (QC) procedures. Adverse findings from a new product pre-approval have led to recalls and other serious regulatory actions for products already in the marketplace.

The PAI process was reinforced in 1988 after several firms were found to be providing false and misleading information to the FDA. This was known in the industry as the "Generic Drug Scandal," where applications were filed fraudulently, although manufacturing capability did not exist for the drug products submitted for market approval, and stability data were falsified. Even though the problems were generated by a small number of companies, the entire pharmaceutical industry and the FDA were shaken by the incidents. Many individuals involved were found guilty of criminal activity and punished. Some individuals were banned from the US drug industry. The FDA maintains a black list of individuals who have been banned from developing or manufacturing drugs for US consumption.

As a result of these tumultuous findings new requirements for PAIs were developed. In the United States, product applications sent to Agency centers may be in perfect order but if significant issues arise at the factory or in the development laboratories during the PAI, the entire approval process can be delayed months or even years. It is, therefore, essential that all branches of manufacturing, including research and development departments prepare carefully for the PAI to assure a timely and successful outcome.

Even the best drug and device firms have demonstrated weaknesses in PAIs. The FDA records objectionable findings on a form known as the FDA-483 Notice of Inspectional Findings. Typical manufacturing and quality system weaknesses encountered during PAIs of the 1990s include:

Validation: Weak, faulty, or non-existent validations supporting the following:

- Processes such as mixing, sterilization, potency adjustments.
- Equipment such as fillers, sealers, processing and packaging equipment.
- Utilities: Medical grade product contact gases such as nitrogen, carbon dioxide, and oil free compressed air.
- Software: Processes controlled by computer; software used in QC calculations.
- Facilities: Floors, wall, and ceilings properly designed and constructed.

Aseptic processing:
- Media runs.
- Training of personnel in aseptic techniques.
- Environmental monitoring program weaknesses.

Specifications:
- Not adequate or present.

Sterility assurance:
- Problems with sterility testing.

- Sterilization process validation weaknesses.
- Closure system integrity.

Training (lack of adequate personnel training or documentation thereof).

GMP violation of bulk drug manufacturing processes including:

- Poor or lacking validation.
- Cleaning processes not validated.
- Impurity profile not understood.
- Deviations from established procedures.
- Sterility assurance and testing.
- Stability (pre-market, post-market, and packaging changes).

The FDA is very serious about enforcement and protection of the public health. FDA investigators who perform PAIs are well armed with Agency-developed guidance documents and training. Some investigators have even received criminal investigator training. In most cases, however, the FDA investigator tries to determine

- that the data filed in the application were gathered under good laboratory and good manufacturing practices;
- if the manufacturing specifications developed thus far are appropriate to control the manufacturing process;
- if the firm is in substantial compliance to current good manufacturing practices;
- if the firm actually can adequately produce the material they have filed for.

2.5. Regulatory Enforcement

The FDA can muster many levels of enforcement activity on the firm under review. In order of severity they can do the following:

- Issue an FDA-483 report, generally considered to be items the FDA investigator feels require attention by the firm. Sometimes these 483s are early warning signs to warrant additional regulatory action.

- Recommend withholding the application approval. This is very painful news to any firm trying to get a new product approved. Usually these recommendations follow a PAI, or result from concerns at FDA Centers reviewing the application, or may be a result of continued unrelated GMP problems at a firm.
- Issue a Warning Letter. This is bad news. A Warning Letter tells the firm's senior management they must develop and provide a responsive plan to the FDA's concerns or further regulatory action may ensue. Boilerplate verbiage in warning letters may say *"These deviations cause drug products manufactured by your firm to be adulterated within the meaning of Section 50 l(a)(2)(B) of the Federal Food, Drug, and Cosmetic Act (the Act)"* and *"You should take prompt action to correct these deviations. Failure to promptly correct these deviations may result in regulatory action without further notice. These actions include, but are not limited to, seizure and injunction."* These actions can be serious indeed.

Warning Letters have cited problems on master batch records, including poor identification of significant steps and failure to comply with the firm's own requirements. Deficiencies in identity testing, labeling violations, complaint handling, stability testing inadequacies, and poor documentation have also been associated with Warning Letters.

FDA-483 observations can also cast doubt on a firm's overall quality programs and the ability to get new products approved quickly. Examples of 483 observations involving injectable products include:

- failure to investigate failures thoroughly, such as exceeding action levels on ingredient water samples, sterility test failures, and stability failures;
- not challenging product stability at upper limits of USP (United States Pharmacopeia) room temperature;
- lack of cleaning validation;
- lack of appropriate change control programs and poor execution of change control;

- reworking, re-inspecting or reprocessing drug substances or products without adequate validation or supporting information;
- testing and reworking materials, ad infinitum, in order to meet a certain specification.
- failure to establish and implement sufficient controls to ensure processes are properly validated prior to drug products being released for sale.
- failure to perform periodic quality evaluations (audits).

3. QUALITY DURING THE PRODUCT DEVELOPMENT PHASE

Many quality professionals in drug and device manufacturing define quality as conformance to specifications. Accordingly, appropriately set specifications are imperative to assure product quality. Researchers and quality professionals alike must assure that the product development process develops specifications that result in effectively monitored processes and process output. Numerous and repetitive objections from industry regulators have focused on inadequacies in specification quality. Regulatory actions have occurred because key quality attributes* are not addressed in specifications or they have been inappropriately set. For the medical device industry the FDA has issued new regulations that give detailed requirements for specification development. Known as the Quality System Regulations,[†] these new regulations focus on the importance of pre-production quality and the specification setting process.

The traditional role of the QC department has been to assure conformance to specifications. If the specifications are set improperly, the QC department will likely not be able to detect a problem, prospectively. The QC department is

---

* The characteristics that impart safety and efficacy to the product.
[†] QSRs, previously known as the Device GMPs, Ref. 21CFR820.

usually not involved in the development or setting of specifications. Instead, the QC department's role is to assure that there is a sound process for specification setting and that product specifications are complied with. Researchers developing and setting specifications should not, therefore, consider the QC department a safety net for bad design or its consequences. Signs of improperly set specifications are high manufacturing loss or scrap rates, excessive laboratory retest rates, stability failures, and customer complaints. Since the design requirements for products typically come from clinical or customer requirements and expectations, the collection of this information is essential in the specification development and setting process. The timing of when specifications should be established and other key activities such as validation and regulatory filing is shown in Fig. 1.

The impact of measurement, raw materials, and processing variation on clinical effectiveness must be addressed. Due to the sheer number of variables involved, statistical tools are commonly used to delineate variables that do or do not impact product performance. The results of these experiments dictate what specifications should be routinely measured. For example, Table 2 shows the results of varying product components with the resultant quality attribute. Other experimentation would be required to understand the relationship between product quality and processing variables in the factory. These relationships should be established and well understood prior to setting final product and process specifications.

3.1. Metrology

For departments generating process and product specifications, it is important not to overlook or underestimate the importance of manufacturing process capability and test method adequacy. Those individuals setting specifications must be aware of manufacturing capability (i.e., the assurance of reliably and consistently operating within developed specifications). Also, as process and product specifications are being established, there must be an assessment on

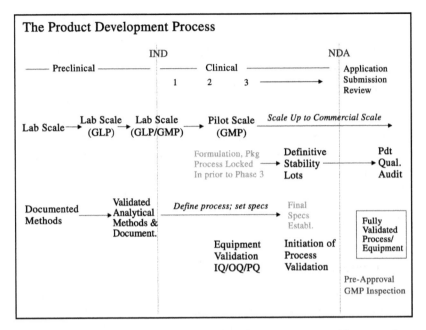

Figure 1 Chronological milestones in injectable product development.

Table 2 Examples of Variation and Effect

Variation in	Will have an impact on
Types and concentrations of oils	Drug solubility and dose
Types and concentrations of phospholipids	Flocculation and coalescence
Types and concentrations of auxiliary emulsifiers	Flocculation and coalescence
Types and concentrations of solubilizers	Solubility and crystal growth
pH and buffering agents	Zeta potential, chemical stability
Types and concentrations of antioxidants	Chemical stability
Type and concentrations of preservatives	Preservative effectiveness

metrology adequacy. Researchers and developers should ask: "Are the test methods that measure compliance to these materials and process specifications adequate to provide confidence levels required?" For example, is the inherent measurement error of the test method understood in relationship to the specification range for the quality attribute being measured? If the test method has a measurement error of $\pm 1\%$ and the specification for the quality attribute is $100 \pm 1\%$, the test method cannot adequately discriminate whether a measurement is or is not in conformance to the specification. Similarly, if researchers set processing specifications tighter than the factory can control or measure, trouble will soon follow.

3.2. Product Quality and Processing

For injectable dispersed drugs, there are numerous routine specifications to be assured prior to batch release or sale (Table 3). Testing these product attributes confirms that the batch was properly formulated, processed, and packaged.

While there are many routine tests required for product batch release in manufacturing, additional tests are required to establish objective evidence that the product works and performs as intended. These tests may be addressed in either the research and development stage or in the marketed product stage, or both, for injectable dispersed products.

3.3. Process Analytical Technology

Due to the advent of new measurement technologies, such as near infrared, Raman, and other spectroscopic techniques and sensor technologies, the pharmaceutical industry and the FDA are moving toward increased in-process and final product control measurements of product quality. Process Analytical Technology (PAT) allows manufacturers the potential to quickly and non-destructively analyze each unit of finished product for certain product parameters such as particulate size, moisture content, oxygen content, content uniformity, and other critical quality features.

Table 3 Typical Marketed Product Batch Release Testing

Physical	Chemical	Microbiological
pH	Active ingredient identification	Sterility
Particulate matter	Active ingredient assay	Endotoxin
Dispersion properties (fat globule size/distribution for emulsions and particle size/distribution for suspensions)	Key component assay (include wetting agents for suspensions)	
Packaging related specifications such as fill volume, labeling, closure system, etc.	Key component identification	
	Heavy metals Single related substances Total related substances	

3.4. Batch Testing

Specifications call for routine QC testing of each batch of finished product. These specifications are intended to demonstrate process control and product fitness for use.

pH: The pH test confirms proper processing. In-process pH measurements may also be required to assure proper ionic conditions for component processing.

Particulate matter: The USP defines acceptable limits for particulate matter in injectables. For products that are essentially particulate in nature, the particulate specifications are intended to control or eliminate unintended foreign particulate matter from the product. Again, special test methods must be developed to distinguish between the product and unintended particulate matter.

Dispersion properties (particle size and size distribution): Size and distribution of drug particles define the dispersed product's clinical effectiveness. Understanding the relationship between these attributes and medical effectiveness should be confirmed in clinical studies. Test methods for sizing can be technically challenging and should be designed to be robust and rugged for use in the QC testing laboratory.

Active ingredient identification: This ID test assures that the proper material was added during manufacturing.

Active ingredient assay: This test assures the correct concentration of the active ingredient.

Key component assay: This test, or series of tests, assures the correct concentrations of other batch ingredients such as excipients.

Sterility test: The product must be sterile when dispensed and must stay sterile upon repeated use, if packaged as a multi-dose formulation. Terminal sterilization using heat is not always possible and aseptic manufacturing must be stringently controlled. Validation of the sterility test method is required. Emulsions and suspensions can provide special challenges due to the techniques used, such as filtration and direct inoculation. See Special Considerations.

Endotoxin test: Injectables must meet USP requirements for pyrogens or bacterial endotoxin. Dispersed products provide special challenges in pyrogen control since these products cannot be depyrogenated using typical methods.* Instead, manufacturers of these products must focus on the prevention of pyrogen introduction into the formulation or from development of pyrogens during manufacturing. Testing for pyrogens is also problematic with these formulations. Due to the physical nature of many suspensions and emulsions, the USP Pyrogen Test (using rabbits) is not always possible. Instead the USP Bacterial Endotoxin Test has to be the logical alternative. Whatever test is selected the absence of potential interference of test sensitivity by the dispersed phase of the product should be addressed in the pyrogen or endotoxin test method validation.

Fill weight/volume: These tests confirm gravimetrically that the individual containers contain the stated mass of material.

---

*Typical methods include rinsing, dilution, distillation, ultrafiltration, reverse osmosis, activated carbon, affinity chromatography, or dry heat.

Package integrity: A representative number of units from the batch are checked to assure proper closure system integrity. This can include destructive testing such as pressure tests (charge container with pressure and look for bubbles underwater) or non-destructive testing such as visual checks, sonic, electrical, or other tests designed to confirm the drug container/closure system has no leaks and will withstand normal handling without breaches in integrity.

3.5. Required Additional Testing

Several other tests are important to the quality of injectable dispersed products. While these tests are not performed on each batch of injectable product, they are conducted during pre-market activities such as clinical material manufacturing and validation studies:

Physical stability testing: The effects of time, storage conditions, packaging, and transportation must be established. Stability programs are designed to gain this understanding. Stability programs should assure a product meets its label claim throughout the stated shelf life (expiration date). There typically are two stages in stability testing; R&D stability and marketed product stability. In the R&D stages material may be stressed to predict real life performance over intended dating. At this stage special storage and handling considerations are confirmed. Once a product is approved and in production a select number of batches per year are placed on stability to monitor product performance in its current configuration. These marketed product stability lots are typically monitored at the storage requirements stated on the label. Each product on the marketed product stability program should have a protocol that dictates storage conditions, test intervals, sampling, and test requirements.

Syringeability and injectability testing: The ease of withdrawal of a product from the container (syringeability) and its subsequent ejection into the desirable site of administration (injectability) must be determined for the final formulation. Syringeability can be affected by the diameter, length, shape of the opening, and surface finish of the syringe needed and, therefore, should be characterized in specification and product labeling development. Injectability denotes the ease with which a dose can be injected. The injection medium must be understood and also characterized during product development.

Preservative effectiveness testing: Multiple-dose injectables contain preservatives to safeguard the product against in-use microbial contamination. The USP preservative effectiveness test method is typically used. The water-insoluble dispersed phase may present special problems in development of a good preservative system. These same product related issues impact the development and qualification of sterility test methods as well. The problems occur because the particles of the product interfere with microbial test methods that rely on turbidity as indicators of microbial growth.

4. RAW MATERIALS

Consistent product performance and manufacturing require quality ingredients. Many key ingredients of dispersed products are biologically provided, meaning variation will be higher than chemically synthesized materials. The natural variation of biologically derived raw materials can cause problems. The quality of complex fats and lipids can vary as well as the composition of ingredients such as soy and safflower oils (Table 4).

Raw materials have stability profiles as do final product formulations. What is the effect of the supplier's manufacturing date, the drug firm's purchase date, and ultimate product performance? The drug development plan should include stability analyses of key component raw materials. Typically, retest or expiration dates are set for raw materials. Retest dates require the raw material be retested against the material specification. Acceptable results allow for material approval status to be extended to the next retest date. Expiration dates are just that (e.g., material has expired). Retesting expired material is not considered an acceptable GMP practice.

Table 4 Some Specific Raw Material Quality Control Issues for Formulation Components of Dispersed Systems

---

Trace quantities of gossypol in oils like cottonseed
Limits on hydrogenated oils, other saturated fatty materials
Limits on unsaponifiable materials such as waxes, steroidal
  components
Contamination with herbicides and pesticides
Vasopressor contaminants in soybean phosphatides
Specifications on lecithin minor components such as cholesterol,
  sphingomyclin, phosphatidic acid, and derivatives

4.1. Variation in Raw Materials

Some suspension and emulsion stability problems have been traced to seasonal variation in raw materials. Small shifts in complex, multi-constituent raw material components such as oils have caused unexpected changes in marketed product stability. A good technical relationship with key material suppliers is important to set sound material specifications and to troubleshoot when necessary.

Many fats and oils used in the manufacture of dispersed products come from natural sources. Accordingly, raw material quality is influenced by mother nature. In one example, shifts in oil fraction components (fatty acids) were detected in high grade soybean oil. The fraction ratio did not meet expectations. Investigation indicated that unseasonably cool and damp conditions in the Western Hemisphere shifted the soy plants production profile of fatty acids. Immediate corrective action was not possible. The manufacturer had to contact the appropriate regulatory body to decide on an acceptable course of action. The regulatory body had to make a quick decision to accept an amendment to the firm's NDA. Working together the FDA and the pharmaceutical firm were able to assure that product quality and clinical efficacy would not be impacted by the shift in the soy oil fatty acid profile.

4.1.1. Raw Material Specifications

Specification quality is key at all stages of manufacturing. Raw material specifications are no exception. The process for establishing specifications is essential to assuring product performance, good supplier relations, and cost.

Raw material specifications should address the following elements:

- List of approved suppliers, by location of manufacturer.
- Key elements of the formula.
- Chemical name and molecular weight.
- Sampling requirements including:

  - special handling considerations (safety, humidity, etc.);
  - sampling plan (quantities, number of samples per container);
  - approved sampling containers/materials;
  - file or reserve sample requirements.

- Specifications for acceptance of material for further processing including but not limited to:

  - receipt quality (any damage during shipment);
  - proper container type and label on receipt;
  - identity;
  - solubility;
  - purity, such as related substances, impurities, degradation products;
  - quality, such as particle size, crystallinity, polymorphic form, etc.;
  - microbial and/or pyrogen quality.

- Testing procedures:

  - compendial;
  - non-compendial, or as required by NDA.

4.1.2. Specific Raw Material Concerns

Research and field experience have provided some insight on potential problems with components of dispersed systems. Impurities and traces of gossypol, an antispermatogenic pigment extracted from cottonseed oil, must be controlled.

Hydrogenated oils and other saturated fatty materials vary seasonally and geographically. Unsaponifiable materials, such as waxes, are also highly variable and should be monitored closely. Volatile organic residues, herbicides, and pesticides are toxic at extremely low levels and difficult to detect. Refined chromatographic and spectrographic procedures are required to achieve low detection levels.

Minor components in lecithin such as cholesterol, sphingomyclin, and phosphatides also must be controlled to below detectable limits.

Microbial, endotoxin, and non-viable particulates are quality attributes that require specifications for injectable products. Assigning specifications for these factors at the raw material stage is important to assure proper QC throughout the manufacturing life cycle.

4.1.3. Toxicological Concerns

Dispersed injectable products provide unique dosage forms for life-saving therapeutic and diagnostic purposes. There are, however, some toxicological considerations. Emulsifiers have been shown to produce hemolytic effects. Lecithin may carry toxic impurities and nearly all emulsifiers possess potential toxic properties. Shifts in free fatty acid content can impact toxicity, stability and clinical effectiveness. Poor control of oil droplet size and size distribution can have untoward clinical implications plus there are clinical hazards associated with injecting these dispersed agents. Hazards include phlebitis, precipitation in the veins, extravasation, emboli, pain and irritation, and interactions with blood cells and plasma proteins.

4.2. Packaging Materials

Selection and qualification of packaging materials are essential to long-lasting quality products. Due to the hydrophobic, non-polar nature of these formulations, there are fewer options for stopper compounds, tubing, and gaskets for manufacturing purposes, and primary containers such as vials, IV bags, or syringes. Packaging Research and Development departments should develop or identify test protocols that will assure that packaging materials are inert and non-reactive for the required period of product contact. Product contact packaging materials such as vials, stoppers, syringe plungers and barrels, and administration tubing must be pyrogen-free when manufactured or rendered pyrogen -free via depyrogenation. In the final product configuration, the "integrity" of the drug delivery system must remain intact from manufacturing assembly to the time of use. Studies should be conducted to assure package integrity remains throughout the product's intended life. Torture tests and challenging the closure system with microbes and/or endotoxin are common when validating the integrity of the container and closure system.

5. SCALE UP AND UNIT PROCESSING

Laboratory batches do not normally translate directly to industrial scale. Well-characterized raw materials, identification of critical process parameters, properly set process and product specifications, and suitable test methods are prerequisites for successful quality scale up operations. When scaling dispersed products, it is important to know where the sources of variability are and to reduce them wherever practical. High degrees of variation in materials, processes, or test methods can mislead researchers, especially when only a limited number of batches or samples can be tested. Bioequivalence and stability should also be considered when significant batch size changes are made. Other in-process manufacturing controls that should be addressed are shown in Table 5.

6. FINISHED PRODUCT CONSIDERATIONS

6.1. Filtration

The physical nature of dispersed products makes submicron filtration a difficult challenge. Globule size is usually greater than the retention requirements for microbes and the particle size of particulates as specified in the USP (10 and 25 μm). Options to the filtration dilemma include filtration

Table 5 Typical In-Process Manufacturing Controls to be Addressed During Scale-Up

---

Agitation (rate, intensity, and duration)
Mixing steps that impact heat gain or loss
Temperature of phases
Rate and order of ingredient additions
Particle size reduction
Emulsification conditions (time, rate, temperature)

---

of components prior to final formulation and/or highly specialized terminal filtration systems. Manufacturers usually rely on component filtration (e.g., oils) prior to homogenization although flow rates can be problematic at the 0.45 μm levels. QC for non-viable and viable particulates may have to be assured at the raw material supplier since submicron filtration may be problematic in many cases. Prefiltration or redundant final filtration methods may be required to achieve acceptable flow rates.

When filtering oily materials hydrophobic filters must be used. The relationship between pore size and oily droplet size must be understood. Special care is required when using microporous filters because dispersed system properties may be disrupted or changed. Oily materials can depress the published bubble point values of the sterilizing filter. Proper filter sizing to assure adequate soil load removal while retaining suitable flow rates can be problematic. Unless significant filtration expertise is available internally, the use of filter vendor services is recommended in these special applications. Bioburden levels should be specified at all stages of manufacturing. Criteria for forward processing should be established for each unit operation.

6.2. Depyrogenation

Pyrogens can be present in the raw materials, as a result of excessive bioburden, or be present on packaging components. Since current approaches to depyrogenation rely primarily on dry heat at temperatures above 250°C, finished dosage forms of injectable dispersed products cannot be depyrogenated. Other techniques such as rinsing or dilution, distillation, ultra-filtration, activated carbon, and chromatographic procedures may be qualified for depyrogenation but all have limitations and may not be suitable for dispersed systems. Accordingly, special care to prevent the introduction of pyrogens should be taken at all stages of manufacturing. Specifications for the absence of pyrogens should be established at applicable stages of manufacturing to assure the finished product meets injectable quality standards. Currently available test methods for emulsions and suspensions include the USP Pyrogen and USP Endotoxin tests. Neither method, however, may work as intended with dispersed products. Limulus amebocyte lysate (LAL) testing is preferred over USP Rabbit Pyrogen testing, but both have limitations. Interferences have been encountered with the LAL test that resulted in potentially suppressed results. Injecting rabbits with certain oily components can result in false positive results.

Packaging materials such as stoppers, plastic containers, vials, bottles, and manufacturing materials such as tubing should be qualified for the absence of pyrogens. The key to pyrogen control is keeping these materials clean and dry at all times. Injection molded materials, blown plastics, molded and extruded vials and medical grade tubing are clean and non-pyrogenic coming off the press or extruders. The challenge is to keep items in dust free environments and especially dry. Many of these materials are shipped in corrugate containers. Wet corrugate containers are an excellent breeding ground for microbes and their resultant pyrogens.

Other considerations in pyrogen control focus on water quality. Any water used to process commodities such as packaging materials (stopper rinsers, bottle washers, etc.) or used as formulation ingredient water must be qualified as non-pyrogenic. Most manufacturers use USP grade Water for Injection for processing materials considered likely for product contact. Wet storage of any in-process materials should be avoided. If, however, there is a need to do so, such as supply moist or siliconized stoppers to a filling line, hold times should be established and validated to assure microbial and pyrogen control.

6.3. Sterility Testing

Sterility testing is typically conducted using the membrane filtration technique. Due to the challenges in filtering suspensions and emulsions, the direct inoculation technique may be required. If the product is aseptically processed, any failures in sterility tests will likely result in rejection of the batch, regardless of whether the positive tubes were determined to be false positives introduced during sterility testing. Isolation technology is now available for sterility testing, greatly reducing the chances of false positive results (Fig. 2). Additives such as dimethylsulfoxide (DMSO) that do not affect the growth promotion characteristics of the media nor damage any microbes in the product may allow the membrane filtration technique to be used for dispersed products. Sterility test methods must be qualified for microbial recovery and growth promotion characteristics.

Sterility testing alone does little to assure sterility of the processed batch. Validation of all sterilization procedures, environmental monitoring and control, and container closure integrity are major components of sterility assurance. Closure system qualification and validation occurs before market entry of a product. Once in routine production, closure system integrity is assured through supplier quality, processing controls, and finished product testing.

6.4. Validation

Validation is required when the results of a process cannot be fully verified by subsequent inspection and testing. Sterility assurance is an excellent example. Testing 20 containers from a batch of aseptically processed material gives little assurance about the sterility of the other 20,000 containers. While US regulators have broadened the scope of expectation to nearly every key process in a manufacturing operation, validation can still result in a business advantage. For injectable dis-

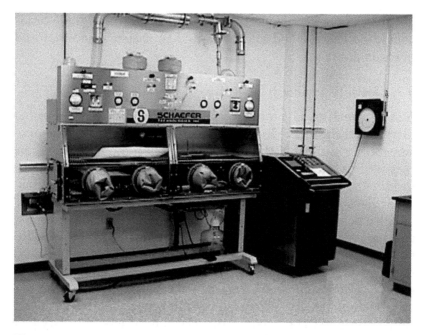

Figure 2   Sterility test isolator. (Courtesy of Baxter Pharmaceutical Solutions.)

persed products the following manufacturing systems should be considered for validation:

- facilities, including construction, floors, walls, ceilings, and lighting;
- barrier isolation systems;
- product contact utilities such as HVAC, water, compressed gases, etc.;
- key component manufacturing equipment such as injection-molding machines;
- in-process and finished product equipment such as filtration equipment, fillers, stoppering/closure machines, lyophilizers, sterilizers, etc.;
- cleaning equipment such as clean-in-place (CIP) or clean-out-of-place (COP) systems;
- test methods for assay, sterility, pyrogen testing, package integrity;

- label verification systems;
- aseptic processing;
- computer systems that are involved in the disposition of materials such as laboratory information management systems (LIMS), test method data handling, algorithms and systems, electronic batch record systems.

Validation must demonstrate that the process in question consistently and reliably produces output meeting pre-determined specifications. Each firm must use validation definitions consistent with current regulatory jargon. Poor or internally developed definitions can cause confusion and even delays in product approval.

While validation protocols may vary due to the subject being covered (i.e., installation qualification of equipment or facilities, operational qualification, performance qualification, or test method validation) validation protocols should be as simple to follow as possible and include, generally, the following elements:

- table of contents;
- purpose and overview;
- description of the product or process;
- listing of test methods used;
- instrument calibration information;
- specifications, acceptance criteria including number or runs/batches and sampling, plans/instructions;
- key document references such as the edition or revision number of operational procedures used.

Documentation must be clear and concise. Many times validation packages do not receive great scrutiny for months to years after they are completed. The protocol initiators such as key scientists or engineers may no longer be available to defend the validation. Therefore, the package must be clear, concise, readily retrievable, and easy to understand years after its completion. Once a process has been validated it has to stay that way. Revalidation must occur when there are changes to the process that could affect its validated status. New product additions, changes to specifications, audits, or relocation of equipment are examples that will trigger a revalidation exercise.

6.5. Cleaning Validation

The processing of formulations containing fats and oils should include a plan for cleaning. Injectables containing fats and or proteins can pose special problems for cleaning. Oily components create slippery conditions, and once dried or denatured by heat, can be extremely difficult to clean. Processing equipment and filling/packaging lines should be designed with special cleaning considerations, especially if the manufacturing equipment is considered "multi-use," meaning more than one product is manufactured using the same equipment. Help in equipment/facility design and cleaning agent selection is available from most major suppliers of cleaning chemical supplies. The pharmaceutical industry can also find assistance* from the dairy industry which is keenly familiar with the equipment and techniques for addressing tough cleaning challenges.

Cleaning validation programs for drugs generally include the following components:

- rationale for selection of cleaning materials;
- listing of materials and surfaces to be cleaned including surface finishes (degree of roughness);
- criteria for "clean."

There are several schools of thought on what constitutes clean. Most regulatory bodies will ask for justification of the approach being used and limits established. Cleanliness must be defined by the user with cleaning limits typically established for:

1. Microbial load after cleaning.
2. Particulates in rinse samples.
3. Post-cleaning residual cleaning compounds or chemicals.

---

* 3A Dairy Standards.

4. Post-cleaning residual product active ingredient(s).
5. Visual determinations for cleanliness.

Sampling methods must be qualified for the recovery of analytes from surfaces or from cleaning rinse water. Special swabs have been developed for different surfaces and analyte types to improve recovery. The approach for analyte recovery from a liquid is similar as for typical assays of drugs in solution formulations.

Some firms use a limit of detection approach for the active ingredient(s). In other words, surfaces or rinse water will be submitted to an assay to detect residual amounts of the product's active ingredient. This approach is frequently used when the manufacturing line or facility is used for multiple products and cross contamination is a concern. Another approach uses toxicological considerations based on the residual active ingredient(s) being no more than a fraction (usually 1/1000) of the therapeutic dose. Some firms use a combination of criteria. In any case, the power of visual inspection cannot be underestimated. While it is not quantitative or qualitative, if it does not look clean, it is not.

The application of total organic carbon (TOC) test methodology is gaining favor in cleaning validation circles. The relationship between carbon-containing analytes of interest, detection limits, and analyte carryover significance must still be established.

6.6. Sterilization and Sterility Assurance

Terminal sterilization using steam heat provides the highest degree of sterility assurance of technologies currently available. Terminal sterilization is also easier to control than aseptic processes. Many dispersed injectables are not stable at the temperatures required for achieving a sterility assurance level of greater that 10^{-6} (probability of non-sterility) and, therefore, aseptic processing provides the only alternative.

6.6.1. Methods

Moist heat sterilization remains the method of choice for sterilization of rubber closures, manufacturing hardware such as filter assemblies, process tubing, tanks, and, if at all possible, the final product. Batch autoclaves remain versatile in achieving steam or dry heat sterilization. Sterilization is achieved by exposing material to high temperatures for various periods of time validated to achieve a sterility assurance level of at least 10^{-6} at the slowest-to-heat location in the batch.

Dry heat sterilization is used for sterilizing items such as glass vials and stainless steel equipment. Temperatures and time cycles are typically higher than those required for steam sterilization due to the kinetics for heat transfer without moisture. Dry heat also is the most effective method of depyrogenation. Since the glass container offers the most surface area for potential contamination, it is fortuitous that glass can be effectively sterilized as well as depyrogenated by dry heat.

Radiation sterilization has gained in popularity and utility. Plastics, paper, clean room gowns, and other non-dense items may be suitable for gamma sterilization. Sterilization is achieved by exposing material either to gamma rays from a radioactive source such as cobalt 60 or accelerated electron beam particles.

Gaseous sterilization due to its handling hazards and human toxicity, is falling from favor as an industrially acceptable method of sterilization. Ethylene oxide has been used to sterilize plastic materials, paper, gowns and medical devices. For barrier isolation systems, internal surfaces are effectively sterilized by gaseous agents such as peracetic acid or vapor-phase hydrogen peroxide.

Aseptic processing: Sterilization using microbially retentive filters (aseptic filtration) is basically the only method of choice for heat labile pharmaceutical products. Most small volume injectable dispersed products are sterilized by aseptic processing. Aseptic processing requires high levels of personnel training and discipline. Sterility assurance levels are lower than what can be achieved with moist heat sterilization.

Filtration becomes an issue again as final filter pore size and physical-chemical properties of the product create technical challenges.

Sterility assurance using aseptic techniques rely on:

- validated sterilization procedures for all manufacturing materials;
- microbiological and particulate control of the facilities;
- certification of air handling systems;
- proper facility design;
- enviromental monitoring program with alert and action limits plus trend analyses;
- training in aseptic technique;
- media fills and operator broth tests;
- control of manufacturing deviations and interventions;
- sterility testing.

6.7. Manufacturing Deviations

Deviation is an alarming word in a regulated industry. Deviations are actions outside of planned or prescribed procedures. Sometimes they are planned in advance, but, in most cases, they occur outside of plan and must be dealt with. In a perfect research and manufacturing environment everything should go according to plan. There should be approved procedures and specifications, and they should be followed without error by trained individuals. Unfortunately, even the best managed operations encounter deviations. They can occur from human failure, poorly written instructions, poor training, mechanical failures, lab errors, and undetected shifts in raw materials or processes. Whenever deviations are encountered they must be documented, explained, and, if necessary, justified. Depending on the nature of the deviation, there could be an impact on validation, material stability, or regulatory compliance. Examples of manufacturing deviations include not taking samples as prescribed, not following procedures exactly, performing steps out of order, failing pumps, and tubing breaks.

6.8. Non-conformances

A non-conformance event is failure to meet specification. Non-conformances are typically associated with the product, but non-conformances can be encountered with processes as well. A manufacturing deviation can cause a non-conformance. Failing to meet specifications is a serious issue. Huge financial losses can be realized, and regulators usually investigate non-conformances of interest. It is sometimes helpful to anticipate the types of non-conformances that may be encountered during the product development stages. Once a product is approved and in the manufacturing environment there are limited options with non-conforming material. A product can be reprocessed, but, in most cases, must be discarded. In the United States, drug products usually cannot be reprocessed unless the reprocessing procedure has been approved by the FDA. In most cases, a product will expire before a reprocessing procedure could be developed and subsequently approved by the FDA. It is therefore prudent to anticipate the likely reprocessing needs of a product and include the procedure and supporting data in the drug regulatory filing.

6.9. Stability

Particle size and size distribution of the dispersed phase are among the key factors controlling the physical stability of a dispersion. A change in these parameters can be indicative of the physical instability of a dispersion. These two parameters also influence therapeutic performance as well as the safety of the product. While many methods are commercially available to measure particle size and distributions many other physical and chemical stability indicating methods are not readily available and must be developed with the particular product. In many cases, the best indicators of product stability are visual with a trained eye. These methods are subjective, due to the human element, but globule size distributions and other physical characteristic methods generally have not proven to be stability-indicating. Accordingly, the quality of the pre-market and after market stability programs are essential in establishing and monitoring the stability of these types of products. Due to the hydrophobic nature of dispersed injectables, the materials used for IV administration should also be considered in product stability programs.

Coalescence and phase separation are the terminal physical instabilities. Coalescence takes place when oil droplets unite to make larger ones. When coalescence progresses, larger and larger droplets are formed, eventually merging and resulting in a separate layer of the dispersed phase. This process is irreversible and renders the product unusable. Sedimentation, caking, creaming, and crystallization are also unwanted outcomes of poor product stability and should be addressed as part of product development and marketed product stability programs.

Peroxidation of unsaturated fatty acids can result in product destabilization resulting in pH decreases. The catalysts for peroxidation can be transition metal ions, oxygen, and light. Peroxidation can produce peroxides that can become a health hazard. From a chemical stability standpoint, measuring known degradants is typically preferred over measuring the potency of the active ingredient although both assays are normally performed in a stability program.

6.10. Facility

A great deal of literature is available on the engineering standards for drug processing facilities and aseptic processing. For the manufacture of injectable dispersed agents special consideration should be given to:

1. Floors, walls, and ceilings are to be readily cleanable. Generally, this means smooth surfaces, but with fats and oils involved, completely smooth floors can become treacherous. Consider cleanable textured floors in areas where there is a possibility of spillage.
2. Make sure there are floor drains in the processing areas and that the floors are adequately sloped to these drains. Engineers have been known to place the drain at the highest part of a floor if not clearly specified.

3. Room air, in areas where product and components are exposed, should be filtered and supplied under positive pressure. Room particle classifications vary depending on the nature of the manufacturing process. Due to the filtration challenges with dispersed products, high air quality (10–10,000 PPCF) is usually required. Humidity and temperature are also a room design requirement that may vary with product design requirements.

Traditional injectable manufacturing clean room facilities will typically have a drug processing area where compounding takes place and a filling and packaging line located separately. These facilities are typically large when one considers the amount of square footage invested per unit of product. Additionally, these traditional designs must be under constant management discipline and monitoring due to the high microbial and particulate standards of injectables.

New technology is becoming available that can reduce manufacturing cost and investment, while increasing the quality assurance of the resulting product(s). Considering that today's marketplace is global and cost control is a major factor, this new technology is a wonderful prospect in today's health care environment. This technology is suitable for working with the components and finished product aspects of injectable dispersed products. Known as isolation technology (Fig. 3), the concept is to protect the product from the greatest potential source of contamination (humans and the factory environment).

While these systems look technologically intimidating, in many respects they are simpler to operate and maintain than traditional clean rooms. Especially noteworthy are the sterility assurance levels that this technology is providing for aseptically manufactured products. The impact of the biologically unclean human is essentially removed from exposure to the product. If isolation technology is chosen for processing or filling suspensions or emulsions, cleanability is of utmost importance. The fats and oils in these formulations can be

External View of Isolator

Internal View of Isolator

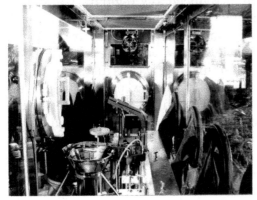

Figure 3   Courtesy of la Calhene, Inc.

challenging from a cleaning standpoint. Clean-in-place /sterilization-in-place (CIP/SIP) systems are recommended and help to qualify the design of the isolator to assure all areas can be adequately reached during clean-up and maintenance.

6.11. Manufacturing Materials

Non-aqueous or non-polar formulations require special precautions in the selection of manufacturing materials such as piping, tubing, gaskets or any plastics that may come into contact with the product. The interaction of the product with these materials should be evaluated. Typical concerns are extractables or the adsorption of the drug product on these plastics and rubbers. Many manufacturers have found the USP reference on Class IV and V biologicals useful in qualifying classes of materials such as nylons, BUNA N, Viton, EPDM, and silastic.

When filters are used in manufacturing, they must also be qualified for their intended use. The concerns of extractables, filter wetting agents, product adsorption, and proper sizing require consideration as well.

7. SUMMARY

In this chapter, we have introduced key regulatory and quality considerations for development and introduction of injectable dispersed agents. Regulatory requirements are stringent and varied amongst the developed countries of the world. Compliance to good manufacturing practices and local regulations are essential to assure timely new product approvals. Product quality is assured through design, not by testing. Raw materials influence product stability and clinical utility. Dispersed products pose special challenges to aseptic processing due to their chemical and physical nature. Special manufacturing and product handling considerations are required in areas such as sub-micron filtration, cleaning, facility design, and final product packaging.

REFERENCES

1. United States Pharmacopeia (http://www.usp.org/USP).

2. 3A Dairy Standards. International Association of Milk, Food and Environmental Sanitariums, Inc (http://www.iamfes.org/).

3. Requirements of Laws and Regulations Enforced by the US Food and Drug Administration (http://www.fda.gov.morechoi-www.fda.gov.morechoices/smallbusiness/bluebook.htm).

4. Portnoff JB, Cohen EM, Henley MW. Development of parenteral and sterile ophthalmic suspensions–the R&D approach. Bull Parenter Drug Assoc 1977; 31:136.

5. Akers MJ, Fites AL, Robison RL. Formulation design and development of parenteral suspensions. J Parenter Sci Tech 1987; 41:88.

6. Boyett JB, Davis CW. Injectable emulsions and suspensions. In: Lieberman HA, Rieger MM, Banker, GS, eds. Pharmaceutical Dosage Forms: Disperse Systems. Vol. 2. New York: Marcel Dekker, Inc., 1989.

7. DeFelippis MR, Akers MJ. Peptides and proteins as parenteral suspensions: an overview of design, development and manufacturing considerations. In: Frokjaer S, Hovgaard L, eds. Pharmaceutical Formulation Development of Peptides and Proteins. Philadelphia: Taylor and Francis, 2000:113–144.

8. Pitkanen OM. Peroxidation of lipid emulsions: a hazard for the premature infant receiving parenteral nutrition? Free Redic Biol Med 1992; 13:239.

9. Scott RR. A practical guide to equipment selection and operating techniques. In: Lieberman HA, Rieger MM, Banker GS, eds. Pharmaceutical Dosage Forms: Disperse Systems. Vol. 2. New York: Marcel Dekker, Inc., 1989.

10. FDA Inspectional References (http://www.FDA.gov/ora/inspect_ref/igs/iglist.html).

11. International Conference on Harmonization of Technical Requirements for Registration of Pharmaceuticals for Human Use (http://www.ifpma.org/ich1.html).

APPENDIX 1. ICH GUIDELINES

Note for Guidance on Stability Testing: Stability Testing of New Drug Substances and Products, November, 2003.

Specifications: Test Procedures and Acceptance Criteria for New Drug Substances and New Drug Products: Chemical Substances, December, 2000.

Guideline on Impurities in New Drug Products; Availability; Notice, November, 2003.

Guidelines for the Photostability Testing of New Drug Substances and Products; Availability; Notice, May 16, 1997.

Availability of Draft Guideline on Quality of Biotechnological/Biological Products: Derivation and Characterization of Cell Substrates Used for Production of Biotechnological/ Biological Products; Notice, May 2, 1997.

Guideline on Impurities: Residual Solvents; Availability; December, 1997.

Guideline on Validation of Analytical Procedures: Methodology, November 6, 1996.

Cells Used for the Production of r-DNA Derived Protein Products, August 21, 1995.

Guideline on Validation of Analytical Procedures: Definitions and Terminology, March 1, 1995.

Key ICH Safety Guidance Documents.

Draft Guideline on the Timing of Nonclinical Studies for the Conduct of Human Clinical Trials for Pharmaceuticals; Notice, May 2, 1997.

Draft Guideline for the Preclinical Testing of Biotechnology-Derived Pharmaceuticals; Availability; Notice, April 4, 1997.

Draft Guideline on Genotoxicity: A Standard Battery for Genotoxicity Testing of Pharmaceuticals; Notice, April 3, 1997.

Draft Guideline on Dose Selection for Carcinogenicity Studies of Pharmaceuticals; Addendum on the Limit Dose; Availability, April 2, 1997.

Reproductive Toxicity Risk Assessment Guidelines; Notice, October 31, 1996.

Single Dose Acute Toxicity Testing for Pharmaceuticals; Revised Guidance; Availability; Notice, August 26, 1996.

Draft Guideline on Testing for Carcinogenicity of Pharmaceuticals; Notice, August 21, 1996.

Guidance on Specific Aspects of Regulatory Genotoxicity Tests for Pharmaceuticals; Availability; Notice, April 24, 1996.

Guideline on Detection of Toxicity to Reproduction for Medicinal Products, April 5, 1996.

Final Guideline on the Need for Long-Term Rodent Carcinogenicity Studies of Pharmaceuticals, March 1, 1996.

Draft Guideline on Conditions Which Require Carcinogenicity Studies for Pharmaceuticals, August 21, 1995.

Draft Guideline on Detection of Toxicity to Reproduction; Addendum on Toxicity to Male Fertility, August 21, 1995.

Guideline on Extent of Population Exposure Required to Assess Clinical Safety for Drugs, March 1, 1995.

Guideline on the Assessment of Systemic Exposure in Toxicity Studies, March 1, 1995.

Guideline on Dose Selection for Carcinogenicity Studies of Pharmaceuticals, March 1, 1995.

20

Regulatory Considerations for Controlled Release Parenteral Drug Products: Liposomes and Microspheres

MEI-LING CHEN

Office of Pharmaceutical Science, Center for
Drug Evaluation and Research, Food and Drug
Administration, Rockville, Maryland, U.S.A.

1. INTRODUCTION

Recent advances in controlled release parenteral drug products have drawn considerable interest and attention from pharmaceutical scientists. These drug products constitute a distinct class of formulations that are designed for sustained

---

The opinions expressed in this chapter are those of the author and do not necessarily reflect the views or policies of the FDA.

release and/or targeted delivery of drugs. The complexity of the delivery systems for these products, in particular, liposomes and microspheres, has presented many unique challenges to scientists in industry, academia, and regulatory agencies. Apart from the multitudes of issues in chemistry, manufacturing, and controls, several questions have arisen in the areas of biopharmaceutics as related to product quality and performance.

This chapter provides an overview of the science- and risk-based regulatory approaches for liposome and microsphere drug products in the field of biopharmaceutics. For a more detailed discussion on the design, formulation, and manufacturing technologies for liposomes and microspheres, the reader is referred to Chapters 8 and 9 in this book.

2. LIPOSOMES

For decades, liposomes have been under extensive investigation as a drug delivery system (1–4). However, pharmaceutical preparations did not become commercially available in the United States until 1995 when the Food and Drug Administration (FDA) approved the first liposome drug product, Doxil®, a doxorubicin HCl liposome injection. In the regulatory environment, nomenclature is an important aspect and, in some cases, can be critical to the registry and approval of a new drug application. In view of the importance of nomenclature, the FDA has proposed the following definitions for liposomes and liposome drug products (5):

- Liposomes are microvesicles composed of one or more bilayers of amphipathic lipid molecules enclosing one or more aqueous compartments.
- Liposome drug products refer to the drug products containing drug substances encapsulated or intercalated in the liposomes.

It is to be noted that based on these definitions, drug-lipid complexes are distinguished from true liposome drug products by the Agency. These complexes are made in such a way that the final product formed does not contain an internal aqueous compartment and thus are not considered as "true" liposomes. Table 1 lists some examples of liposome-associated drug products currently in the US marketplace (6). Doxil provides an example where the drug substance, doxorubicin, is encapsulated in the aqueous space of the liposome. By contrast, AmBisome® is a liposome product with the drug substance amphotericin B intercalated within the lipid bilayers. (See Chapters 14 and 15 of this book.)

From a regulatory perspective, a liposome drug product consists of the drug substance, lipid(s), and other inactive ingredients. All liposome drug products approved to date are formulated using phospholipids. The lipids in a liposomal formulation are considered "functional" excipients. The pharmacological and toxicological properties, as well as the quality of a liposome drug product, can vary significantly with changes in the formulation, including the lipid composition. Unlike conventional dosage forms, the physicochemical characteristics of a liposome drug product are critical to establishing the identity of the product. These properties are also important for setting specifications and evaluation

Table 1 Examples of Approved Liposome-Associated Drug Products

Trade Name	Generic Name	Year of Approval in U.S.
Doxil®	Doxorubicin HCl Liposome Injection	1995
DaunoXome®	Daunorubicin Citrate Liposome Injection	1996
AmBisome®	Amphotericin B Liposome for Injection	1997
Depocyt®	Cytarabin Liposome Injection	1999
Abelcet®	Amphotericin B Lipid Complex Injection	1995
Amphotec®	Amphotericin B Cholesteryl Sulfate Complex for Injection	1997

of manufacturing changes. Liposome dosage forms are extremely sensitive to changes in manufacturing conditions, including changes in scale. As such, the FDA currently recommends complete characterization of the liposome drug product should any changes occur to the critical manufacturing parameters (5). To reduce the industry and regulatory burden, there is a need for better understanding of the formulation and manufacturing variables for liposome products.

Liposome drug products can be broadly categorized into different types in many ways (1). For example, they may be classified based on liposome size or lamellarity, such as multilamellar large vesicles (MLVs), small unilamellar vesicles (SUVs), and large unilamellar vesicles (LUVs). They may also be distinguished based on coatings on the liposome surface, such as stealth liposomes vs. conventional liposomes. Of particular relevance to in vivo performance and clinical outcome may be the classification on the basis of pharmacological behavior of liposomes towards the reticuloendothelial system (RES) in the body. The RES, also often referred to as mononuclear phagocyte system (MPS), can be found in the liver, spleen, and bone marrow. Most conventional liposomes are easily taken up by the RES macrophages and thus have a relatively short residence time in the bloodstream. In contrast, with the advent of modern technology, liposomes can be designed to shy away from uptake by RES, and circulate in the blood for a long period of time. In addition, these liposomes can be made small enough that they eventually extravasate into the tissues through the "leaky" vascular membranes where the permeability has been compromised due to the underlying disease (7).

2.1. Pharmacokinetic Studies

The regulatory requirement to provide human pharmacokinetic data for submission of a new drug application can be found in a series of FDA regulations (8). Most drug applications submitted for liposomes have been based on an approved drug product in the conventional dosage form given by the same route of administration. Liposomes are generally used to improve the therapeutic index of drugs by increasing efficacy and/or reducing toxicity. Since liposome and non-liposome preparations have the same active moiety, it is vital to compare the product performance in terms of their pharmacokinetic profiles. In such circumstances, the FDA has suggested (5) that the following studies be conducted:

1. A comparative single-dose, pharmacokinetic study to evaluate the absorption, distribution, metabolism and excretion (ADME) of the drug between the liposome and non-liposome drug product.
2. A comparative mass-balance study to assess the differences in systemic exposure, excretion, and elimination of the drug between the liposome and non-liposome drug product.

The pharmacokinetic information will be useful in determining the dose–(concentration–) response relationship and establishing dosage/dosing regimen for the liposome drug product. Table 2 illustrates a side-by-side comparison of pharmacokinetic characteristics between Doxil (doxorubicin HCl liposome injection) and Adriamycin (doxorubicin HCl injection) (6). As expected, there are distinct differences in the pharmacokinetic parameters between a liposome product and a non-liposome product. Doxil has a much smaller volume of distribution at steady state compared to Adriamycin

Table 2 Pharmacokinetics of Doxil vs. Adriamycin HCl injection

		Drug product	
Pharmacokinetic parameter[a]	Unit	Doxil	Adriamycin
$V_{d,ss}$	L/m^2	2.7–2.8	700–1100
CL_p	$L/h/m^2$	0.04–0.06	24–35
$t_{1/2}$ (1st phase)	hr	4.7–5.2	0.08
$t_{1/2;}$(2nd phase)	hr	52–55	20–48

The plasma pharmacokinetics of Doxil was evaluated in 42 patients with AIDS-related Kaposi's sarcoma who received single doses of 10 or 20 mg/m^2 administered by a 30-min infusion. The pharmacokinetics of Adriamycin was determined in patients with various types of tumors at similar doses.
[a]$V_{d,ss}$, volume of distribution at steady state; CL_p, plasma clearance; $t_{1/2}$, half-life.

injection. The liposome injection is mostly confined in the vascular fluid volume while Adriamycin injection has an extensive tissue uptake. The plasma clearance for Doxil is also much slower than for Adriamycin injection. As a result, Doxil circulates in the bloodstream longer than the Adriamycin injection.

In addition to the pharmacokinetic studies mentioned above, the FDA recommends (5) that a new drug application for a liposome product include a multiple-dose study, and a dose-proportionality study for the product under investigation. Pending the results of these studies, additional studies such as drug–drug interaction studies or studies in special populations may be needed to refine the dose or dosage regimen under different conditions.

Liposomes can be destabilized by interacting with lipoproteins and/or other proteins in the blood (9). Therefore, the possible effect of protein binding should be taken into consideration when evaluating the pharmacokinetics of a liposomal formulation. If a protein-binding effect is confirmed, determination of the protein binding of both drug substance and drug product is recommended over the expected therapeutic concentration range.

2.2. Analytical Methods

To evaluate the pharmacokinetics of a liposomal formulation, it is pertinent to develop a sensitive and selective analytical method that can differentiate the encapsulated drug from unencapsulated drug. The development of such an analytical method may not be an easy task, but is not impossible given the current science and technology. Sponsors of new drug applications on liposomes are always encouraged by the FDA to develop an analytical method with accuracy, specificity, sensitivity, precision, and reproducibility.

The choice of moieties to be measured in a pharmacokinetic study will depend on the integrity of the liposome product in vivo. The in vivo integrity can be evaluated by conducting a single-dose study and determining the ratio of unencapsulated to encapsulated drug. Presumably, the liposome product can be considered stable in vivo if the drug substance remains in the circulation substantially in the encapsulated form and the ratio of unencapsulated to encapsulated drug is constant over the time course of the study. In such circumstances, measurement of total drug concentration would be adequate. Conversely, if the product is unstable in vivo, separate measurement of encapsulated and unencapsulated drug is necessary to allow for proper interpretation of the pharmacokinetic data.

2.3. Assessment of Bioavailability and Bioequivalence

Establishment of bioavailability and/or bioequivalence constitutes an integral part in the development of new drugs and their generic equivalents. For both, bioavailability and bioequivalence studies are also vital in the presence of manufacturing changes during the post-approval period. The assessment of bioavailability and/or bioequivalence can generally be achieved by considering the following three questions (10,11):

1. What is the primary question of the study?
2. What are the tests that can be used to address the question?
3. What degree of confidence is needed for the test outcome?

The primary question in bioavailability and bioequivalence studies can be considered in the context of regulatory definitions for these terms. In the Federal Food, Drug and Cosmetic Act, bioavailability is defined as (12):

The rate and extent to which the active ingredient or active moiety is absorbed from a drug product and becomes available at the site of action. For drug products that are not intended to be absorbed into the bloodstream, bioavailability may be assessed by measurements intended to reflect the rate and extent to which the active ingredient or active moiety becomes available at the site of action.

Similarly, bioequivalence is defined as (12):

The absence of a significant difference in the rate and extent to which the active ingredient or active moiety in pharmaceutical equivalents or pharmaceutical alternatives becomes available at the site of drug action when administered at the same molar dose under similar conditions in an appropriately designed study.

Based on these definitions, therefore, it is essential to consider two key factors when assessing bioavailability and bioequivalence. The first factor to be considered is the release of the drug substance from the drug product and the second factor is the availability of the drug at the site of action.

The second question focuses on the test procedures that are deemed adequate to address the primary question of bioavailability and bioequivalence. An important principle that prevails in the US regulations is the reliance on the most accurate, sensitive, and reproducible method to measure bioavailability and demonstrate bioequivalence. In this regard, the US regulations (13) include the following methods, in descending order of preference, for purposes of establishing bioavailability and bioequivalence:

1. Comparative pharmacokinetic studies.
2. Comparative pharmacodynamic studies.
3. Comparative clinical trials.
4. Comparative in vitro tests.
5. Other approaches deemed adequate by the FDA.

As can be seen, from the regulatory standpoint, a pharmacokinetic approach is the preferred method for assessment of bioavailability and bioequivalence, whenever feasible. In the case of liposome drug products, however, it is unknown at this time if the measurement of drug concentration in the blood/plasma/serum can be used to determine bioavailability or bioequivalence. Each liposomal formulation has its own unique characteristics and currently there is a lack of a clear understanding of the disposition of a liposome product in the body. Since uncertainty exists with regard to when and where the drug is released from liposomes, it remains an open question as to whether the drug concentration in the blood will reflect the drug concentration at the site of action.

Retrospectively, the issue of whether a pharmacokinetic approach is appropriate for determination of bioavailability or bioequivalence of liposome products has been discussed on several occasions (14,15). A proposal was once made to use this approach in conjunction with the classification scheme of liposomes based on uptake by the RES. It was theorized that the feasibility of using a pharmacokinetic approach for these products might rely on the type of liposomes as follows.

For liposome drug products designed to target the RES, the liposomes would be taken up by the RES macrophages immediately after administration in vivo. Following uptake, the RES could act as a depot and drug could be released slowly back to the systemic circulation. Under such conditions, the drug concentrations in the blood might be used to estimate the bioavailability of the liposome drug product. This theory, however, has been challenged on two accounts: (a) the accumulation of drug in RES may not be an instantaneous process, and (b) all of the drug may not be released from liposomes following RES uptake.

Conversely, for liposome drug products designed to avoid uptake by the RES, the liposome-encapsulated drug would be circulating in the blood for a long time, and thus it was speculated that measurement of drug levels in the blood might provide a tool for determination of bioavailability and bioequivalence. This theory may hold on the grounds that the liposomes are fairly stable in the blood and all drugs eventually become available at the site of action. In reality, however, the assumptions may not be true for the liposome products currently available. Moreover, even if all drugs are eventually released to the tissues, it is still uncertain if they will be directed to the specific site of action.

Finally, the two classes of liposomes in terms of RES uptake described above may only represent the extreme scenarios whereas most liposome drug products fall in between the two categories.

Based on the above considerations, it appears that the conventional pharmacokinetic approach may not be suitable for assessment of bioavailability or bioequivalence. It has been suggested that radiolabeled studies be used for these purposes. However, further research is needed to explore this possibility.

2.4. In Vitro Release Tests

In vitro dissolution testing is widely used for solid oral dosage forms in drug development and during the regulatory approval process (16,17). It is commonly employed to guide drug development and select the appropriate formulations for further in vivo studies. It is also used to ensure batch-to-batch consistency in product quality and performance. In the regulatory setting, in vitro dissolution testing may be suitable for assessing bioavailability and bioequivalence when a minor change occurs in formulation or manufacturing (18–21). When an in vitro–in vivo correlation or association is available, the in vitro test can serve not only as a quality control check for the manufacturing process, but also as an indicator of product performance in vivo (22). Under such circumstances, bioequivalence can be documented using in vitro dissolution alone (16).

Just as for in vitro dissolution testing of solid oral dosage forms, development of an in vitro release test is essential for controlled release parenteral products such as liposomes. For liposome products, currently the in vitro release test is mainly used for assurance of product quality and process controls. In rare situations, the in vitro release test can be used as a substitute for in vivo testing. This is primarily attributed to the difficulty in developing an appropriate in vitro release test that is correlated to in vivo performance of a liposome product. Ideally, an in vitro release test may be developed depending on the mechanism of drug release from the liposome product under investigation. For example, if a liposome product is intended for systemic drug delivery, the conventional dissolution method may be adequate for in vitro release testing. However, if a liposome product is designed for targeted delivery, it may be more appropriate to use a cell-based model for in vitro release. The ultimate goal is to link in vitro and in vivo performance such that the in vitro release test can be used as a tool to monitor liposome stability in vivo and further serve as a surrogate for in vivo studies in the presence of changes in formulation or manufacturing.

3. MICROSPHERES

Microspheres are solid, spherical drug carrier systems usually prepared from polymeric materials with particle sizes in the micron region (see Chapter 9). Based on morphology, microspheres can be classified into two types, microcapsules and micromatrices (23). Microcapsules have a distinct capsule wall with the drug substance entrapped in the polymer matrix within the wall. Micromatrices have no walls and drug is just dispersed throughout the carrier. Table 3 provides some examples of controlled release microsphere products marketed in the United States for parenteral use (6). All of these drugs are either proteins or peptides. They were all developed using the biodegradable polymer poly-lactic-co-glycolic acid (PLGA) based on an approved conventional dosage form.

Microsphere technology has been widely used for controlled and prolonged release of drugs (23–31). The microsphere product incorporates the drug in a polymer matrix that subsequently hydrolyzes in vivo, releasing the drug in the body at a constant rate. The polymer matrices can be formulated for drug release up to several weeks or months, depending on the physicochemical properties of the specific drug to be encapsulated and the specific polymer that will

Table 3 Examples of Approved Microsphere Products

Trade name	Generic name	Year of approval in US
Leupron Depot®	Leuprolide acetate for depot suspension	1989
Sandostatin LAR® Depot	Octreotide acetate for injectable suspension	1998
Nutropin Depot®	Somatropin (rDNA origin) for injectable suspension	1999
Trelstar Depot®	Triptorelin pamoate for depot suspension	2000

be used. For PLGA, the polymer degrades in the body through non-enzymatic hydrolysis, resulting in lactic acid and glycolic acid that are further broken down into carbon dioxide and water (32). The release mechanism of drugs from most PLGA microspheres is through polymer erosion, and perhaps accompanied by drug diffusion (32).

As with the liposome dosage form, there are several regulatory concerns about the chemistry, manufacturing and controls (CMC) for microsphere drug products, which are beyond the scope of this chapter. Among others, in addition to the general CMC requirements for conventional dosage forms, special attention should be given to the safety of polymer materials used for manufacture of microsphere products. Demonstration of biocompatibility is necessary for these products. Also, CMC specifications must be established for both the drug substance and the drug product.

3.1. Microspheres vs. Conventional Dosage Forms

In general, the application of microsphere technology prolongs the retention time of the drug in the body and reduces the frequency of drug administration required for achieving clinical efficacy. The availability of these formulations offers the opportunity for greater patient convenience and compliance as compared to conventional dosage forms. For illustration purposes, provided below are some examples of microsphere products that are available in the United States.

Leuprolide Acetate. Leuprolide acetate, an agonist of luteinizing hormone-releasing hormone (LH-RH), acts as a potent inhibitor of gonadotropin secretion when given continuously and in therapeutic doses (33). The original non-microsphere formulation of leuprolide acetate (Lupron® Injection) is administered daily by subcutaneous injection. As with other drugs given chronically by this route, the injection site has to be varied periodically. In contrast, with appropriate doses, the microsphere products (Leupron Depot ®) can be administered once every month, 3 months or 4 months, yielding similar therapeutic outcomes as the original, non-microsphere, daily dosage form (6). For both microsphere and non-microsphere formulations, the steady-state concentrations of leuprolide are maintained over the intended therapeutic dosing interval.

Somatropin. Somatropin (rDNA origin) for injection is a human growth hormone (hGH) produced by recombinant DNA technology (6). Somatropin has 191 amino acid residues and a molecular weight of 22,125 Da. The amino acid sequence of the product is identical to that of pituitary-derived hGH. The original formulation, Nutropin®, is required for daily subcutaneous injection whereas Nutropin Depot® is administered once or twice monthly. Nutropin Depot consists of micronized particles of recombinant human growth hormone (rhGH) embedded in biocompatible and biodegradable PLGA microspheres.

In a clinical study (34), 56 pre-pubertal children were treated with Nutropin Depot at 1.5 mg/kg once monthly (1×/mon) or 0.75 mg/kg twice monthly (2×/mon) for 24 months. The mean pre-study growth rate was 5.0 ± 2.4 cm/yr. The 0–12 month growth rate was 8.3 ± 1.5 cm/yr in the 1×/mon group and 8.2 ± 2.0 cm/yr in the 2×/mon group. The corresponding 12–24 month growth rate was 7.2 ± 2.0 and 6.9 ± 1.5 cm/yr, respectively. Although the microsphere product (Nutropin Depot) has been shown to be effective in the clinical trials, it is noteworthy that experience is limited in patients who were treated with daily growth hormone and switched to Nutropin Depot (6).

Octreotide Acetate. Octreotide acetate is a synthetic cyclic peptide that exerts pharmacologic actions similar to the natural hormone, somatostatin (33). Compared with somatostatin, octreotide is highly resistant to enzymatic degradation and has a prolonged plasma half-life of about 100 min in humans, allowing its use in the long-term treatment of various pathological conditions (35).

The original formulation of octreotide acetate, Sandostatin®, is prepared as a clear sterile solution for administration by deep subcutaneous or intravenous injection (6). Octreotide is indicated in the treatment of patients with acromegaly, an adjunct to surgery and radiotherapy. The goal is to achieve normalized levels of growth hormone and insulin-like growth factor-1 (IGF-I), also known as somatomedin C. In patients with acromegaly, Sandostatin (octreotide acetate) reduces growth hormone to within normal ranges in 50% of patients and reduces IGF-I to within normal ranges in 50–60% of patients (6). Octreotide has also been used to treat the symptoms associated with metastatic carcinoid tumors (flushing and diarrhea), and Vasoactive Intestinal Peptide (VIP) secreting adenomas (watery diarrhea). Subcutaneous injection is the usual route of administration of Sandostatin for control of symptoms. As with most drugs given chronically, frequent injections of the non-microsphere formulation at the same site within short periods of time cause pain and thus injection sites must be rotated in a systematic manner.

On the contrary, the microsphere product Sandostatin LAR® Depot is a long-acting injectable suspension to be given intramuscularly (intragluteally) once every 4 weeks. It maintains the clinical characteristics of the immediate-release dosage form Sandostatin Injection with the added feature of slow release of the drug from the injection site while reducing the need for frequent administration.

3.2. Pharmacokinetics of Microsphere Drug Products

In general, the magnitude and duration of drug concentrations in the plasma after a subcutaneous or intramuscular injection of a long-acting microsphere formulation reflect the release of drug from the microsphere polymer matrix. Drug release is governed by slow degradation of the microspheres at the injection site, but once present in the systemic circulation, the drug will be distributed and eliminated in a manner similar to that from the immediate-release formulation.

Most plasma profiles of microsphere products can be characterized by an initial burst of drug followed by the onset of steady state levels within days or weeks, and then the drug levels decline gradually throughout several weeks. Encapsulated drugs are released over an extended period of time, depending on several factors associated with the drug and microspheres. The extent of the initial burst also depends on the type of microsphere drug products. The initial burst may be a result of release of microsphere-surface associated drug. However, it is unclear if a high level of drug shortly after administration contributes to therapeutic effects or adverse reactions. This question may have to be addressed on a case-by-case basis.

In a study (6) of pediatric patients with growth hormone deficiency (GHD), Nutropin Depot exhibited an appreciable burst of drug immediately after injection, with $AUC_{0-2\,days}$ constituting 50–60% of total $AUC_{0-28\,days}$ following 0.75–1.5 mg/kg doses. The serum hGH concentrations were found to decrease thereafter, but persisted at a concentration of greater than 1 mcg/L for 11–14 days after dosing. Table 4 summarizes the pharmacokinetics of somatropin (rDNA origin) between Nutropin and Nutropin Depot through different routes of administration or in different populations (6).

As expected, a great deal of fluctuation was also observed in the serum profiles produced by Sandostatin LAR Depot. After a single intramuscular injection of Sandostatin LAR Depot in healthy subjects, the serum octreotide concentration reached a transient initial peak of about 0.03 ng/mL/mg dose within 1 h. However, the drug level progressively declined over the following 3–5 days to a nadir of < 0.01 ng/mL/mg dose, then slowly increased to reach a plateau of 0.07 ng/mL/mg dose at 2–3 weeks post-administration. After about 6 weeks post-injection, octreotide levels further decreased to < 0.01 ng/mL/mg dose by weeks 12–13, which was concomitant with the terminal degradation phase of the polymer matrix of the dosage form (6).

3.3. Bioavailability and Bioequivalence

Since the drug is readily available once the microspheres degrade in the body, measurement of drug concentrations in the blood has been commonly used for the assessment of bioavailability and bioequivalence. In general, the drug encapsulated in a microsphere formulation is less bioavailable than that from an immediate release dosage form. For

Table 4 Mean Pharmacokinetic Parameters (Standard Deviation) of Somatropin[a]

Pharmacokinetic parameter[a]	Unit	Nutropin		Nutropin Depot	
		0.02 mg/kg, IV (n = 19)	0.1 mg/kg, SC (n = 36)	0.75 mg/kg, SC (n = 12)	1.5 mg/kg, SC (n = 8)
C_{max}	mcg/L	—	67 (19)	48 (26)	90 (23)
T_{max}	hr	—	6 (2)	12–13	12–13
CL/F	mL/hr·kg	116–174	158 (19)	—	—
$t_{1/2}$	min	20 (3)	126 (26)	—	—
AUC_{0-inf}	mcg·hr/L	—	643 (77)	—	—
$AUC_{0-28\ days}$	mcg·day/L	—	—	83 (49)	140 (34)
$AUC_{0-2\ days}/AUC_{0-28\ days}$	%	—	—	52 (16)	61 (10)

[a]Growth hormone data for Nutropin were obtained from healthy adult males, while those for Nutropin Depot were from pediatric patients with GHD; IV: intravenous; SC: subcutaneous.

[a]C_{max}: maximum concentration; T_{max}: peak time; CL/F: systemic clearance; F, bioavailability (not determined); $t_{1/2}$, half life; AUC_{0-inf}, area under the curve to time infinity.

example, it has been reported (6) that after a single dose, the relative bioavailability of Nutropin Depot in GHD children was about 33–38% when compared to Nutropin AQ® in healthy adults, and 48–55% when compared to Protropin® in GHD children. Similarly, the relative bioavailability of Sandostatin LAR Depot was 60–63% relative to the immediate-release Sandostatin injection given subcutaneously (6).

As noted, the plasma profiles generated by microspheres such as PLGA consist of an initial burst followed by a relatively slow and prolonged release of the drug. Because of these unique characteristics, questions have been raised as to what would be the optimal measures for evaluation of bioavailability or bioequivalence in these drug products. Traditionally, the maximum concentration (C_{max}) and peak time (T_{max}) obtained from the plasma/serum/blood curves are employed as measures for rate of absorption in an orally administered product. However, these measures may not be a sensible index for a microsphere dosage form in view of its peculiar plasma profiles. Several proposals have been suggested for a better characterization of these profiles, such as plateau height, plateau duration, and exposure measures.

Among others, the exposure measures have been proposed in an FDA guidance document for orally administered drug products (16). The FDA recommends a change in focus from the measures of rate and extent of absorption to measures of systemic exposure based on the rationale that "rate" is a continuous and varying function, and cannot be denoted by a single number (36). In contrast, systemic exposure is well known to often correlate with the efficacy and/or safety of a drug. Accordingly, to achieve the regulatory goal, it is proposed that a plasma concentration–time profile be categorized in terms of three fundamental exposure attributes, namely, total exposure, peak exposure, and early exposure. Systemic exposure can then be estimated by the plasma concentration–time profile, which in turn will reflect the rate and extent of drug absorption. Presumably, these measures can be extended to controlled release parenteral dosage forms such as microspheres.

3.4. In Vitro Release Testing

From a regulatory perspective, an appropriate in vitro release test method should be capable of discriminating between "acceptable" and "unacceptable" batches so that it can be used for batch release and quality control. As a further step, if an in vitro–in vivo correlation or association is available, the in vitro test can serve not only as a quality control for manufacturing process, but also as an indicator of product performance in vivo. Therefore, the in vitro release method is best developed to simulate the physiological conditions. As described (16), under specified conditions, the in vitro release test data may also be utilized to support waiver of bioavailability and/or bioequivalence studies.

Since microsphere products are designed to release drug over a long period of time, it is essential to have both long- and short-term in vitro release tests in place for quality control. The long-term release test, sometimes referred to as a real-time test, can be employed to monitor product release over the dosing interval. This test is preferably developed during the early stage of drug development. The short-term release test, also called an accelerated test, can be used for setting specifications for batch release after manufacturing. A logical approach to devising in vitro release testing for microsphere dosage forms is to first develop a real-time test using experimental conditions that simulate the in vivo environment, and then develop a short-term release test based on its relevance to the real-time test.

When developing a bio-relevant accelerated release test method for a microsphere drug product, it is particularly important to maintain the release mechanism designed for the product. A number of means have been employed to accelerate drug release for short-term in vitro testing, including the use of organic solvents, pH change, temperature adjustment, and agitation, etc. Ideally, to develop an appropriate test, investigation should be conducted to determine if these various factors alter the release mechanism of the formulation under study. It is suspected that organic solvents and alkaline pH may solubilize PLGA instead of speeding up its breakdown. Also, high temperature and rapid agitation may cause microsphere agglomeration (37).

Another point to consider is the possibility of drug degradation when conducting in vitro release testing. This is particularly important for proteins and peptides during long-term release testing or under certain conditions of accelerated testing. In addition, the acid release upon breakdown of PLGA may also cause drug degradation.

One of the concerns about conducting a pharmacokinetic or bioavailability/bioequivalence study for microsphere products is that it usually takes a long time to complete in view of the prolonged release of the drug from the dosage form. In this regard, it is particularly advantageous if the in vitro release test can be correlated with the in vivo performance of the drug product. Admittedly, a meaningful in vitro–in vivo correlation or association could be difficult to obtain for microsphere formulations because of the unique characteristics inherent in this dosage form. To facilitate the development of such relationships, the in vivo measurement may not be limited to the plasma concentration of the drug, a conventional compartment for obtaining in vivo data. Alternative measurements may be made through tissue concentrations, biomarkers, surrogate endpoints, or clinical endpoints for safety/efficacy. In the case of microsphere products that are designed for systemic delivery, it may be appropriate to measure drug concentrations in the blood or plasma. However, tissue concentrations may be more relevant for local or targeted delivery.

Animal models are currently not used for the purposes of regulatory approval of drug products in the United States. Nonetheless, they can be employed to assess if an in vitro release method is discriminating. Animal models can also serve as a valuable tool in initial research for development of a possible in vitro–in vivo correlation or association. This is especially useful for controlled release dosage forms since in vivo human studies can be difficult and are generally time consuming for these products. To understand the general principles of in vitro/in vivo correlation, readers are strongly encouraged to review the FDA guidance for industry on

"Extended Release Oral Dosage Forms: Development, Evaluation and Application of In vitro/In vivo Correlation" (22).
Although the guidance was developed mainly for extended
release oral dosage forms, the same principles apply to
controlled release parenteral drug products. For further information on in vitro/in vivo correlation of controlled release
parenterals, the reader is referred to Chapter 5 of this book.

REFERENCES

1. Storm G, Crommelin DJA. Liposomes: quo vadis? PSIT 1998;
    1:19–31.

2. Drummond DC, Meyer O, Hong K, Kirpotin DB, Papahadjo-
    poulos D. Optimizing liposomes for delivery of chemotherapeu-
    tic agents to solid tumors. Pharmacol Rev 1999; 51:691–743.

3. Lian T, Ho RJ. Trends and developments in liposome drug
    delivery systems. J Pharm Sci 2001; 90:667–680.

4. Janoff AS, ed. Liposomes: Rational Design. New York: Marcel
    Dekker, Inc., 1999.

5. US Department of Health and Human Services, Food and
    Drug Administration, Center for Drug Evaluation and Rese-
    arch. *Draft Guidance for Industry: Liposome Drug Products—
    Chemistry, Manufacturing and Controls; Human Pharma-
    cokinetics and Bioavailability; and Labeling Documentation.*
    Division of Drug Information, Office of Training and Commu-
    nication, Center for Drug Evaluation and Research, Food and
    Drug Administration
    (http://www.fda.gov/cder/guidance/index.htm).

6. Physicians' Desk Reference. 58th ed. Montvale, NJ: Medical
    Economics Company, 2004.

7. Martin F. Liposome drug products – Product evolution and
    influence of formulation on pharmaceutical properties and phar-
    macology. Advisory Committee for Pharmaceutical Science
    Meeting, US Food and Drug Administration, July 20, 2001
    (http://www.fda.gov/ohrms/dockets/ac/01/slides/3763s2.htm).

8. US Food and Drug Administration, Title 21, Code of Federal
    Regulations, Part 314.50, 320.21, and 320.29. Office of the

Federal Register, National Archives and Records Administration, 2004.

9. Scherpho G, Damen J, Hoekstra D. Interactions of liposomes with plasma proteins and components of the immune system. In: Knight G, ed. Liposomes—From Physical Structure to Therapeutic Applications. Amsterdam: Elsevier, 1981:299–322.

10. Sheiner LB. Learning versus confirming in clinical drug development. Clin Pharmacol Ther 1997; 61:275–291.

11. Chen M-L, Shah V, Patnaik R, et al. Bioavailability and bioequivalence: an FDA regulatory overview. Pharm Res 2001; 18:1645–1650.

12. US Food and Drug Administration, Title 21, Code of Federal Regulations, Part 320.1. Office of the Federal Register, National Archives and Records Administration, 2004.

13. US Food and Drug Administration, Title 21, Code of Federal Regulations, Part 320.24. Office of the Federal Register, National Archives and Records Administration, 2004.

14. Burgess DJ, Hussain AS, Ingallinera TS, Chen M-L. Assuring quality and performance of sustained and controlled release parenterals. AAPSPharmSci 2002; 4(2): 7 (http://www.aapsph armsci.org/scientificjournals/pharmsci/journal/040207.htm); Pharm Res 2002; 19(11):1761–1768.

15. US Department of Health and Human Services, Food and Drug Administration, Center for Drug Evaluation and Research. Advisory Committee for Pharmaceutical Science Meeting, Complex Drug Substances—Liposome Drug Products, July 20, 2001. (http://www.fda.gov/ohrms /dockets/ac/01/slides/3763s2.htm).

16. US Department of Health and Human Services, Food and Drug Administration, Center for Drug Evaluation and Research. *Guidance for Industry: Bioavailability and Bioequivalence Studies for Orally Administered Drug Products—General Considerations.* March 2003. Division of Drug Information, HFD-240, Center for Drug Evaluation and Research, Food and Drug Administration. (http://www.fda.gov/cder/ guidance/index.htm).

17. US Department of Health and Human Services, Food and Drug Administration, Center for Drug Evaluation and Research. *Guidance for Industry: Waiver of In Vivo Bioavailability and Bioequivalence Studies for Immediate-Release Solid Oral Dosage Forms Based on a Biopharmaceutics Classification System.* August 2000. Office of Training and Communications, Division of Communications Management, Drug Information Branch, HFD-210, Rockville, MD (http://www.fda.gov/cder/guidance/index.htm).

18. US Department of Health and Human Services, Food and Drug Administration, Center for Drug Evaluation and Research. *Guidance for Industry: Immediate Release Solid Oral Dosage Forms. Scale-Up and Post-approval Changes: Chemistry, Manufacturing and Controls, In Vitro Dissolution Testing, and In Vivo Bioequivalence Documentation.* November 1995. Office of Training and Communications, Division of Communications Management, Drug Information Branch, HFD-210, Rockville, MD (http://www.fda.gov/cder/guidance/index.htm).

19. US Department of Health and Human Services, Food and Drug Administration, Center for Drug Evaluation and Research. *Guidance for Industry: Nonsterile Semisolid Dosage Forms. Scale-Up and Postapproval Changes: Chemistry, Manufacturing and Controls, In Vitro Release Testing, and In Vivo Bioequivalence Documentation,* May 1997. Office of Training and Communications, Division of Communications Management, Drug Information Branch, HFD-210, Rockville, MD (http://www.fda.gov/cder/guidance/index.htm).

20. US Department of Health and Human Services, Food and Drug Administration, Center for Drug Evaluation and Research. *Guidance for Industry: Dissolution Testing of Immediate Release Solid Oral Dosage Forms,* August 1997. Office of Training and Communications, Division of Communications Management, Drug Information Branch, HFD-210, Rockville, MD (http://www.fda.gov/cder/guidance/index.htm).

21. US Department of Health and Human Services, Food and Drug Administration, Center for Drug Evaluation and Research. *Guidance for Industry: Modified Release Solid Oral Dosage Forms. Scale-Up and Postapproval Changes: Chemistry, Manufacturing and Controls, In Vitro Dissolution Testing, and In*

Vivo Bioequivalence Documentation. September 1997. Office of Training and Communications, Division of Communications Management, Drug Information Branch, HFD-210, Rockville, MD (http://www.fda.gov/cder/guidance/index.htm).

22. US Department of Health and Human Services, Food and Drug Administration, Center for Drug Evaluation and Research. *Guidance for Industry: Extended Release Oral Dosage Forms: Development, Evaluation and Application of In Vitro/In Vivo Correlation.* September 1997. Office of Training and Communications, Division of Communication Management, Drug Information Branch, HFD-210, Rockville, MD (http://www.fda.gov/cder/guidance/index.htm).

23. Davis SS, Illum L. Microspheres as drug carriers. Roerdink FHD, Kroon AM, eds. Drug Carrier System. New York: John Wiley & Sons Ltd, 1989:131–153.

24. Cowsar DR, Tice TR, Gilley RM, English JP. Poly(lactide-co-glycolide) microspheres for controlled release of steroids. Methods Enzymol 1985; 112:101–116.

25. Hora MS, Rana RK, Nunberg JH, Tice TR, Gilley RM, Hudson ME. Release of human serum albumin from poly(lactide-co-glycolide) microspheres. Pharm Res 1990; 7:1190–1194.

26. Jacobs E, Setterstrom JA, Bach DE, Heath JR, McNiesh LM, Cierny IG. Evaluation of biodegradable ampicillin anhydrate microspheres for local treatment of experimental staphylococcal osteomyalitis. Clin Orthop Relat Res 1991; 267:237–244.

27. Mathiowitz E, Jacobs JS, Jong YS, Carino GP, Chickering DE, Chaturvedi P, Santos CA, Morrell C, Bassett M, Vijayaraghaven K. Biologically erodable microspheres as potential oral delivery systems. Nature 1997; 386:410–414.

28. Barrow ELW, Winchester GA, Staas JK, Quennelle DC, Barrow WW. Use of microsphere technology for targeted delivery of rifampin to *Mycobacterium tuberculosis*-infected macrophages. Antimicrob Agent Chemotherapy 1998; 42:2682–2689.

29. Johansen P, Men Y, Merkle HP, Gander B. Revisiting PLA/PLGA microspheres: an analysis of their potential in parenteral vaccination. Eur J Pharm Biopharm 2000; 50:129–146.

30. Vasir JK, Tambwekar K, Garg S. Bioadhesive microspheres as a controlled drug delivery system. Intern J Pharmaceutic 2003; 255:13–32.

31. Varde NK, Pack DW. Microspheres for controlled release drug delivery. Expert Opin Biol Ther 2004; 4:35–51.

32. Tice TR, Cowsar DR. Biodegradable controlled-release parenteral systems. J Pharm Technol 1984; 8:26–36.

33. Hardman JG, Limbird LZ, ed. Goodman & Gilman's. The Pharmacological Basis of Therapeutics. 9th ed. The McGraw-Hill Co., Inc., 1996.

34. Silverman BL, Blethen SL, Reiter EO, Attie KM, Neuwirth RB, Ford KM. A long-acting human growth hormone (Nutropin Depot): Efficacy and safety following two years of treatment in children with growth hormone deficiency. J Pediatr Endocrinol Metab 2002; 15(suppl 2):715–722.

35. Chanson P, Timsit J, Harris AG. Clinical pharmacokinetics of octreotide: therapeutic applications in patients with pituitary tumors. Clin Pharmacokinet 1993; 25:375–391.

36. Chen M-L, Lesko LJ, Williams RL. Measures of exposure versus measures of rate and extent of absorption. Clin Pharmacokinet 2001; 40:565–572.

37. Zolnik BS, Asandei AD, Raton J-L, Chen M-L, Hussain AS, Burgess DJ. In vitro testing methods of dexamethasone release from PLGA microspheres. AAPS PharmSci 5(4), Abstract W4216 (2003).

Index